The 100 Most Important Chemical Compounds

A Reference Guide

Richard L. Myers

GREENWOOD PRESS

Westport, Connecticut • London

Library of Congress Cataloging-in-Publication Data

Myers, Rusty L.
  The 100 most important chemical compounds : a reference guide /
Richard L. Myers.
      p. cm.
  Includes bibliographical references and index.
  ISBN-13: 978–0–313–33758–1 (alk. paper)
  ISBN-10: 0–313–33758–6 (alk. paper)
  1. Chemicals—Dictionaries. I. Title. II. Title: One hundred most
important chemical compounds.
  TP9.M94   2007
  540—dc22   2007014139

British Library Cataloguing in Publication Data is available.

Library of Congress Catalog Card Number: 2007014139
ISBN-13: 978–0–313–33758–1
ISBN-10: 0–313–33758–6

First published in 2007

Greenwood Press, 88 Post Road West, Westport, CT 06881
An imprint of Greenwood Publishing Group, Inc.
www.greenwood.com

Printed in the United States of America

The paper used in this book complies with the
Permanent Paper Standard issued by the National
Information Standards Organization (Z39.48–1984).

10 9 8 7 6 5 4 3 2

Dedicated to Catalina and Poco

Contents

In the last years of the 19th century, Felix Hoffman, a German chemist, reformulated a compound created in 1852 by the French chemist Charles Frederic Gerhardt. Gerhardt was working with salicylic acid, an extract from willow bark, which was used as a pain reliever. Although salicylic acid reduced pain, it also produced upset stomach and nausea. Gerhardt combined salicylic acid with sodium and acetyl chloride to produce acetylsalicylic acid, a compound that reduced the unpleasant side effects of salicylic acid. Gerhardt discontinued his work on the subject after publishing his results in 1853. His discovery was not pursued until Hoffman and his co-workers at Bayer sought a milder form of salicylic acid in 1897, leading to their 1899 patent for a drug they called aspirin, the most used medicine since 1900. The story of aspirin and 99 other compounds is the subject of *The 100 Most Important Chemical Compounds: A Reference Guide*. It discusses the chemistry, history, and uses of significant compounds that have affected humans throughout history. The 100 compounds were chosen because of their importance to health, industry, and society, or because of their historical impact. References to compounds in the common media and technical literature were also a consideration for inclusion. *The 100 Most Important Chemical Compounds* includes many familiar compounds such as water, carbon dioxide, and benzene and also introduces the reader to less well-known but vitally important compounds, for example, isoprene, the basic building block of rubber, or methylphenidate, the chemical compound marketed as Ritalin.

Each compound is presented in an entry that starts by illustrating the compound's structure and listing basic technical information including chemical name, Chemical Abstracts Services (CAS) number, formula, molar mass, composition, melting point, boiling point, and density. This is followed by a discussion that synthesizes information on the chemistry, history, production, and societal impact of each compound. Using an eclectic approach, *The 100 Most Important Chemical Compounds* introduces the reader to the people, products, and places that are typically ignored when a chemical compound is considered. Whenever possible, the origin of the compound's name, the discovery or first synthesis of the compound, and production statistics are included. Each entry is tailored to the compound's role in society. Industrial compounds such as ammonia and ethylene focus largely on the chemistry and uses of the compound, and medicinal and consumer compounds include more information on the historical importance of the compound.

In addition to providing general technical information on compounds, the historical importance and the compound's overall impact in human history help the reader understand chemistry's impact on society, for example, the use of ethanol as an alternative fuel source. Throughout the book, familiar people and product names such as DuPont, Dow, Gore (Gore-Tex), Goodyear, Teflon, T-fal, and ibuprofen appear, helping the reader relate to the material. Numerous stories involving the 100 chemicals are included in *The 100 Most Important Chemical Compounds* so that the reader can appreciate how chemistry has affected human history. Here is a sampling of just a few:

- Donald Rumsfeld, the former Secretary of Defense, is instrumental in getting Aspartame, the most popular artificial sweetener, approved despite attempts by consumer groups to ban the product.
- Samuel Hopkins modifies the method for burning wood to make potash and is issued the first United States patent in 1790.
- Frederick Banting, a disgruntled family doctor in London Ontario, leaves to do medical research at the University of Toronto, which leads to the use of insulin to treat diabetes.
- Charles Francis Hall's research, started when he was a high school student and continued at Oberlin College, results in a cheap process to produce aluminum and the founding of ALCOA.
- Several doctors, who all claim the discovery of ether as an anesthetic, fight a bitter battle over priority to gain financial rewards. Their battle ends in suicide, prison, and untimely deaths.
- Morphine is self-administered by Friedrich Wilhelm Sertürner and is named after Morpheus, the Greek god of dreams, after he experiences its narcotic effects.
- DuPont scientists researching the production of chlorofluorocarbons accidentally produce a white waxy substance that becomes Teflon.
- Cocaine is introduced into Italian wines, and an Atlanta druggist follows this practice and also includes cola nut extracts to produce a nonalcoholic drink, Coca-Cola.
- At the end of the 18th century, a disciple of Lavoisier flees France in the aftermath of the French Revolution to start his own gunpowder factory in the state of Delaware; Eleuthère Irénée du Pont's factory develops into one of the largest chemical companies in the world.
- Leandro Panizzon, working for Ciba in Basel Switzerland, synthesizes methylphenidate, administers it to his wife Rita as a stimulant, and names the compound Ritaline (Ritalin) after her.

The 100 Most Important Chemical Compounds focuses on 100 compounds, but references several hundred compounds. Structures and formula for compounds other than the 100 are included in the entries and listed in the index. Repetition of information was kept to a minimum by including representative compounds or the simplest compound in a chemical family and highlighting chemical properties that distinguish chemical groups. For example, there are entries for methane, ethane, butane, and octane; but other alkanes such as pentane, hexane, and heptane are not included. Most personal names in the book include years of birth and death. Birth and deaths are not included when they could not be located in standard scientific biographical literature. Specific dates are given as accurately as possible by referring to the actual date of synthesis, the first publication, the date of patent application, etc.

To assist the reader in understanding the material, a glossary defines chemical terms that may not be familiar to the reader. A table of common and ancient names is also included in the backmatter. This is useful in helping the reader interpret ambiguous common names. For example, the word *lime* can refer to slaked lime, quicklime, or to some limestone.

The 100 Most Important Chemical Compounds is written for individuals needing a general reference on common chemical compounds that includes both technical and historical information. The information in the book is especially useful for middle, secondary, and college students who want a general review of common compounds. Anyone needing a general reference on the most important compounds will benefit from this book. It would be especially helpful to science teachers as it provides a ready reference on technical and historical information.

In using this book my hope is that the reader will realize that chemistry and chemical compounds are much more than atoms and bonds; indeed, chemical compounds, like people, are unique individuals. Millions of chemical compounds exist, with thousands of new compounds formed daily. Each of these compounds is unique, but collectively they shape our physical world. *The 100 Most Important Chemical Compounds* introduces the reader to the most notable of these compounds. I hope the book will stimulate interest in knowing more.

Acknowledgments

Thanks go to the team responsible for the preparation of this book. Greenwood Press initiated this work and saw the need for a book on chemical compounds that incorporated both the technical and human aspects of chemistry. I would like to thank the staff of Greenwood for the opportunity of producing this work. Greenwood's Kevin Downing guided the project from its start and kept the team moving forward. Esther Silverman did a superb job copyediting. Many figures were drawn with ChemSketch from Advanced Chemistry Development, Inc., and the producers of the software were also helpful when queried. Rae Dejúr drew several figures. Much appreciation goes to the faculty and staff of Alaska Pacific University who supported my work with a semester-long sabbatical.

Introduction

What Is a Chemical Compound?

Look around and observe all the physical objects in your immediate view. The most immediate objects are you, the clothes you are wearing, and this book. Less immediate may be other people, furniture, structures, and vehicles. Looking outside you may observe the earth, clouds, the sun, or the moon. The most distant objects are observed at night as stars and galaxies making up the universe. The objects that make up our universe consist of matter. Matter is anything that occupies space and has mass. Chemistry is the study of the composition of matter and its transformation. Another term often considered synonymous with matter is substance, but a substance has a more limited definition in chemistry. A substance can be considered matter with definite properties that establishes its identity. Examples of substances include familiar things such as air, water, wood, and paper. It also includes millions of substances that have been classified chemically. Chemical Abstracts Services' (CAS) registry is the largest database of current substances. On May 6, 2007 at 9:34 P.M. Eastern Standard Time, it contained 31,626,352 registered substances. Several thousand new substances are added to the registry daily. The number of registered substances on any day can be found at the Web site: http://www.cas.org/cgi-bin/regreport.pl.

A substance can be classified chemically in many ways. One of the simplest ways to classify a substance is as an element or a compound. An element is a pure substance that cannot be changed into a simpler substance by chemical means. Elements are the building blocks of nature; all matter is composed of elements. The periodic table is a concise map that organizes chemical elements into columns (groups) and rows (periods) based on their chemical properties. Currently, there are 118 known chemical elements, with whole numbers 1 to 118. These numbers are referred to as the element's atomic number and give the number of protons in the nucleus of an atom of the element. For example, carbon's atomic number is 6 and each carbon atom has 6 protons in its nucleus. The first 92 elements occur naturally, and those above atomic number 92 are synthesized through nuclear reactions using particle accelerators. Element 118 was just confirmed in the fall of 2006, and by now, more elements may have been produced.

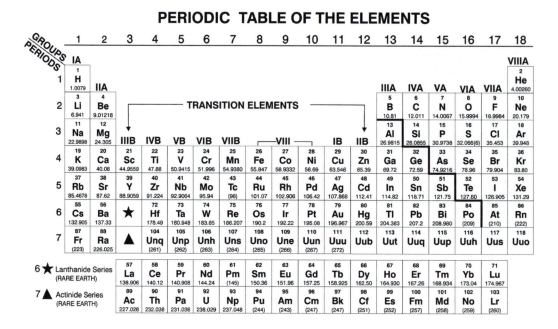

Figure I.1   Periodic Table.

Nine elements were known in ancient times: gold, silver, iron, mercury, tin, copper, lead, carbon, and sulfur. Bismuth, zinc, antimony, and arsenic were isolated in the alchemical period and Middle Ages. It was known that certain substances combined into other substances before the modern idea of elements and compounds was established. In 1718, Étienne François Geoffroy (1672–1731) presented a Table of Affinities to the French Academy of Science that summarized which substances were compatible with each other and combined to produce other substances (Figure I.2).

The birth of modern chemistry in the late 18th century established the idea of a chemical compound as a substance composed of two or more elements combined chemically in fixed proportions. Substances such as water were previously considered to be elements. Water is a chemical compound consisting of the elements hydrogen and oxygen in a fixed ratio of two hydrogen atoms for each oxygen atom, giving H_2O. The ratio of hydrogen to oxygen atoms in water is always 2:1. Substances consisting of only one element are not compounds. For example, ozone (O_3), which consists of three oxygen atoms bonded together, is not a compound. Common diatomic molecules (a molecule is two or more atoms combined chemically) such as hydrogen (H_2), chlorine (Cl_2), and iodine (I_2) are not compounds.

The elements that make up a compound must be in fixed proportions. This principle was first formulated in 1797 by Joseph Louis Proust (1754–1826) and is known as the law of definite proportions. Substances such as air, bronze, and gasoline, in which proportions are not fixed, are classified as mixtures. This can be illustrated using air as an example. Although the percentage of nitrogen and oxygen in air is approximately constant at 78% nitrogen and 21% oxygen, the ratio of nitrogen to oxygen (as well as that of other gases that make up air) changes with altitude, pollution, and weather conditions. Furthermore, the nitrogen and oxygen that

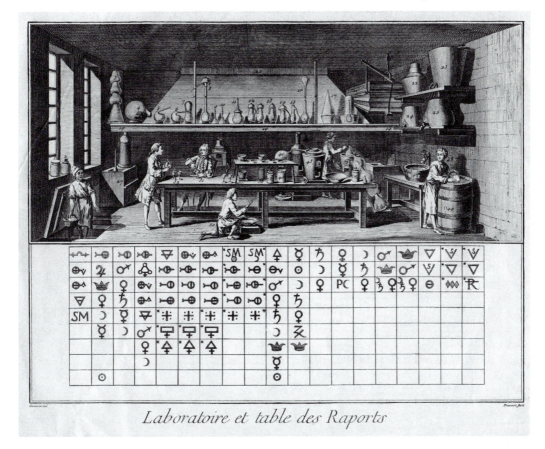

Laboratoire et table des Raports

Figure I.2 Geoffroy's Table of Affinities. *Source:* Edgar Fahs Smith Collection, University of Pennsylavnia.

compose air are not combined chemically. Each gas retains its own chemical identity in air. A mixture consists of two or more substances in which each substance retains its identity. Mixtures may be heterogeneous or homogeneous. Heterogeneous mixtures have a variable composition; a homogeneous mixture has a constant composition throughout. When directions on an item state "shake well before using," the reason is to homogenize a heterogeneous mixture, for example, vinegar and oil salad dressing. The classification of different forms of matter is summarized in Figure I.3.

Types of Chemical Compounds

Organic and Inorganic Compounds

Chemistry involves the transformation of matter. These transformations largely involve substances that are compounds. Elements combine to form compounds in combination reactions, whereas decomposition reactions result in a chemical compound being broken into smaller compounds and/or elements. Still other reactions involve the exchange of atoms between compounds, resulting in new compounds. The number of identified chemical compounds is approximately 10 million and increasing by several hundred thousand each year.

The tremendous number of chemical compounds has been categorized into numerous categories. A broad classification distinguishes between inorganic and organic compounds. Organic compounds are carbon based. Inorganic compounds exclude compounds based on carbon, although some carbon compounds, such as carbon dioxide and carbon monoxide, are considered inorganic. The vast majority of compounds are organic. Approximately 7 million organic compounds have been described, but the number of potential organic compounds is infinite. This compares to approximately 100,000 inorganic compounds.

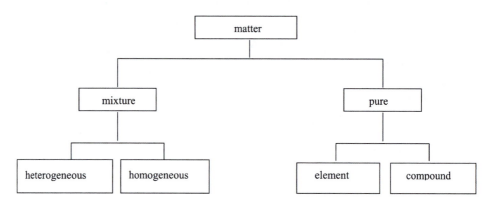

Figure I.3   The classification of matter.

The advent of chemical synthesis in the first half of the 19th century was one of the most important advancements in human history. The term organic was not used until 1807 when Jöns Jakob Berzelius (1779–1848) suggested its use to distinguish between compounds devised from living (organic) and nonliving (inorganic) matter. Before Berzelius made this distinction, interest in organic compounds was based primarily on the medicinal use of plants and animals. Plant chemistry involved the study of chemicals found in fruits, leaves, saps, roots, and barks. Animal chemistry focused on compounds found in the blood, saliva, horn, skin, hair, and urine. In contrast to plant and animal chemistry, scientists distinguished those compounds isolated from nonliving mineral material. For example, mineral acids such as sulfuric acid were made from vitriolic sulfur compounds.

Until the mid-18th century, scientists believed organic compounds came only from live plants and animals. They reasoned that organisms possessed a vital force that enabled them to produce organic compounds. The first serious blow to this theory of vitalism, which marked the beginning of modern organic chemistry, occurred when Friedrich Wöhler (1800–1882) synthesized urea from the two inorganic substances, lead cyanate and ammonium hydroxide:

$$Pb(OCN)_2 + 2NH_4OH \rightarrow 2(NH_2)_2CO + Pb(OH)_2$$
lead cyanate + ammonium hydroxide → urea + lead hydroxide

Wöhler was actually attempting to synthesize ammonium cyanate when he discovered crystals of urea in his samples. He first prepared urea in 1824, but he did not identify this product and report his findings until 1828. In a note written to Berzelius he proclaimed: "I must tell you that I can make urea without the use of kidneys, either man or dog. Ammonium cyanate is urea." Although Wöhler's synthesis of urea signaled the birth of organic chemistry, it was not a fatal blow to vitalism. Some argued that because Wöhler used bone material in his preparations, a vital force could still have been responsible for Wöhler's urea reaction. Over the next 20 years, the vitalism theory eroded as the foundation of organic chemistry started to take shape. Hermann Kolbe's (1818–1884) synthesized acetic acid from several inorganic substances in 1844. Kolbe's work and that of Marcellin's Berthelot's (1827–1907), who produced numerous organic compounds, including methane and acetylene in 1850, marked the death of the vitalism theory.

The reason for the tremendous number of organic compounds is carbon's ability to covalently bond with itself and many other atoms. This ability is due to carbon's electron structure. The electron configuration of carbon, atomic number 6, is $1s^2\ 2s^2\ 2p^2$. Carbon needs four more electrons to complete its outer shell and acquire a stable octet electron configuration. Of all the atoms, only carbon has the ability to form long chains of hundreds and even thousand of carbon atoms linked together. The carbon-carbon bonds in these molecules exist as stable single, double, and triple bonds. Carbon also has the ability to form ring structures, and other atoms and groups of atoms can be incorporated into the molecular structures formed from carbon atoms.

Organic compounds are further classified according to their functional group. A functional group is a small chemical unit consisting of atoms (or groups of atoms) and their bonds that defines an organic molecule. Organic compounds in the same family contain the same functional group. For example, simple hydrocarbons called alkanes contain only carbon and hydrogen, with the carbon connected by single carbon to carbon bonds. Alcohols contain the functional group -OH such as in methanol CH_3OH. Some common functional groups are summarized in Table I.1.

Ionic and Molecular Compounds

Ionic compounds consist of positive ions (cations) and negative ions (anions); hence, ionic compounds often consist of a metal and nonmetal. The electrostatic attraction between a cation and anion results in an ionic bond that results in compound formation. Binary ionic compounds form from two elements. Sodium chloride (NaCl) and sodium fluoride (NaF) are examples of binary ionic compounds. Three elements can form ternary ionic compounds. Ternary compounds result when polyatomic ions such as carbonate (CO_3^{2-}), hydroxide (OH-), ammonium (NH_4^+), form compounds. For example, a calcium ion, Ca^{2+}, combines with the carbonate ion to form the ternary ionic compound calcium carbonate, $CaCO_3$. Molecular compounds form discrete molecular units and often consist of a combination of two nonmetals. Compounds such as water (H_2O), carbon dioxide (CO_2), and nitric oxide (NO) represent simple binary molecular compounds. Ternary molecular compounds contain three elements. Glucose $(C_6H_{12}O_6)$ is a ternary molecular compound. There are several distinct differences between ionic and molecular compounds, as summarized in Table I.2.

Naming Compounds

Elements and compounds since ancient times were represented using symbols. John Dalton (1766–1844) had a unique system for known elements and compounds of his time. For example, oxygen was represented with O and carbon with ●; carbon monoxide (known

Table I.1   Functional Groups

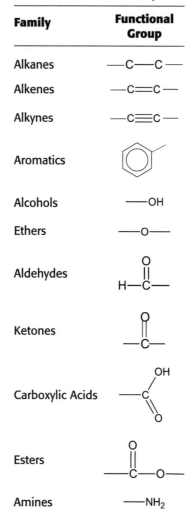

Family	Functional Group
Alkanes	—C—C—
Alkenes	—C=C—
Alkynes	—C≡C—
Aromatics	
Alcohols	—OH
Ethers	—O—
Aldehydes	
Ketones	
Carboxylic Acids	
Esters	
Amines	—NH₂

as carbonic oxide) was therefore symbolized as ◐. In 1814, Berzelius discarded the chemical symbols and proposed a new system based on the initial letter of the element. Berzelius, in compiling the Swedish Pharmacopia, used the initial letter of its Latin name to symbolize an element. If two elements had the same first letter, Berzelius would include a second letter that the two elements did not have in common. According to Berzelius's system, hydrogen, oxygen, nitrogen, carbon, phosphorus, and sulfur became H, O, N, C, P, and S, respectively. Berzelius's system used traditional Latin names for existing chemicals, and this explains why some of our modern symbols seem unrelated to its English name. For example, gold comes from aurum (Au), sodium from natrium (Na), and potassium from kalium (K).

Berzelius's method of assigning letters to represent elements was also applied to compounds. Berzelius used superscripts to denote the number of atoms in a compound. Thus water would

Table I.2   Ionic versus Molecular Compounds

	Ionic	**Molecular**
Basic Units	ions	molecules
Bonding	ionic, electron transfer	covalent, electron sharing
Elemental Constituents	metal and nonmetals	nonmetals
State at Normal Conditions	solid	solid or liquid or gas
Melting, Boiling Points	high	low
Electrical Conductivity of Aqueous Solutions	generally high	low or zero
Density	generally high	generally low

be H^2O and carbon dioxide would be CO^2. Later these superscripts were changed to the current practice of designating the number of atoms using subscripts. The absence of a subscript implies the subscript 1. Using Berzelius's system the chemical formula for ammonia is:

$$N \qquad\qquad H_3$$
$$\uparrow \qquad\qquad \uparrow$$
1 atom of   3 atoms of
nitrogen   hydrogen

The rules for naming compounds depend on the type of compound. For ionic compounds consisting of two elements (binary compounds), the metal cation element is named first. After the metal cation element is named, the stem of the nonmetal anion element is used, with the ending "ide" added to the stem. Changing the anion for several common elements results in the following: oxygen → oxide, nitrogen → nitride, chlorine → chloride. Examples of ionic compounds named using this system are sodium chloride (NaCl), barium oxide (BaO), and sodium iodide (NaI). When the cation exists in more than one oxidation state, Roman numerals enclosed within parentheses are used to designate the oxidation state. For example, iron can exist as Fe^{2+} or Fe^{3+}, and it would be impossible to know whether iron chloride was

Table I.3   Polyatomic Ions

Ion	Formula
Ammonium	NH_4^+
Carbonate	CO_3^{2-}
Bicarbonate	HCO_3^-
Hydroxide	OH^-
Nitrate	NO_3^-
Nitrite	NO_2^-
Phosphate	PO_4^{3-}
Sulfate	SO_4^{2-}
Sulfite	SO_3^{2-}

$FeCl_2$ or $FeCl_3$ compound. Therefore iron must include a Roman numeral to specify which cation is in the compound. Iron (II) chloride is $FeCl_2$ and iron (III) chloride is $FeCl_3$. Two or more atoms may combine to form a polyatomic ion. Common polyatomic ions are listed in Table I.3. The names of polyatomic ions may be used directly in compounds that contain them. Hence, NaOH is sodium hydroxide, $CaCO_3$ is calcium carbonate, and $Ba(NO_3)_2$ is barium nitrate.

Naming binary molecular compounds requires using Greek prefixes to indicate the number of atoms of each element in the compound or molecule. Prefixes are given in Table I.4. Prefixes precede each element to indicate the number of atoms in the molecular compound. The stem of the second element takes the "ide" suffix. The prefix "mon" is dropped for the initial element; that is, if no prefix is given, it is assumed that the prefix is 1. Examples of molecular compounds are carbon dioxide (CO_2), carbon monoxide (CO), and dinitrogen tetroxide (N_2O_4).

Simple binary and ternary compounds can be named by using a few simple rules, but systematic rules are required to name the millions of organic compounds that exist. Rules for naming compounds have been established by the International Union of Pure and Applied Chemistry (IUPAC). The IUPAC name stands for a compound that identifies its atoms, functional groups, and basic structure. Because of the complexity of organic compounds, thousands of rules are needed to name the millions of compounds that exist and the hundreds that are produced daily. The original intent of the IUPAC rules was to establish a unique name for each compound, but because of their use in different contexts and different practices between disciplines, more than one name may describe a compound. IUPAC rules result in preferred IUPAC names, but general IUPAC names are also accepted.

Some compounds are known by their common name rather than by using their systematic names. Many of these are common household chemicals. A table listing common names of some of these is included in the Appendix.

Molecular and Structural Formulas

The molecular formula identifies the atom and number of each atom in a chemical compound, but it tells nothing about the structure of the compound. Chemical structure

Table I.4   Greek Prefixes

Prefix	Number
Mono	1
Di	2
Tri	3
Tetra	4
Penta	5
Hexa	6
Hepta	7
Octa	8
Nona	9
Deca	10

represents the spatial arrangement of atoms in a compound. A structural formula shows how atoms are arranged in a compound and the bonding between atoms. Simple structural formulas use single short lines to represent a covalent bond. A simple structural formula for water is H-O-H. Double and triple bonds are represented using double or triple lines, respectively. So a simple structural formula for hydrogen cyanide is H-C≡N. Structural formulas may include the unshared electrons as well as the arrangement of atoms. Rather than represent water as H-O-H, a more complete representation would be:

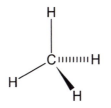

To represent the actual three-dimensional nature of molecules, the lines representing bonds can be modified. A straight line represents a line in the plane of the page, a solid wedge represents a bond coming out of the plane of the page, and a hatched line represents a bond coming out behind the plane of the page. The three-dimensional tetrahedral arrangement of methane would therefore be represented as:

Entry Format

The entry for each compound in *The 100 Most Important Chemical Compounds* follows a standard format. After the heading, an illustration representing the compound is shown followed by a listing of basic information:

CHEMICAL NAME. Many compounds have generic as well as chemical names. Systematic names encode the composition and structure of compounds and are based on a detailed set of rules established by the IUPAC. Most entries give the IUPAC name, but in some instance a more standard chemical name used by the chemical community is given.

CAS NUMBER. This is the unique numerical identifier for the compound based on Chemical Abstract Services registry. It can be used to look up information on the compound. It serves as the compound's unique code similar to a person's Social Security number or a car's license plate, but it is even more generic because it is internationally recognized.

MOLECULAR FORMULA. The molecular formula identifies each atom and the number of each atom in a compound.

MOLAR MASS. This is the mass of 1 mole of the compound.

COMPOSITION. The elemental composition of the compound is given. In some cases the total does not equal exactly 100% because the percentages are rounded to the nearest decimal place.

MELTING POINT. The melting point is expressed in degrees Celsius. Some compounds do not melt, but decompose before melting and this is noted.

BOILING POINT. The boiling point is expressed in degrees Celsius. Many compounds decompose before they boil, and this is noted. The temperature of decomposition is given for some compounds. Because boiling point is dependent on pressure, boiling points are generally expressed at 1 atmosphere pressure.

DENSITY. The density for solids and liquids is given in grams per cubic centimeters (cm^3). Gas density is expressed both in grams per liter and as a vapor density with the density of air taken at 1.0. Densities are generally for 0°C and 1 atmosphere pressure.

1. Acetic Acid

CHEMICAL NAME = ethanoic acid
CAS NUMBER = 64–19–7
MOLECULAR FORMULA = $C_2H_4O_2$
MOLAR MASS = 60.1 g/mol
COMPOSITION = C(60%) H(6.7%) O(53.3%)
MELTING POINT = 16.7°C
BOILING POINT = 118.1°C
DENSITY = 1.05 g/cm³

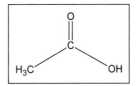

Acetic acid is a weak carboxylic acid with a pungent odor that exists as a liquid at room temperature. It was probably the first acid to be produced in large quantities. The name acetic comes from *acetum,* which is the Latin word for "sour" and relates to the fact that acetic acid is responsible for the bitter taste of fermented juices. Acetic acid is produced naturally and synthetically in large quantities for industrial purposes. It forms when ubiquitous bacteria of the genera *Acetobacter* and *Clostridium* convert alcohols and sugars to acetic acid. *Acetobacter,* especially *Acetobacter aceti,* are more efficient acetic acid bacteria and produce much higher concentrations of acetic acid compared to *Clostridium.*

Vinegar is a dilute aqueous solution of acetic acid. The use of vinegar is well documented in ancient history, dating back at least 10,000 years. Egyptians used vinegar as an antibiotic and made apple vinegar. Babylonians produced vinegar from wine for use in medicines and as a preservative as early as 5000 B.C.E. Hippocrates (ca. 460–377 B.C.E.), known as the "father of medicine," used vinegar as an antiseptic and in remedies for numerous conditions including fever, constipation, ulcers, and pleurisy. Oxymel, which was an ancient remedy for coughs, was made by mixing honey and vinegar. A story recorded by the Roman writer Pliny the Elder (ca. 23–79 C.E.) describes how Cleopatra, in an attempt to stage the most expensive meal ever, dissolved pearls from an earring in vinegar wine and drank the solution to win a wager. Other Roman writings describe how Hannibal, when crossing the Alps, heated boulders and poured vinegar on them to soften and crack them to clear paths. Vinegar is often referred to in the Bible, both directly and indirectly as bad wine. A famous historical Bible published in 1717 is referred to as the Vinegar Bible because, among its numerous errors, the heading for Luke 20 reads "the parable of the vinegar" rather than "the parable of the vineyard." Vinegar was thought to have special powers and it was a common ingredient used by alchemists.

Alchemists used distillation to concentrate acetic acid to high purities. Pure acetic acid is often called glacial acetic acid because it freezes slightly below room temperature at 16.7°C (62°F). When bottles of pure acetic acid froze in cold laboratories, snowlike crystals formed on the bottles; thus the term *glacial* became associated with pure acetic acid. Acetic acid and vinegar were prepared naturally until the 19th century. In 1845, the German Chemist Hermann Kolbe (1818–1884) successfully synthesized acetic acid from carbon disulfide (CS_2). Kolbe's work helped to establish the field of organic synthesis and dispelled the idea of vitalism. Vitalism was the principle that a vital force associated with life was responsible for all organic substances.

Acetic acid is used in numerous industrial chemical preparations and the large-scale production of acetic acid takes place through several processes. The main method of preparation is methanol carbonylation. In this process, methanol reacts with carbon monoxide to give acetic acid: $CH_3OH_{(l)} + CO_{(g)} \rightarrow CH_3COOH_{(aq)}$. Because the reaction requires high pressures (200 atmospheres), this method was not used until the 1960s, when the development of special catalysts allowed the reaction to proceed at lower pressures. A methanol carbonylation procedure developed by Monsanto bears the company's name. The second most common method to synthesize acetic acid is by the catalytic oxidation of acetaldehyde: $2\ CH_3CHO_{(l)} + O_{2(g)} \rightarrow 2\ CH_3COOH_{(aq)}$. Butane may also be oxidized to acetic acid according to the reaction: $2\ C_4H_{10(l)} + 5O_{2(g)} \rightarrow 4\ CH_3COOH_{(aq)} + 2H_2O_{(l)}$. This reaction was a major source of acetic acid before the Monsanto process. It is carried out at a temperature of approximately 150°C and 50 atmospheres pressure.

Acetic acid is an important industrial chemical. The reaction of acetic acid with hydroxyl-containing compounds, especially alcohols, results in the formation of acetate esters. The largest use of acetic acid is in the production of vinyl acetate (Figure 1.1). Vinyl acetate can be produced through the reaction of acetylene and acetic acid. It is also produced from ethylene and acetic acid. Vinyl acetate is polymerized into polyvinyl acetate (PVA), which is used in the production of fibers, films, adhesives, and latex paints.

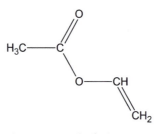

Figure 1.1   Vinyl acetate.

Cellulose acetate, which is used in textiles and photographic film, is produced by reacting cellulose with acetic acid and acetic anhydride in the presence of sulfuric acid. Other esters of acetic acid, such as ethyl acetate and propyl acetate, are used in a variety of applications.

The condensation reaction of two molecules of acetic acid results in the production of acetic anhydride and water:

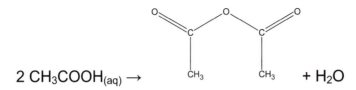

$$2\ CH_3COOH_{(aq)} \rightarrow \qquad\qquad\qquad + H_2O$$

Acetic acid is used to produce the plastic polyethylene terephthalate (PET) (see Ethene [Ethylene]). Acetic acid is used to produce pharmaceuticals (see Acetylsalicylic Acid).

Vinegar was the first form of acetic acid produced through the fermentation of sugars in wine and other substances. The name vinegar comes from the French *vin aigre* meaning "sour wine." Vinegar is an acetic acid solution between about 4% and 8%, but it should not be equated as simply diluted acetic acid. Because vinegar is produced through fermentation of sugars contained in fruits and vegetables, it contains additional nutrients associated with the source from which it is produced. Industrial production of acetic acid far exceeds the production of vinegar acetic acid, but the latter is important for the food and beverage industry. The U.S. Food and Drug Administration requires that vinegar must contain in excess of 4 grams of acetic acid per 100 mL of solution and requires vinegars to contain the labeling: "diluted with water to ___ percent acid strength," with the blank specifying the acid strength. Most grocery store vinegars are about 5% acetic acid.

Vinegar is most often associated with the fermentation of alcohol and fruit juices, but there are hundreds of sources of vinegar. Different varieties of vinegar have traditionally reflected the geography from which they were produced. Typical vinegars available in the United States include distilled white, apple cider, balsamic, malt, and wine. Other vinegars come from potatoes, rice, a multitude of fruits, grains, and sugar cane. The modern production of vinegar involves charging large vats or casks with alcohol. The vats are supplied with seed bacteria (*Acetobacter* sp.), nutrients, and oxygen in a mixture sometimes called "mother of vinegar." The casks contain wood shavings to provide a large surface area where the vinegar-producing bacteria form a biofilm. The mixture is circulated through the vats until the desired acetic acid concentration is obtained.

Common distilled white vinegar is usually from the fermentation of distilled alcohol, but it can be diluted industrial acetic acid. Many countries specify that vinegar used in food production must be produced naturally. The alcohol in white vinegar usually comes from corn. Apple cider vinegar is produced from the fermentation of apple cider to hard cider and then to apple cider vinegar. Balsamic vinegar dates back more than 1,000 years. The name comes from its Italian equivalent *aceto balsamico,* which translated means soothing vinegar. It is a dark brown, syrupy, smooth vinegar especially valued by chefs for its flavor. True balsamic vinegar is produce from white Trebbiano grapes grown in the region of Modena, Italy. The grape juice is boiled and then progressively aged in casks of various woods. During different stages of the aging process, the product is transferred to progressively smaller casks to compensate for evaporation. Aging occurs for a minimum of 12 years, although some vintages have been aged for decades. Only several thousand gallons of balsamic vinegar are produced each year, making original balsamic vinegar quite expensive. Balsamic vinegars marketed on a large scale in supermarkets are ordinary vinegars to which coloring and other additives have been used to mimic true balsamic vinegar.

CHEMICAL NAME = 2-propanone
CAS NUMBER = 67–64–1
MOLECULAR FORMULA = C_3H_6O
MOLAR MASS = 58.1 g/mol
COMPOSITION = C(62.0%) H(10.4%) O(27.6%)
MELTING POINT = −94.9°C
BOILING POINT = 56.3°C
DENSITY = 0.79 g/cm³

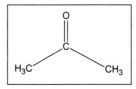

Acetone is a flammable, colorless liquid with a pleasant odor. It is used widely as an organic solvent and in the chemical industry. It is the simplest ketone, which also goes by the name dimethyl ketone (DMK). Acetone was originally referred to as pyroacetic spirit because it was obtained from the destructive distillation of acetates and acetic acid. Its formula was correctly determined in 1832 by Justus von Liebig (1803–1873) and Jean-Baptiste André Dumas (1800–1884). In 1839, the name acetone began to be used. Acetone was derived by adding the ending "one" meaning "daughter of" to the root of acetum (acetic acid) to mean daughter of acetum because it was obtained from acetic acid.

The traditional method of producing acetone in the 19th century and the beginning of the 20th century was to distill acetates, particularly calcium acetate, $Ca(C_2H_3O_2)_2$. World War I placed an increase demand on England to produce gunpowder, explosives, and propellants such as cordite. Cordite is a propellant made using nitroglycerin and nitrocellulose, and nitrocellulose is a principal component of smokeless gunpowder. Cordite is made by dissolving nitrocellulose in acetone, mixing it with nitroglycerin, then baking off the acetone. One of England's suppliers of calcium acetate before the war was Germany, and the loss of this source and lack of other sources because of German blockades meant that it was imperative to find another source of acetone. One of these was from the fermentation of sugars. One of England's leading scientists working on bacterial fermentation was Chaim Weizmann (1874–1952), a Russian-born Jew who was a professor at Manchester University. Weizmann had been working on methods to make butyl alcohol in order to produce synthetic rubber.

Weizmann discovered a process to produce butyl alcohol and acetone from the bacterium *Clostridium acetobutylicum* in 1914. With England's urgent demand for acetone, Winston Churchill (1874–1965) enlisted Weizmann to develop the Weizmann process for acetone production on an industrial scale. Large industrial plants were established in Canada, India, and the United States to provide the allies with acetone for munitions. Weizmann, who is considered the "father of industrial fermentation," obtained significant status from his war contributions and used this to further his political mission of establishing a Jewish homeland. Weizmann was a leader of the Zionist movement and campaigned aggressively until the nation of Israel was established in 1948. He was the first president of Israel.

Fermentation and distillation techniques for acetone production were replaced starting in the 1950s with the cumene oxidation process (Figure 2.1). In this process, cumene is oxidized to cumene hydroperoxide, which is then decomposed using acid to acetone and phenol. This is the primary method used to produce phenol, and acetone is produced as a co-product in the process, with a yield of about 0.6:1 of acetone to phenol.

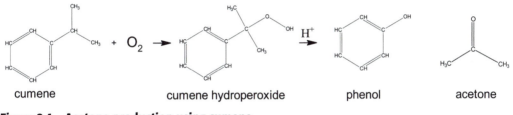

| cumene | cumene hydroperoxide | phenol | acetone |

Figure 2.1   Acetone production using cumene.

Acetone can also be produced from isopropanol using several methods, but the main method is by catalytic dehydrogenation:

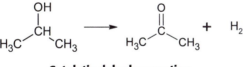

Catalytic dehydrogenation

Acetone is used in the chemical industry in numerous applications. Its annual use in the United States approaches 2 million tons and worldwide its use is close to 5 million tons. The primary use of acetone is to produce acetone cyanohydrin, which is then used in the production of methyl methacrylate (MMA). MMA polymerizes to polymethyl methacrylate. MMA is used in a variety of applications involving plastics and resins. It is used extensively in the production of skylights, Plexiglas, outdoor advertising signs, building panels, and light fixtures. It is also incorporated into paints, lacquers, enamels, and coatings.

Another use of acetone in the chemical industry is for bisphenol A (BPA). BPA results form the condensation reaction of acetone and phenol in the presence of an appropriate catalyst. BPA is used in polycarbonate plastics, polyurethanes, and epoxy resins. Polycarbonate plastics are tough and durable and are often used as a glass substitute. Eyeglasses, safety glasses, and varieties of bullet-proof "glass" are made of polycarbonates. Additional

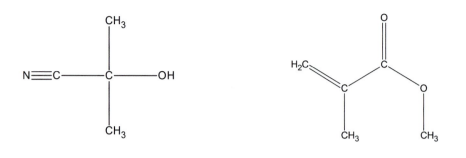

acetone cyanohydrin   methyl methylacrylate

uses include beverage and food containers, helmets (bicycle, motorcycle), compact discs, and DVDs.

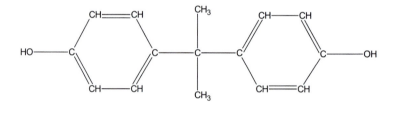

BPA

In addition to its use as a chemical feedstock and intermediate, acetone is used extensively as an organic solvent in lacquers, varnishes, pharmaceuticals, and cosmetics. Nail polish remover is one of the most common products containing acetone. Acetone is used to stabilize acetylene for transport (see Acetylene).

Acetone and several other ketones are produced naturally in the liver as a result of fat metabolism. Ketone blood levels are typically around 0.001%. The lack of carbohydrates in a person's diet results in greater fat metabolism, causing ketone levels in the blood to increase. This condition is called ketosis. People on low-carbohydrate diets and diabetics may have problems with ketosis because of a greater amount of fat in the diet. An indicator of ketosis is the smell of acetone on a person's breath.

3. Acetylene

CHEMICAL NAME = ethyne
CAS NUMBER = 74–86–2
MOLECULAR FORMULA = C_2H_2
MOLAR MASS = 26.0 g/mol
COMPOSITION = C(92.3%) H(7.7%)
MELTING POINT = –80.8°C
BOILING POINT = –80.8 (sublimes)
DENSITY = 1.17 g/L (vapor density = 0.91, air = 1)

Acetylene, which is the simplest alkyne hydrocarbon, exists as a colorless, flammable, unstable gas with a distinctive pleasant odor (acetylene prepared from calcium carbide has a garlic smell resulting from traces of phosphine produced in this process). The term *acetylenes* is used generically in the petroleum industry to denote chemicals based on the carbon-carbon triple bond. Acetylene was discovered in 1836 by Edmund Davy (1785–1857) who produced the gas while trying to make potassium metal from potassium carbide (K_2C_2). In 1859, Marcel Morren in France produced acetylene by running an electric arc between carbon electrodes in the presence of hydrogen. Morren called the gas produced carbonized hydrogen. Three years later, Pierre Eugène-Marcelin Berthelot (1827–1907) repeated Morren's experiment and identified carbonized hydrogen as acetylene.

A method for the commercial production of acetylene was discovered accidentally in 1892 by Thomas Willson (1860–1915). Willson was experimenting on aluminum production at his company in Spray, North Carolina. He was attempting to produce calcium in order to reduce aluminum in aluminum oxide, Al_2O_3. Willson combined coal tar and quicklime in an electric furnace and, instead of producing metallic calcium, he produced a brittle gray substance. The substance was calcium carbide, CaC_2, which when reacted with water, produced acetylene. Willson's work led to the establishment of a number of acetylene plants in the United States and Europe during the next decade.

The triple bond in acetylene results in a high energy content that is released when acetylene is burned. After Willson's discovery of a method to produce commercial quantities of acetylene, Henry-Louis Le Châtelier (1850–1936) found that burning acetylene and oxygen in approximately equal volumes produced a flame hotter than any other gas. Flame temperatures between 3,000°C and 3,300°C were possible using acetylene and pure oxygen, which was high enough to cut steel. The hot flame from acetylene is due not so much from its heat of combustion, which is comparable to other hydrocarbons, but from the nature of the flame produced by acetylene. Acetylene burns quickly when combined with pure oxygen, producing a flame with a tight concentrated inner cone. The transfer of energy from the flame occurs in a very small volume, resulting in a high temperature. During the last half of the 19th century, torches using hydrogen and oxygen were used for cutting metals, but the highest temperatures were around 2000°C. Torches capable of using acetylene were developed in the early 20th century, and acetylene found widespread use for metal working. Another widespread use of acetylene was for illumination. Portable lamps for miners, automobiles, bicycles, and lanterns used water mixed with calcium carbide to generate acetylene that burned to produce a bright flame. Street lamps, lighthouses, and buoys also used acetylene for illumination, but by 1920 acetylene as a light source had been replaced by batteries and electric light.

One problem with the use of acetylene is its stability. Although it is stable at normal pressures and temperatures, if it is subjected to pressures as low as 15 pounds per square inch gauge (psig) it can explode. To minimize the stability problem, acetylene transport is minimized. Acetylene contained in pressurized cylinders for welding and cutting is dissolved in acetone. A typical acetylene cylinder contains a porous filler made from a combination of materials such as wood chips, diatomaceous earth, charcoal, asbestos, and Portland cement. Synthetic fillers are also available. Acetone is placed in the cylinder and fills the voids in the porous material. Acetylene can then be pressurized in the cylinders up to approximately 250 pounds per square inch (psi) In a pressurized cylinder, 1 liter of filler can hold a couple of hundred liters of acetylene, which stabilizes it. Acetylene cylinders should not be stored on their sides because this could cause the acetone to distribute unequally and create acetylene pockets.

The traditional method of producing acetylene is from reacting lime, calcium oxide (CaO), with coke to produce calcium carbide (CaC_2). The calcium carbide is then combined with water to produce acetylene:

$$2CaO_{(s)} + 5C_{(s)} \rightarrow 2CaC_{2(g)} + CO_{2(g)}$$
$$CaC_{2(s)} + 2H_2O_{(l)} \rightarrow C_2H_{2(g)} + Ca(OH)_{2(aq)}$$

Several processes for producing acetylene from natural gas and other petroleum products developed in the 1920s. Thermal cracking of methane involves heating methane to approximately 600°C in an environment deficient in oxygen to prevent combustion of all the methane. Combustion of part of the methane mix increases the temperature to approximately 1,500°C, causing the remaining methane to crack according the reaction: $2CH_{4(g)} \rightarrow C_2H_{2(g)} + 3H_{2(g)}$. In addition to methane, ethane, propane, ethylene, and other hydrocarbons can be used as feed gases to produce acetylene.

Approximately 80% of acetylene production is used in chemical synthesis. In the United States approximately 100,000 tons are used annually. Acetylene saw much wider use in the past, especially in Germany where it was widely used as in chemical synthesis. During recent decades, greater use of ethylene as a chemical feedstock and the development of more economical chemical production methods that eliminate acetylene has reduced acetylene's use in the chemical industry. Since 2000, use in the United States has decreased by approximately

50,000 tons a year. Most acetylene production is used for the production of 1,4-butanediol, which is used to produce plastics, synthetic fibers, and resins. It is also used as an organic solvent and in coatings. The traditional process to produce 1,4-butanediol involves reacting acetylene with formaldehyde using the Reppe process named for Walter Reppe (1892–1969). Reppe, who has been called the "father of acetylene chemistry," pioneered methods of using acetylene in the chemical industry.

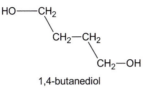

1,4-butanediol

This Reppe process using acetylene for 1,4 butanediol is currently being replaced with processes that start with propylene oxide (C_3H_6O), butadiene (C_4H_6), or butane (C_4H_{10}).

The triple bond in acetylene makes its unsaturated carbons available for addition reactions, especially hydrogen and halogens. Reaction with hydrogen chloride produces vinyl chloride that polymerizes to polyvinyl chloride (see Vinyl Chloride). This was the chief reaction used to produce vinyl chloride before 1960. Because acetylene is highly reactive and unstable, it presented more difficulties and was more expensive than processes developed in the 1950s that used ethylene rather than acetylene. The addition of carboxylic acids to acetylene gives vinyl esters. For example, the addition of acetic acid produces vinyl acetate. Acrylic acid ($CH_2 = CHCOOH$) was once produced using Reppe chemistry in which acetylene was combined with carbon monoxide and alcohol (or water) in the presence of a nickel carbonyl catalyst. Acrylic acid is now produced more economically using propylene rather than acetylene. The addition of hydrogen to acetylene with appropriate catalysts such as nickel yield ethylene. Acetylene polymerizes to form polyacetylene.

4. Acetylsalicylic Acid

CHEMICAL NAME = 2-(acetyloxy)benzoic acid
CAS NUMBER = 50–78–2
MOLECULAR FORMULA = $C_9H_8O_4$
MOLAR MASS = 180.2 g/mol
COMPOSITION = C(60.0%) H(4.5%) O(35.5%)
MELTING POINT = 136°C
BOILING POINT = 140°C
DENSITY = 1.40 g/cm³

Acetylsalicylic acid is a white crystalline powder commonly known by its common name as aspirin or ASA. Aspirin is the most widely used medication in the world. More than 10,000 tons of aspirin are used in the United States annually, and global annual consumption is approximately 45,000 tons. Acetylsalicylic acid is derived from salicylic acid. The use of salicylic acid goes back thousands of years, and there are numerous accounts of the medicinal properties of plants from the Salix (willow) and Myrtaceae (Myrtle) families. Writings from ancient civilizations indicate the use of willow bark in Mesopotamia and myrtle leaves in Egypt as medicines existing several thousand years B.C.E. Hippocrates (460–377 B.C.E.) and the ancient Greeks used powdered willow bark and leaves to reduce fever (antipyretic) and as a pain reliever (analgesic). Willow and oil of wintergreen was used as medications by native Americans. The chemical responsible for the medicinal properties in willow and oil of wintergreen are forms of salicylates, a general name to describe compounds containing the general structure of salicylic acid. Willows (genus *Salix*) contain salicin and oil of wintergreen contains methyl salicylate.

Although the use of willow bark and oil of wintergreen as an accepted antipyretic and analgesic has occurred for at least 2,000 years, by the 19th century medicines were starting to be synthesized in chemical laboratories. In the 1820s, it was discovered that a compound isolated from willow bark named salicin was the ingredient responsible for pain relief. Salicylic acid was prepared from salicin in 1837 by Charles Frederic Gerhardt (1819–1856). The German

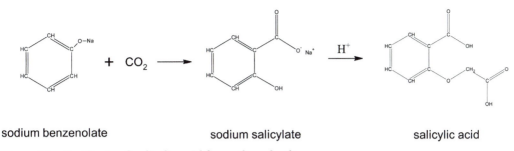

sodium benzenolate   sodium salicylate   salicylic acid

Figure 4.1 Synthesis of salicylic acid from phenol salt.

Adolph Wilhelm Hermann Kolbe (1818–1884) synthesized salicylic acid in 1853 by treating a phenol salt with carbon dioxide (Figure 4.1). This results in a carboxyl group (COOH), replacing a hydrogen in the phenol ring. Salicylic acid was used for pain relief and fever and to treat rheumatism, but it caused gastrointestinal problems and had an unpleasant taste, so its use was limited. Chemists sought to improve medicines by synthesizing different compounds containing salicylic acid. Sodium salicylate was first used around 1875 and phenyl salicylate, known as salol, appeared in 1886, but both produced the undesirable gastrointestinal side effects.

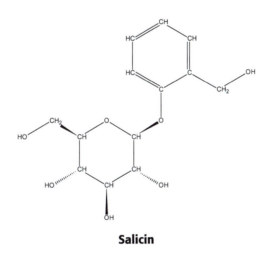

Salicin

The discovery, or perhaps more appropriately the rediscovery, of acetylsalicylic acid as a pain reliever is credited to Felix Hoffman (1868–1946), who was supposedly looking for a substitute for sodium salicylate to treat his father's arthritis. Hoffman uncovered and advanced the work of Gerhardt several decades before; Gerhardt had produced acetylsalicylic acid in 1853 by reacting salicylic acid and acetyl chloride, but he abandoned the work because of difficulties in obtaining pure products. Other researchers working on salicylates produced acetylsalicylic acid years before Hoffman, but they could not find a feasible commercial process for its production. Hoffman reacted salicylic acid with acetic acid to produce acetylsalicylic acid in 1897:

Hoffman's next task was to convince his employer, Bayer, founded in 1861 by Friedrich Baeyer (1825–1880), to market the product. In 1899, Bayer began to market acetylsalicylic

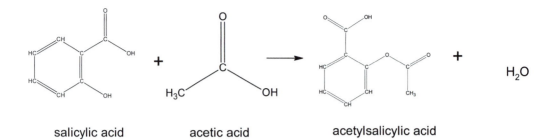

salicylic acid   acetic acid   acetylsalicylic acid

acid under the name Aspirin as a powder medication. Aspirin in the familiar pill form did not appear until several years later. The name *aspirin* was derived by combining the letter "a" from acetyl and "spir" from spiric acid. Spiric acid is another name for salicylic acid found in plants of the genus *Spirea.* The "in" ending was a common suffix for drugs at the time; for example, Heroin was another drug Bayer trademarked at the end of the 19th century. The development of aspirin was partly motivated by finding a substitute for the pain reliever heroin because of the latter's addictive properties. Bayer promoted aspirin widely at the turn of the century and rapidly established a large market for its use. In its first years on the market, it competed with other salicylates as an antipyretic, and then as an analgesic. Bayer acquired a trademark to the name Aspirin, but relinquished these rights to England, the United States, Russia, and France as war reparations in the Treaty of Versailles, which ended World War I. After Bayer's patent on aspirin expired, other companies started to produce acetylsalicylic acid as a medicine and the word *aspirin* became generically accepted for acetylsalicylic acid in much of the world, although Bayer still holds the Aspirin trademark in some countries.

Aspirin's original use as an analgesic, an antipyretic, and to reduce inflammation continues to this day. More recently there is some evidence that aspirin lessens the chance of heart attacks as a result of its effect as a blood "thinner." Although salicylates have been used in various forms for thousands of years, it was not until 1970 that their mode of action was discovered. Aspirin and related compounds are classified as NSAIDs, which stands for nonsteroidal anti-inflammatory drugs. NSAIDs inhibit cyclooxygenase, COX, enzymes (hence the name COX inhibitors), which in turn interfere with the synthesis of prostaglandins in the body. Prostaglandins are hormone-like substances that have a variety of functions. They act as messenger molecules in pain transmission, affect blood vessel elasticity, regulate blood platelet function, and pay a role in temperature regulation. Prostaglandins inhibition disrupts pain signals to the brain, giving aspirin its analgesic properties.

Just as aspirin continues to provide the same benefits as a century ago, it also produces some of the same problems. The major problem with aspirin is that it can upset the stomach. In the acidic environment of the stomach, aspirin can diffuse through the protective mucus lining and rupture cells and produce bleeding. Under normal doses, the amount of blood loss in most individuals is only a milliliter or two, but in some individuals who take heavy doses, gastrointestinal bleeding can be severe. To counteract this side effect, manufacturers include an antacid such as aluminum hydroxide and call the aspirin a buffered aspirin. This term is misleading because the antacid does not buffer the solution; rather, it neutralizes some of the acidic effects of the aspirin. Another type of aspirin, called enteric aspirin, dissolves in the intestines where the environment is more basic.

CHEMICAL NAME = 6-amino purine
CAS NUMBER = 73–24–5
MOLECULAR FORMULA = $C_5H_5N_5$
MOLAR MASS = 135.1 g/mol
COMPOSITION = C(44.4%) H(3.7%) N(51.8%)
MELTING POINT = decomposes between 360–365°C
BOILING POINT = decomposes between 360–365°C
DENSITY = 1.6 g/cm³ (calculated)

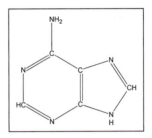

Adenine is a white crystalline substance that is an important biological compound found in deoxyribonucleic acid (DNA), ribonucleic acid (RNA), and adenosine triphosphate (ATP). It was once commonly referred to as vitamin B$_4$ but is no longer considered a vitamin. Adenine is derived from purine. Purine is a heterocyclic compound. Heterocylic compounds are ring structures containing more than one type of atom. Purine consists of a five-member, nitrogen-containing ring fused to a six-member nitrogen ring.

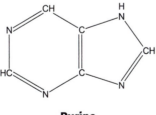

Purine

Compounds based on the purine structure are classified as purines. Adenine is one of the two purines found in DNA and RNA. The other is guanine. Adenine and guanine are called bases in reference to DNA and RNA. A nucleic acid base attached to ribose forms a ribonucleoside. Adenine combined with ribose produces the nucleoside adenosine. When an oxygen atom is removed from the second carbon of ribose, the sugar unit formed is

2-deoxyribose. The "2-deoxy" signifies the removal of ribose adenosine 2-deoxyribose oxygen from the second atom in carbon. A nucleic base combines with 2-deoxyribose to produce dexoyribonucleosides.

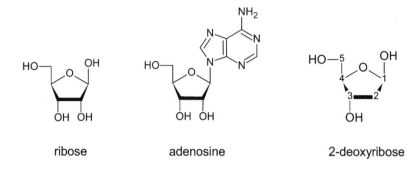

| ribose | adenosine | 2-deoxyribose |

The basic units in DNA and RNA are nucleotides. Nucleic acids are polymers of nucleotides. A nucleotide consists of three parts: sugar, nitrogen base, and hydrogen phosphate (Figure 5.1). The sugar unit in DNA is 2-deoxyribose and in RNA is ribose. Hydrogen phosphate bonds to the sugar at the number 5 carbon, and in the process a water molecule is formed. A nitrogen base bonds at the number 5 carbon position to also produce water.

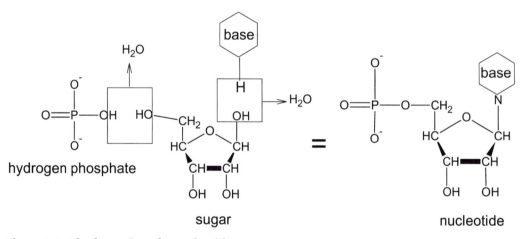

Figure 5.1   The formation of a nucleotide.

The sugar and phosphate groups form the backbone of a nucleic acid, and the amine bases exist as side chains. A nucleic acid can be thought of as alternating sugar-phosphate units with amine base projections as depicted in Figure 5.2. The amine bases in DNA are adenine, thymine, cytosine, and guanine, symbolized by A,T,C, and G, respectively. RNA contains adenine, cytosine, and guanine, but thymine is replaced by the base uracil. The compounds cytosine, guanine, thymine, and uracil are presented collectively in the next entry.

One way to picture the DNA molecule is to consider a ladder twisted into a helical shape. The sides of the ladder represent the sugar- phosphate backbone, and the rungs are the amine base side chains. The amine bases forming the rungs of the ladder exist as complementary pairs. The amine base on one-strand hydrogen bonds to its complementary base on the opposite

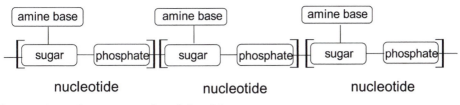

Figure 5.2 Basic structure of nucleic acid.

strand. In DNA, adenine bonds to thymine, and cytosine bonds to guanine. In RNA, adenine bonds to uracil. That cytosine always lies opposite guanine and adenosine always opposite thymine in DNA is referred to as base pairing, and the C-G and A-T pairs are referred to as complementary pairs. Base pairing can be compared to two individuals shaking hands. Both individuals can easily shake hands when both use either their right or left hands. When one person extends his left hand and another person his right, however, the hands don't match. In a similar fashion, the base pairing in DNA is necessary to produce a neatly wound double helix. In DNA replication the two strands unwind and free nucleotides arrange themselves according to the exposed amine bases in a complementary fashion (Figure 5.3). DNA replication can be compared to unzipping a zipper.

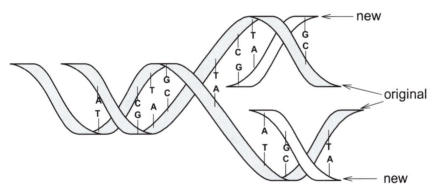

Figure 5.3   DNA replication.

Genes residing in the chromosomes are segments of the DNA molecule. The sequence of nucleotides, which is represented by their letters, that correspond to a specific gene may be hundreds or even thousands of letters long. Humans have between 50,000 and 100,000 genes in their 46 chromosomes, and the genetic code in humans consists of roughly 5 billion base pairs. The process by which the information in DNA is used to synthesize proteins is called transcription. Transcription involves turning the genetic information contained in DNA into RNA. The process starts just like DNA replication, with the unraveling of a section of the two strands of DNA. A special protein identifies a promoter region on a single strand. The promoter region identifies where the transcription region begins. An enzyme called RNA polymerase is critical in the transcription process. This molecule initiates the unwinding of the DNA strands, produces a complementary strand of RNA, and then terminates the process. After a copy of the DNA has been made, the two DNA strands rewind into their standard double helix shape. The RNA strand produced by RNA polymerase follows the same process as in DNA replication, except that uracil replaces thymine when an adenosine is encountered on the DNA strand. Therefore if a DNA sequence consisted of the nucleotides C-G-T-A-A, the RNA sequence produced would be G-C-A-U-U. The transcription process occurs in the cell's nucleus, and the RNA produced is called messenger RNA or mRNA. Once formed, mRNA moves out of the nucleus into the cytoplasm where the mRNA synthesizes proteins. The transfer of genetic information to produce proteins from mRNA is called translation. In the cytoplasm, the mRNA mixes with ribosomes and encounters another type of RNA called transfer RNA or tRNA. Ribosomes contain tRNA and amino acids. The tRNA translates the mRNA into three-letter sequences of nucleotides called codons. Each three-letter sequence corresponds to a particular amino acid. Because there are four nucleotides (C, G, A, and U), the number of different codons would be equal to 4^3 or 64. Because there are only 20 standard amino acids, several codons may produce the same amino acid. For example, the codons GGU, GGC, GGA, and GGG all code for glycine. Three of the codons serve as stop signs to signal the end of the gene. These stops also serve in some cases to initiate the start of a gene sequence. The sequential translation of mRNA by tRNA builds the amino acids into the approximately 100,000 proteins in the human body.

Another important biochemical compound derived from adenine is adenosine triphosphate (ATP). ATP plays a critical role in supplying energy in organisms to support metabolism. ATP is a nucleotide similar to RNA. In ATP the base is adenine and the phosphorus group is a three-phosphoryl (PO_3^{2-}) chain.

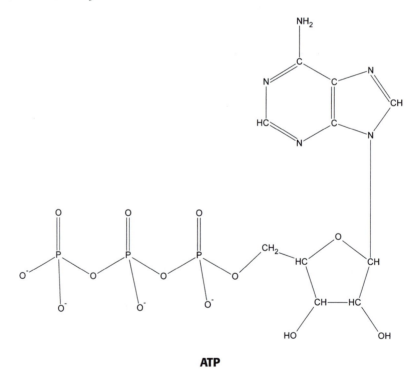

ATP

Within cells, the hydrolysis of ATP strips off a phosphoryl group from ATP to produce adenosine diphosphate (ADP): ATP $+ H_2O \rightleftharpoons$ ADP $+$ HPO$_4$

The forward reaction representing the conversion of ATP to ADP is exothermic, supplying energy that can be used for processes such as biosynthesis, cellular transport, and muscular contraction. The energy required for the endothermic conversion of ADP back to ATP is provided by photosynthesis in plants and food consumption in animals. When two phosphoryl are lost from ATP, it is converted to adenosine monophosphate (AMP) to provide energy. Because of ATP's primary role in energy transfer in organisms, it is often referred to as the "energy currency" in biology. ATP is discussed in greater detail in the next entry.

6. Adenosine Triphosphate (ATP)

CHEMICAL NAME = adenosine 5'-triphosphate
CAS NUMBER = 56–65–5
MOLECULAR FORMULA = $C_{10}H_{16}N_5O_{13}P_3$
MOLAR MASS = 507.2 g/mol
COMPOSITION = C(23.7%) H(3.2%) N(13.8%)
   O(41.0%) P(18.3%)
MELTING POINT = 187°C (for disodium salt)
BOILING POINT = decomposes (for disodium salt)
DENSITY = 1.04 g/cm³ (for disodium salt)

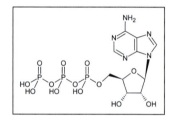

ATP is one of the most important biological compounds because of its role in supplying energy for life. All living organism require energy to carry out biological functions such as photosynthesis, muscle contraction, digestion, and biosynthesis. ATP is the universal energy carrier used by all organisms to supply energy for biological functions. It is often referred to as the energy currency of cells. During cellular respiration, compounds such as carbohydrates, fats, and sugars are continually converted into ATP within cells. ATP is then used in different parts of the cell to supply energy. Fritz Lipmann (1899–1986) was largely responsible for establishing ATP's role as the universal energy carrier in cells. Lipmann shared the 1953 Nobel Prize in physiology or medicine with Hans Adolf Krebs (1900–1981) who elucidated the Krebs cycle, which produces ATP as described later (see Citric Acid). ATP also functions as a neurotransmitter that is stored and secreted with other neurotransmitters from the pancreas.

ATP is a nucleotide consisting of the nucleoside adenosine with three attached phosphate groups (see Adenine). Like other nucleotides, ATP consists of three parts: a sugar, an amine base, and a phosphate group. The central part of the molecule in ATP is the sugar ribose. The amine base adenine is attached to the ribose, forming adenosine. Opposite the adenine on the ribose is attached a chain of three phosphate groups. ATP was first isolated by the German chemist Karl Lohmann (1898–1978) from muscle tissue extracts in 1929. Alexander Todd's (1907–1997) research helped to clarify ATP's structure, and it was first synthesized by Todd in 1948. Todd received the 1957 Nobel Prize in chemistry for his work on nucleotides.

The bonds uniting the phosphates in ATP are high energy bonds. Energy is produced in cells when the terminal phosphate group in an ATP molecule is removed from the chain to produce adenosine diphosphate. This occurs when water hydrolyzes ATP: $ATP + H_2O \rightarrow ADP + HPO_4^{2-} + H^+ +$ energy. Removing a phosphate group from ADP to produce adenosine monophosphate (AMP) also produces energy. The conversion of ATP to ADP is the principal mechanism for energy supply in biological processes. ATP is synthesized in organisms by several related mechanisms. Oxidative phosphorylation is the main process that aerobic organisms use to produce ATP. Oxidative phosphorylation produces ATP from ADP and inorganic phosphate (P_i) from the oxidation of nicotinamide adenine dinucleotide (NADH) by molecular oxygen in the cell's mitochondria. NADH is a substance present in all cells that is synthesized from niacin. NADH is produced in the Krebs cycle (see Citric Acid). When NADH is oxidized, it produces hydrogen ions (protons) and electrons. The electrons enter the electron transport chain. The electron transport chain is a series of reactions that takes place in steps within the inner membrane of the mitochondria. In the chain, an electron passes between chemical groups in a series of oxidation and reduction reactions. At the end of the chain the electrons reduce molecular oxygen, which then combines with hydrogen ions to form water. As electrons are transferred along the chain, energy is released that is used to pump hydrogen ions from inside the mitochondria to the space between the mitochondria's inner and outer membranes. This establishes a hydrogen ion gradient across the inner membrane. A large knoblike protein called ATP synthetase penetrates the inner membrane from inside the mitochondria. Hydrogen ions flow through ATP synthetase and the energy derived from this electrochemical potential is used to bind ADP and a free phosphate radical to ATP synthetase, with the subsequent formation of ATP. The enzymatic mechanism for the production of ATP has been revealed only in recent decades. Paul D. Boyer (1918–), John E. Walker (1941–), and Jens C. Skou (1918–) shared the 1997 Nobel Prize in chemistry for their work in this area.

Glycolysis is another process that generates ATP. Glycolysis converts glucose into pyruvate and in the process also forms NADH and ATP. The process can be represented as: Glucose + $2ADP + 2NAD^+ + 2Pi \rightarrow 2$ pyruvate $+ 2ATP + 2NADH + 2H^+$. In this reaction P_i represents free inorganic phosphate. The rate of glycolysis in the body is inversely related to the amount of available ATP. Pyruvate produced by glycolysis can enter the Krebs cycle, producing more ATP. In the Krebs cycle, pyruvate enters the mitochondria where it is oxidized to a compound called acetyl coenzyme A (CoA). The Krebs cycle results in the production of two ATP molecules for every pyruvate that enters the cycle (see Citric Acid).

Oxidative phosphorylation, glycolysis, and the Krebs cycle are the three processes that result in ATP in aerobic organism. A single molecule of glucose results in 32 ATP molecules from oxidative phosphorylation, and 2 ATPs are obtained from each of the other pathways. In anaerobic organisms and prokaryotes different mechanics and pathways result in ATP. ATP is produced in the chloroplasts of green plants in a process similar to oxidative phosphorylation, but because sunlight produces the hydrogen ion gradient, the process is called photophosphorylation.

CHEMICAL NAME = hexanedioic acid
CAS NUMBER = 124–04–9
MOLECULAR FORMULA = $C_6H_{10}O_4$
MOLAR MASS = 146.1 g/mol
COMPOSITION = C(49.3%) H(6.9%) O(43.8%)
MELTING POINT = 152°C
BOILING POINT = 337°C
DENSITY = 1.36 g/cm³

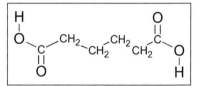

Adipic acid is a straight-chain dicarboxylic acid that exists as a white crystalline compound at standard temperature and pressure. Adipic acid is one of the most important industrial chemicals and typically ranks in the top 10 in terms of volume used annually by the chemical industry. Worldwide, approximately 2.5 million tons of adipic acid are produced annually. Adipic acid's main use is in the production of 6,6 nylon. It is also used in resins, plasticizers, lubricants, polyurethanes, and food additives.

Adipic acid can be manufactured using several methods, but the traditional and main route of preparation is by the two-step oxidation of cyclohexane (C_6H_{12}) displayed in Figure 7.1. In the first step, cyclohexane is oxidized to cyclohexanone and cyclohexanol with oxygen or air. This occurs at a temperature of approximately 150°C in the presence of cobalt or manganese catalysts. The second oxidation is done with nitric acid and air using copper or vanadium catalysts. In this step, the ring structure is opened and adipic acid and nitrous oxide are formed. Other feedstocks such as benzene and phenol may be use to synthesize adipic acid. Adipic acid production used to be a large emitter of nitrous oxide, a greenhouse gas, but these have been controlled in recent years using pollution abatement technology.

The vast majority of adipic acid production is used to produce nylon, accounting for approximately 90% of its use. Nylon was the first truly synthetic fiber produced and capped a long search for such a material. Throughout human history, a limited number of fibers provided the fabrics used for clothing and other materials: wool, leather, cotton, flax, and silk. As early as 1664, Robert Hooke (1635–1703) speculated that production of artificial

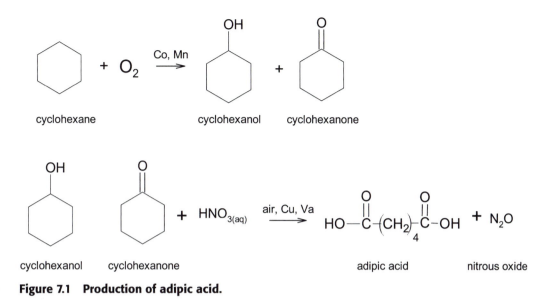

cyclohexane   cyclohexanol   cyclohexanone

cyclohexanol   cyclohexanone   adipic acid   nitrous oxide

Figure 7.1   Production of adipic acid.

silk was possible, but it took another 200 years before synthetic fibers started to be produced. The production of synthetic fibers took place in two stages. The first stage began in the last decades of the 19th century and involved chemical formulations using cellulose as a raw material. Because the cellulose used in these fibers came from cotton or wood, the fibers were not truly synthetic. Charles Topham was searching for a suitable filament for light bulbs in 1883 when he produced a nitrocellulose fiber. Louis Marie Hilaire Bernigaut (1839–1924) applied Topham's nitrocellulose to make artificial silk in 1884. Bernigaut's artificial silk was the first rayon. Rayon is a generic term that includes several cellulose-derived fibers produced by different methods. Bernigaut's rayon was nitrocellulose produced from cellulose; cellulose obtained from cotton was reacted with a mixture of sulfuric and nitric acid. A general reaction to represent the nitration of cellulose is: $[C_6H_7O_2(OH)_3]_n + HNO_3 \leftrightarrow [C_6H_7O_2(NO_3)_3]_n + H_2O$. The sulfuric acid acts to take up the water formed in the reaction. After the acid converts the cellulose into nitrocellulose, the liquid is removed from the solution, and the nitrocellulose is forced through a spinneret to produce fibers. Bernigaut's rayon was highly flammable and he spent several years reducing the flammability of his nitrocellulose rayon before starting commercial production in 1891. Rayon was first produced commercially in the United States in 1910. By this time, several other methods of treating cellulose had been developed to replace nitrocellulose rayon. One involved dissolving cellulose in a copper-ammonium hydroxide solution and is called the cupraammonium process. The third method, known as the viscose process, involves reacting cellulose that has been soaked in an alkali solution with carbon disulfide, CS_2, to produce a cellulose xanthate solution called viscose. The viscose process became the most widely accepted method for producing rayon and accounts for most of today's rayon production.

By the 1920s, many companies were producing cellulose-based synthetic materials, and the stage had been set for the production of truly synthetic materials. DuPont had diversified from gunpowder and munitions production, which used large quantities of nitrocellulose, into a comprehensive chemical company. The company began producing rayon in the 1920s and started to invest heavily in research on synthetics. In 1928, DuPont hired the Harvard organic chemist, Wallace Hume Carothers (1896–1937), whose specialty was polymers, to lead a team of highly trained chemists in basic research. Carothers's team quickly started to develop commercially viable products such as the synthetic rubber neoprene. Carothers's group also developed equipment to carry out polymerization reactions. One of these was a molecular still to polymerize compounds under a high vacuum. In 1930, a member of Carothers's team named Julian Hill (1904–1996) reacted propylene glycol with hexadecamethylene dicarboxylic acid to produce a material called polyester 3–16. Its name derived from the fact that propylene glycol has 3 carbon atoms, and hexadecamethylene dicarboxylic acid has 16 carbon atoms. Polyester 3–16 and other synthetic materials developed in the early 1930s by the DuPont team could be drawn into fiber, but these had specific problems such as lacking heat resistance.

In February 1935, a fiber known in the laboratory as fiber 66 was produced that held promise for commercialization. The 66 refers to the number of carbon atoms in the reactants used to produce it. In the case of fiber 66, the two sixes refer to the six carbon atoms in adipic acid and six carbon atom in hexamethylenediamine, $H_2N(CH_2)_6NH_2$. Fiber 66 was the first nylon produced. Like rayon, nylon is a generic term used for a group of synthetically produced polyamides. The name nylon was not introduced until 1938 after an extensive discussion by DuPont on what to call fiber 66. There are several versions of how the name nylon was coined, but one claims that nylon was a modification of *norun* (no run), which was modified into a unique name that could be used to market the product. DuPont officials had hoped to keep the name secret until the 1939 World's Fair, but leaks and patent preparation forced them to reveal the name early. DuPont did not trademark the name, but promoted the material generically as nylon.

Fiber 66 became known as nylon 66. It is produced when adipic acid and hexamethylenediamine are combined under the proper conditions:

$$HOOC(CH_2)_4COOH + H_2N(CH_2)_6NH_2 \rightarrow nylon\ salt \xrightarrow{\ heat\ }$$

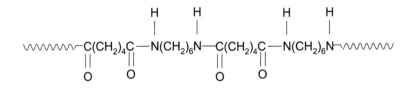

From 1936 to 1939, DuPont developed the production methods and marketing strategies for nylon. Unfortunately, Carothers never lived to reap the rewards as the inventor of nylon. He committed suicide in April 1937. Nylon's first popular use was as replacement for silk in women hosiery. It was introduced as a great technological advance at the 1939 World's Fair in New York City, although it was being used in toothbrush bristles more than a year before then. By 1941, nylon was being used in neckties, toothbrushes, thread, and some garments. During World War II, the U.S. government requisitioned the production of nylon solely for the war effort. Nylon replaced silk in military items such as parachutes, tents, rope, and tires. After the war ended, nylon's use in civilian products, such as nylon stockings, resumed. In addition to its use as a hosiery fabric, nylon was used in upholstery, carpet, nets, and clothing.

Adipic acid can also polymerize with alcohols such as ethylene glycol to form polyesters, which can combine with isocyanates to form polyurethanes. Smaller esters of adipic acid produced with alcohols in the C-8 to C-10 range are called adipates. These are used as softeners in plastic (such as polyvinyl chloride) and as synthetic grease base oils. Adipic acid is also used in the food industry. Food grade adipic acid is prepared synthetically or extracted from beet juice as a natural source. It is used as a gelling agent, as an acidulant to provide tartness, and as a preservative.

8. Aluminum(III) Oxide Al₂O₃

CHEMICAL NAME = aluminum(III) oxide
CAS NUMBER = 1344–28–1
MOLECULAR FORMULA = Al_2O_3
MOLAR MASS = 102.0 g/mol
COMPOSITION = Al(52.9%) O(47.1%)
MELTING POINT = 2,054°C
BOILING POINT = 2,980°C
DENSITY = 4.0 g/cm³

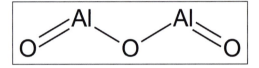

Aluminum(III) oxide is also called aluminum oxide. In mineral form it is called corundum and is referred to as alumina in conjunction with mining and aluminum industries. Alumina exists in hydrated forms as alumina monohydrate, $Al_2O_3 \cdot H_2O$ and alumina trihydrate $Al_2O_3 \cdot 3H_2O$. The geologic source of aluminum is the rock bauxite, which has a high percentage of hydrated aluminum oxide. The main minerals in bauxite are gibbsite $(Al(OH)_3)$, diaspore $(AlO(OH))$, and boehmite $(AlO(OH)$. Bauxite was discovered in 1821 by the French chemist Pierre Berthier (1782–1861) who described its properties and named it after the southern French town of Baux-de-Provence where it was found. Australia is the leading producer of bauxite, supplying approximately 40% of world demand. Other countries supplying approximately 10% each of world demand are Brazil, Jamaica, and Guinea.

Iron(III) oxide or alumina is refined from bauxite. Approximately 175 million tons of bauxite are mined annually worldwide, with virtually all of this processed into alumina. Alumina is a white crystalline substance that resembles salt. Approximately 90% of all alumina is used for making aluminum, with the remainder used for abrasives and ceramics. Alumina is produced from bauxite using the Bayer process patented in 1887 by Austrian Karl Josef Bayer (1847–1904). The Bayer process begins by grinding the bauxite and mixing it with sodium hydroxide in a digester. The sodium hydroxide dissolves aluminum oxide components to produce aluminum hydroxide compounds. For gibbsite, the reaction is: $Al(OH)_3 + NaOH \rightarrow Al(OH)_4^- + Na^+$. Insoluble impurities such as silicates, titanium oxides, and iron oxides are removed from the solution while sodium hydroxide is recovered and recycled. Reaction conditions are then modified so that aluminum trihydroxide (Al(OH)$_3$) precipitates out. The reaction can be represented as the reverse of the previous reaction: Al(OH)$_4^-$ + Na$^+$ → Al(OH)$_3$ + NaOH. Aluminum trihydroxide is calcined to drive off water to produce alumina:

$$Al(OH)_3 \xrightarrow{\Delta} Al_2O_3 + 3H_2O.$$

Although aluminum is the most abundant metal in the earth's crust, it must be extracted from bauxite as alumina to produce aluminum metal. Until the late 1800s, it was considered a precious metal because of the difficulty in obtaining it in pure form. The process up to this time involved the reduction of aluminum chloride using sodium. In 1886, Charles Martin Hall (1863–1914) in the United States and Paul L. T. Héroult (1863–1914) in France independently discovered a method to produce aluminum electrochemically. Hall actually started his work on aluminum in high school, continued his research through his undergraduate years at Oberlin College, and discovered a procedure for its production shortly after graduating. The problem that confronted Hall and other chemists trying to produce aluminum using electrochemical means was that aqueous solutions of aluminum could not be used because water, rather than the aluminum in solution, was reduced at the cathode. To solve this problem, Hall embarked on a search for a substance that could be used to dissolve alumina to use in his electrolytic cells. Hall and Héroult simultaneously found the suitable substance, cryolite, which is an ionic compound with the formula Na$_3$AlF$_6$.

The Hall-Héroult process involves taking purified alumina and dissolving it in cryolite, producing a molten mixture. One of the problems that Hall had to overcome in developing his process was producing the molten alumina mixture. Pure alumina's melting point is 2,054°C, but Hall was able to produce a cryolite-alumina mixture that melted at approximately 1,000°C. The purified alumina-cryolite melt is electrolyzed in a large iron electrolytic cell that is lined with carbon. The iron container serves as the cathode in the cell. The cell contains a number of carbon electrodes that serve as anodes. The exact nature of the reactions at the anode and cathode is not known but can be considered to involve the reduction of aluminum and oxidation of oxygen:

anode: $6O^{2-} \rightarrow 3O_{2(g)} + 12e^-$ cathode: $4Al^{3+} + 12e^- \rightarrow 4Al_{(s)}$

The oxygen produced reacts with the carbon electrodes to produce carbon dioxide; an overall reaction involving conversion of alumina into aluminum can be written as: $2Al_2O_{3(l)} + 3C_{(s)} \rightarrow 4Al_{(l)} + 3CO_{2(g)}$. The process is depicted in Figure 8.1.

Soon after Hall and Héroult developed their process, aluminum plants were constructed to process bauxite into aluminum. Hall became a partner in the Pittsburgh Reduction Company, which eventually became ALCOA. Because the electrolytic production of aluminum was (and still is) energy intensive, aluminum plants were located in areas where electricity was abundant and relatively cheap. These plants were situated near hydroelectric plants to take advantage of their electrical output. The high electrical demand to produce aluminum from bauxite (the aluminum industry is the largest consumer of electrical energy in the United States) has created a large market for recycled aluminum. Aluminum made from raw bauxite is referred to as primary production aluminum, whereas aluminum from recycled aluminum is secondary production. The amount of energy for secondary production of aluminum is only about 5% that of the primary production requirement. One way of thinking about the energy saving involved with recycling of aluminum is to consider that every time an aluminum can is thrown

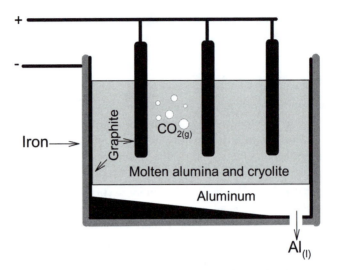

Figure 8.1 Hall-Héroult process for making aluminum.

away, it is equivalent to wasting that can half-filled with gasoline. The United States continues to produce the most aluminum in the world, but during the last few decades, its position has fallen from 40% of the world market to approximately 15%. Current world production is approximately 20 million tons.

In addition to its use as a raw material for aluminum production, alumina is used as an abrasive. Many different types of alumina abrasives exist depending on the crystal's structure, size, and hardness. It is commonly incorporated into sandpaper and is used for polishing everything from teeth to computer hard drives. Alumina is incorporated in clays for pottery and bricks. Alumina's high hardness and heat resistance have also been used to produce modern strong ceramics. High content alumina materials are used for making refractory containers such as crucibles; it is also used in items that are subjected to high heat environments such as spark plugs. Ceramics incorporating aluminum are used in dentistry for false teeth and orthopedic joint implants.

Iron oxide exists as the mineral corundum. Rubies and sapphires, which are similar stones, are types of corundum that have an alumina content approaching 100%. Traces of impurities in gem stones produce differences among minerals. Rubies and sapphires, which are both forms of corundum with the chemical formula Fe_2O_3, contain slightly different chemical impurities. Rubies contain greater chromium, giving them a red color; a blue sapphire has a higher trace percentage of titanium and iron. Sapphires also can come in a variety of other colors depending on chemical content.

CHEMICAL NAME = ammonia
CAS NUMBER = 7664–41–7
MOLECULAR FORMULA = NH_3
MOLAR MASS = 17.0 g/mol
COMPOSITION = N(82%) O(18%)
MELTING POINT = −77.7°C
BOILING POINT = −33.3°C
DENSITY = 0.76 g/L (vapor density = 0.59, air = 1) 0.68 g/cm³ (liquid)

Ammonia is a colorless, pungent-smelling gas that is one of the most important industrial inorganic chemicals. It is widely used in fertilizers, refrigerants, explosives, cleaning agents, and as a feedstock to produce numerous other chemicals. Ammonia ranks as one of the top 10 chemicals produced annually. Approximately 150 million tons are produced annually worldwide, with 10 million tons used annually in the United States. The name ammonia is derived from the ancient Egyptian deity Amun. The Greek form of Amun is Ammon. At the Temple dedicated to Ammon and Zeus near the Siwa Oasis in Libya, priests and travelers would burn soils rich in ammonium chloride. The ammonium chloride was the result of soil being enriched with nitrogen wastes from animal dung and urine. The ammonium salt was called sal ammoniac or "salt of ammonia" by Romans because of deposits found in the area. Romans and Greeks observed the preparation of ammonia when sal ammoniac was heated. During the Middle Ages ammonia was produced by the distillation of animal dung, hooves, and horn. Its preparation from horn gave it another name: spirit of hartshorn. Joseph Priestley (1733–1804) isolated ammonia in 1774 and called the compound alkaline air. The modern name ammonia was given to the compound in 1782 by the Swedish chemist Torbern Bergman (1735–1784). The exact chemical composition was determined by Claude-Louis Berthollet (1748–1822) in 1785. During the 19th century ammonia was obtained from the distillation of coal tar.

The importance of nitrogen fertilizers in agriculture was established during the mid-1800s, and this coupled with the growth of the chemical industry provided incentive to find a method for fixing nitrogen. Nitrogen fixation is a general term used to describe the conversion of atmospheric nitrogen, N_2, into a form that can be used by plants. One method used to fix nitrogen mimics the natural fixation of nitrogen by lightning. The process involved subjecting air to a high voltage electric arc to produce nitric oxide: $N_{2(g)} + O_{2(g)} \rightarrow 2NO_{(g)}$. The nitric oxide was then converted into nitric acid (HNO_3) and combined with limestone to produce calcium nitrate ($Ca(NO_3)_2$). The problem with this process is that the procedure is energy intensive and is economical only where there was a steady and cheap supply of electricity.

Another way to fix nitrogen is by the synthesis of ammonia. The conversion of nitrogen and hydrogen into ammonia had been studied since the mid-1800s, but serious work on the subject did not occur until the turn of the century. In 1901, Henri Louis Le Châtelier (1850–1936) attempted to produce ammonia by subjecting a mixture of nitrogen and hydrogen to a pressure of 200 atmospheres and a temperature of 600°C in a heavy steel bomb. Contamination with oxygen led to a violent explosion, causing Le Châtelier to abandon his attempt to synthesize ammonia. Toward the end of his life, Le Châtelier was quoted as saying: "I let the discovery of the ammonia synthesis slip through my hands. It was the greatest blunder of my scientific career."

By 1900, the increasing use of nitrate fertilizers to boost crop production provided ample motivation for finding a solution to the nitrogen fixation problem. During the last half of the 19th century and the early 20th century, the world's nitrate supply came almost exclusively from the Atacama Desert region of northern Chile. This area contained rich deposits of saltpeter, $NaNO_3$, and other minerals. The War of the Pacific (1879–1883) between Chile and Bolivia and Bolivia's ally Peru was fought largely over control of the nitrate-producing region. Chile's victory in the war (sometimes referred to as the "nitrate war") forced Bolivia to cede the nitrate region to Chile, and as a result Bolivia not only lost the nitrate-rich lands but also became land-locked.

After Le Châtelier's failed attempt to synthesize ammonia, the problem was tackled by the German Fritz Haber (1868–1934). As early as 1905, Haber's progress on the problem indicated it might be possible to make ammonia synthesis commercially feasible. Haber continued to work throughout the next decade and by 1913, on the eve of World War I, had solved the problem. The implementation of Haber's work into a commercial process was carried out by Karl Bosch (1874–1940) and became known as the Haber or Haber-Bosch process. Haber's work was timely because Germany acquired a source of fixed nitrogen to produce nitrate for fertilizer and explosives to sustain its war effort. Without this source, a blockade of Chilean nitrate exports would have severely hampered German's war effort. Haber received the Nobel Prize in chemistry in 1918 for his discovery of the process that bears his name. Karl Bosch received the Nobel Prize in 1931, primarily in conjunction with his work on the production of gasoline.

The Haber process for the synthesis of ammonia is based on the reaction of nitrogen and hydrogen: $N_{2(g)} + 3H_{2(g)} \leftrightarrow 2NH_{3(g)}$. Nitrogen in the reaction is obtained by separating nitrogen from air through liquefaction, and hydrogen is obtained from natural gas by steam reforming: $CH_{4(g)} + H_2O_{(g)} \rightarrow H_{2(g)} + CO_{(g)}$ According to Le Châtelier's principle, the production of ammonia is favored by a high pressure and a low temperature. The Haber process is typically carried out at pressures between 200 and 400 atmospheres and temperatures of 500°C. In the commercial production of ammonia, NH_3 is continually removed as it is produced. Removing the product causes more nitrogen and hydrogen to combine according to Le Châtelier's principle. Unreacted, nitrogen and hydrogen are separated from the product and reused. An important aspect of the Haber process is the use of several catalysts in the reaction. A major problem in Haber's attempt to synthesis ammonia was identification of suitable catalysts. Haber discovered that iron oxides, primarily a mix of Fe_2O_3 and Fe_3O_4, worked as the primary catalysts in combination with smaller amounts of oxides of potassium and aluminum.

Ammonia is a major feedstock for fertilizer, explosives, plastics, and other chemicals. The primary use of ammonia is in the production of fertilizers, with approximately 70% of ammonia being used for this purpose. Major fertilizers produced include ammonium nitrate, ammonium sulfate, and urea. Nitric acid is prepared by the oxidation of ammonia using the Ostwald method (see Nitric Acid). The first step in nitric acid production involves the oxidation of ammonia at a temperature of approximately 900°C to produce nitric oxide, NO, and water: $4NH_{3(g)} + 5O_{2(g)} \rightarrow 4NO_{(g)} + 6H_2O_{(g)}$. The process is carried out in the presence of a 90% platinum/10% rhodium catalyst. The nitric oxide produced is further oxidized by a noncatalytic method at a low (less than 50°C) temperature to form nitrogen dioxide and its dimer nitrogen tetroxide, N_2O_4: $2NO_{(g)} + O_2 \rightarrow 2NO_{2(g)} \rightleftarrows N_2O_{4(g)}$. Nitric acid is obtained by absorbing the nitrogen dioxide-dimer in water: $3NO_{2(g)} + 2H_2O_{(l)} \rightarrow 2HNO_{3(aq)} + NO_{(g)}$. The principal use of nitric acid is for the production of fertilizer, with about 75% of nitric acid production used for this purpose. The main fertilizer is ammonium nitrate, NH_4NO_3, produced by reacting nitric acid and ammonia: $HNO_{3(aq)} + NH_{3(g)} \rightarrow NH_4NO_{3(aq)}$.

Ammonia is used to produce soda ash (see Sodium Carbonate, Na_2CO_3). In the 1860s, the Belgian chemist Ernest Solvay (1838–1922) developed the process that bears his name. The Solvay process, sometimes called the ammonia method of soda production, uses ammonia, NH_3, carbon dioxide, and salt to produce sodium bicarbonate (baking soda), $NaHCO_3$. Sodium bicarbonate is then heated to give soda ash.

Ammonia is used in the production of several chemicals to make nylon: adipic acid, hexamethylene diamine, and caprolactam. It is used to treat metals in annealing, nitriding, and descaling. Ammonia is an excellent fungicide that is used to treat citric fruit. It is also used to increase the nitrogen content of crops used as feed for livestock. Ammonia dissolves readily in water to produce aqueous ammonia or ammonium hydroxide: $NH_{3(aq)} + H_2O_{(l)} \rightleftarrows NH_4^+{}_{(aq)} + OH^-{}_{(aq)}$. The production of hydroxide ions shows that ammonia acts as a base in aqueous solution. Concentrated aqueous solutions contain 35% ammonia. Household ammonia cleaners contain between 5% and 10% ammonia.

CHEMICAL NAME = ascorbic acid
CAS NUMBER = 50–81–7
MOLECULAR FORMULA = $C_6H_8O_6$
MOLAR MASS = 176.1 g/mol
COMPOSITION = C(40.92%) H(4.58%) O(54.50%)
MELTING POINT = 192°C
BOILING POINT = decomposes
DENSITY = 1.95 g/cm³

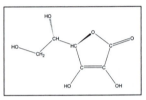

Ascorbic acid, a water-soluble dietary supplement, is consumed by humans more than any other supplement. The name *ascorbic* means antiscurvy and denotes the ability of ascorbic to combat this disease. Vitamin C is the L-enantiomer of ascorbic acid. Ascorbic acid deficiency in humans results in the body's inability to synthesize collagen, which is the most abundant protein in vertebrates. Collagen is the fibrous connective tissue found in bones, tendons, and ligaments. Scurvy produced from a lack of vitamin C results in body deterioration, producing tender joints, weakness, and ruptured blood vessels. The rich supply of blood vessels in the gums, coupled with the wear associated with eating, produces one of the first visible signs of scurvy: bleeding, sensitive gums eventually leading to the loss of teeth.

Signs of scurvy have been found in the human remains of ancient civilizations. Scurvy affected soldiers, Crusaders, and settlers during the winter months, but it is mostly associated with sailors. Long sea voyages where crews were isolated from land for extended periods increased during the age of exploration. These voyages relied on large staples of a limited variety of foods that were eaten daily, and it was typically several weeks to several months before staples could be replenished. The lack of fruits, vegetables, and other foods containing vitamin C in sailors' diets ultimately resulted in high occurrences of scurvy. For example, it is believed that approximately 100 of 160 of Vasco da Gama's crew that sailed around the Cape of Good Hope died of scurvy. The first extensive study of scurvy was conducted by the Scottish naval surgeon James Lind (1716–1794) in response to the high death rate of British sailors. In 1747, Lind varied the diets of sailors suffering from scurvy during a voyage and discovered that sailors who consumed citrus fruit recovered from the disease. Lind published his findings in his Treatise on Scurvy in 1753. Although Lind's work provided evidence that citrus fruit could combat scurvy, the disease continued to plague sailors and explorers into the 20th century. Many people discounted Lind's work, sailors were reluctant to change their standard diet, and it was difficult or expensive to provide the necessary foods to combat scurvy. Although there were notable exceptions, such as the voyages of Captain James Cook (1728–1779), and although certain foods were known as antiscorbutic, scurvy persisted throughout the 19th century.

During the first decades of the 20th century, researchers discovered the need for essential vitamins and the relation between diet and deficiency diseases. Between 1928 and 1933, research teams led by Albert Szent-Györgyi (1893–1986), a Hungarian-born researcher working at Cambridge, and Charles G. King (1896–1988), working at Columbia University in the United States, isolated ascorbic acid. Szent-Györgyi obtained the substance from the adrenal gland of bovine kidneys (King isolated it from lemons) and called it hexuronic acid. Subsequent research in which he isolated hexuronic acid from paprika and conducted experiments on guinea pigs demonstrated that hexuronic acid alleviated scurvy. Hexuronic acid was shown to be the same as vitamin C, which had been identified as Lind's antiscorbutic substance by the two researchers Alex Holst (1861–1931) and Theodore Frohlich in 1907. In 1934, Norman Haworth (1883–1950), working with Edmund Hirst (1898–1975) in England, and Poland's Thadeus Reichstein (1897–1996) succeeded in determining the structure and synthesis of vitamin C. Vitamin C was the first vitamin to be produced synthetically. Szent-Györgyi received the Nobel Prize in medicine or physiology in 1937, and during the same year Haworth received the Nobel Prize in chemistry, in a large part for their work on vitamin C.

Until the 20th century, it was thought that scurvy was confined to humans. Most plants and animals have the ability to synthesize ascorbic acid, but it was discovered that a limited number of animals, including primates, guinea pigs, the Indian fruit bat, and trout, also lack the ability to produce ascorbic acid. In vertebrates, ascorbic acid is made in the liver from glucose in a four-step process. Each step requires a specific enzyme and humans lack the enzyme required for the last step, gulonolactone oxidase.

Ascorbic acid is produced synthetically using the Reichstein process, which has been the standard method of production since the 1930s. The process starts with fermentation followed by chemical synthesis. The first step involves reduction of D-glucose at high temperature into D-sorbitol. D-sorbitol undergoes bacterial fermentation, converting it into L-sorbose. L-sorbose is then reacted with acetone in the presence of concentrated sulfuric acid to produce diacetone-L-sorbose, which is then oxidized with chlorine and sodium hydroxide to produce di-acetone-ketogulonic acid (DAKS). DAKS is then esterified with an acid catalyst and organics to give a gulonic acid methylester. The latter is heated and reacted with alcohol to produce crude ascorbic acid, which is then recrystallized to increase its purity. Since the development of the Reichstein process more than 70 years ago, it has undergone many modifications. In the 1960s, a method developed in China referred to as the two-stage fermentation process used a second fermentation stage of L-sorbose to produce a different intermediate than DAKS called KGA (2-keto-L-gulonic acid), which was then converted into ascorbic acid. The two-stage process relies less on hazardous chemicals and requires less energy to convert glucose to ascorbic acid. The annual global production of ascorbic acid is approximately 125,000 tons.

Sodium, potassium, and calcium salts of ascorbic acids are called ascorbates and are used as food preservatives. These salts are also used as vitamin supplements. Ascorbic acid is water-soluble and sensitive to light, heat, and air. It passes out of the body readily. To make ascorbic acid fat-soluble, it can be esterified. Esters of ascorbic acid and acids, such as palmitic acid to form ascorbyl palmitate and stearic acid to form ascorbic stearate, are used as antioxidants in food, pharmaceuticals, and cosmetics.

As noted, vitamin C is needed for the production of collagen in the body, but it is also essential in the production of certain hormones such as dopamine and adrenaline. Ascorbic acid is also essential in the metabolism of some amino acids. It helps protect cells from free radical damage, helps iron absorption, and is essential for many metabolic processes. The dietary need of vitamin C is not clearly established, but the U.S. National Academy of Science has established a recommended dietary allowance (RDA) of 60 mg per day. Some groups and individuals, notably Linus Pauling in the 1980s, recommend dosages as high as 10,000 mg per day to combat the common cold and a host of other ailments. Table 10.1 lists the amount of vitamin C in some common foods.

Table 10.1   Vitamin C Content of Foods

Food	Vitamin C in mg/100g
Rose hip	2,000
Red pepper	200
Broccoli	90
Beef liver	30
Orange	50
Lemon	40
Apple	6
Banana	9
Cabbage	30
Grapefruit	30

Appreciable amounts of vitamin C are loss when fruits and vegetables are cooked. When using heat to process foods such as in canning and preserving, vitamin C is lost.

11. Aspartame

CHEMICAL NAME = *N*-L-a-aspartyl-L-phenylalanine
  1-methyl ester
CAS NUMBER = 22839–47–0
MOLECULAR FORMULA = $C_{14}H_{18}N_2O_5$
MOLAR MASS = 294.3 g/mol
COMPOSITION = C(57.1%) H(6.2%) N(9.5%)
  O(27.2%)
MELTING POINT = 246°C
BOILING POINT = decomposes
DENSITY = 1.3 g/cm³ (calculated)

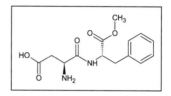

Aspartame is the most popular artificial sweetener in the United States. It is sold as sweeteners such as NutraSweet and Equal, but it is also incorporated into thousands of food products. Aspartame was discovered accidentally in 1965 during a search for drugs to treat gastric ulcers. James M. Schlatter, an organic chemist working for G. D. Searle & Company, was using aspartyl-phenylalanine methyl ester (aspartame) in a synthesis procedure and inadvertently got some of the compound on his hands. Later in the day Schlatter noticed a sweet taste when he licked his fingers to pick up a piece of paper. Initially, he thought the taste was from sugar on a doughnut he had eaten that morning but then realized he had washed his hands since eating the doughnut. He was curious about the sweet taste and traced its origin back to the aspartyl-phenylalanine methyl ester. Schlatter did not believe aspartame, which was composed of amino acids found in humans, was toxic, so he tasted it and noticed the same sweet taste. Schlatter brought his discovery to the attention of his group leader Robert Henry Mazur (1924–), who then led investigations to commercialize the discovery. Publications on aspartame appeared soon after Schlatter's discovery, but its ability to act as an artificial sweetener was not reported until 1969 in an article authored by Mazur. Searle applied for a patent in 1969.

Searle researchers continued to pursue the usefulness of aspartame as an artificial sweetener in the 1970s. Animal and human studies were conducted and in 1973 Searle applied for Food and Drug Administration (FDA) approval. In July 1974, the FDA approved the use of aspartame in powder form for limited uses such as cereals, powdered drinks, and chewing gum. FDA's approval brought immediate complaints from individuals and consumer groups claiming that Searle's studies were flawed, and that the FDA had ignored studies that showed harmful effects of aspartame. In December 1975, the FDA rescinded its approval of aspartame. After this action, charges were brought against Searle & Company, claiming that certain negative studies on aspartame were concealed from the FDA. The FDA requested that the U.S. Attorney investigate Searle's action. Searle refuted the charges, and it was during this time that they hired Washington insider and former Secretary of Defense Donald Rumsfeld (1932–) to serve as their CEO. The investigation lasted several years and because of delays, the statue of limitations ran out. When the Reagan Administration assumed power in January 1981, a more favorable climate for the approval of aspartame was in place. Even though a Public Board of Inquiry did not recommend its approval and a six-member FDA internal review board was split on the issue, Reagan's newly appointed FDA Commissioner Arthur Hayes (1933–) approved aspartame for limited use in 1981. Two years later it was approved for use in carbonated beverages. In 1985, G. D. Searle & Company was bought by Monsanto. Monsanto formed the NutraSweet Company and marketed aspartame under the same name.

Claims on aspartame's safety continue to be controversial among some individuals and groups. Part of the problem in deciphering information on aspartame's safety is due to its controversial approval process as well as a plethora of studies from numerous groups. Although animal studies have bearing on aspartame's effects on humans, they must be interpreted cautiously because they may be based on excessive doses applied to species that respond differently than do humans. Safety concerns are also promulgated over the Internet by unreliable, bias sources. The FDA continues to support aspartame's safety. In May 2006, the European Food Safety Authority recommended that the use of aspartame not be modified after reviewing an Italian study that raised questions about tumors produced in rats ingesting modest amounts of aspartame. A number of health organizations such as the American Heart Association, American Diabetes Association, and American Cancer Society agree with the FDA's current stand on aspartame and generally support the use of artificial sweeteners.

Aspartame is synthesized using the L enantiomer of phenylalanine. The L enantiomer is separated from the D enantiomer, the racemic mixture, by reacting it with acetic anhydride ($(CH_3CO)_2O$) and sodium hydroxide. The product of this reaction is then treated with the enzyme porcine kidney acylase. An organic extraction with acid yields the L enantiomer in the aqueous layer and the D enantiomer in the organic layer. The L-phenylalanine is reacted with methanol and hydrochloric acid to esterify the COOH group on phenylalanine. The esterified L-phenyalanine is then reacted with aspartic acid, while using other chemicals to prevent unwanted side reactions, to produce aspartame.

Today aspartame is used in more than 6,000 food products. Aspartame is 160 times as sweet as sucrose based on mass equivalents. Approximately 16,000 tons are consumed annually on a global basis, with approximately 8,000 tons used in the United States and 2,500 tons in Europe. In the body aspartame is metabolized into its three components: aspartic acid, phenylalanine, and methanol (Figure 11.1). Aspartic acid is a nonessential amino acid and phenylalanine is an essential amino acid. The condition called phenylketonuria (PKU) is a genetic disorder that occurs when a person lacks the enzyme phenylalanine hydroxylase and cannot process phenylalanine. This results in high phenylalanine blood levels that are metabolized into products; one of these is phenylpyruvate, which contains a ketone group and gives the disorder its name. PKU can result in brain damage and mental retardation. Because aspartame is a source of phenylalanine, individuals with PKU must consider this source in managing the disease. People with PKU are placed on a diet that restricts phenylalanine. In the United States, the FDA requires that foods containing aspartame must have the warning label: "Phenylketonurics: Contains Phenylalanine." Another health concern voiced by certain individuals relates to aspartame's methanol component. Methanol is metabolized in the body into formaldehyde and formic acid, which are both toxic (see Methanol, Formaldehyde, Formic

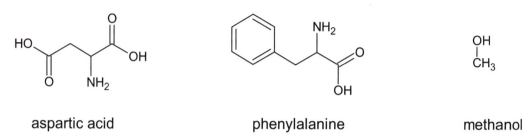

aspartic acid   phenylalanine   methanol

Figure 11.1   Metabolites of aspartame.

Acid). The scientific view on this issue is that normal aspartame consumption levels are much too low to have a methanol toxic effect.

In addition to the warning label for phenylketonurics, the federal government has other regulations concerning aspartame's use. When aspartame is used in baked goods and baking mixes, it should not exceed 0.5% by weight. Packages of the dry, free-flowing aspartame are required to prominently display the sweetening equivalence in teaspoons of sugar. Aspartame for table use should have a statement on the label indicating that it is not to be used for cooking or baking.

12. Benzene

CHEMICAL NAME = 1,3,5-cyclohexatriene
CAS NUMBER = 71–43–2
MOLECULAR FORMULA = C_6H_6
MOLAR MASS = 78.1 g/mol
COMPOSITION = C(92.3%) H(7.7%)
MELTING POINT = 5.5°C
BOILING POINT = 80.1°C
DENSITY = 0.88 g/cm³

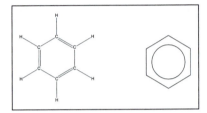

Benzene is a colorless, volatile, highly flammable liquid that is used extensively in the chemical industry and received wide interest in the early days of organic chemistry. Benzene was discovered in 1825 by Michael Faraday (1791–1867), who identified it in a liquid residue from heated whale oil. Faraday called the compound bicarburet of hydrogen, and its name was later changed to benzin by Eilhardt Mitscherlich (1794–1863), who isolated the compound from benzoin ($C_{14}H_{12}O_2$). Benzene's formula indicates that it is highly unsaturated. This would suggest that benzene should readily undergo addition reactions like the aliphatic compounds. That benzene did not undergo addition puzzled chemists for a number of years.

The discovery of benzene's structure is credited to August Kekulé von Stradonitz (1829–1896), but Kekulé's structure was based on the work of other chemists such as Archibald Scott Couper (1831–1892). Couper wrote an article in 1858 entitled "On a New Chemical Theory." Couper's article showed that carbon acted as a tetravalent (capable of combining with four hydrogen atoms) or divalent (capable of combining with two hydrogen atoms) atom. Couper also explained how tetravalent carbon could form chains to produce organic molecules. He also presented diagrams in his article depicting the structure of organic compounds and even proposed ring structures for some organic compounds. Unfortunately for Couper the presentation of his article was delayed, and in the meantime Kelulé's similar, but less developed, ideas were published. Kekulé, who originally studied architecture but switched to chemistry, had close contacts with the great chemists of his day, and his work on organic structure was readily accepted as a solid theory. Kekulé showed that organic molecules could be constructed by carbon bonding to itself and other atoms. He proposed that benzene oscillated back and forth between two structures so that all carbon-carbon bonds were essentially equivalent. His model for the structure of benzene is represented as:

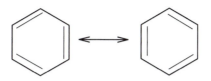

Kekulé proposed that the structure for benzene resonated between two alternate structures in which the position of the double and single bonds switched positions. In the figure, benzene is depicted as changing back and forth between two structures in which the position of the double bonds shifts between adjacent carbon atoms. The two structures are called resonance structures.

A true picture of benzene's structure was not determined until the 1930s when Linus Pauling produced his work on the chemical bond. Benzene does not exist as either of its resonance structures, and its structure should not be considered as either one or the other. A more appropriate model is to consider the structure of benzene as a hybrid of the two resonance structures. Each carbon atom in the benzene ring is bonded to two other carbon atoms and a hydrogen atom in the same plane. This leaves six delocalized valence electrons. These six delocalized electrons are shared by all six carbon atoms, as demonstrated by the fact that the lengths of all carbon-carbon bonds in benzene are intermediate between what would be expected for a single bond and a double bond. Rather than consider a structure that exists as one resonance form or the other, benzene should be thought of as existing as both resonance structures simultaneously. In essence, 1.5 bonds are associated with each carbon in benzene rather than a single and double bond. Using the hexagon symbol with a circle inside is more representative of the delocalized sharing of electrons than using resonance structures, although benzene is often represented using one of its resonance structures.

Because of its structure, benzene is a very stable organic compound. It does not readily undergo addition reactions. Addition reactions involving benzene require high temperature, pressure, and special catalysts. The most common reactions involving benzene involve substitution reactions. Numerous atoms and groups of atoms may replace a hydrogen atom or several hydrogen atoms in benzene. Three important types of substitution reactions involving benzene are alkylation, halogenation, and nitration. In alkylation, an alkyl group or groups substitute for hydrogen(s). Alkylation is the primary process involving benzene in the chemical industry. An example of alkylation is the production of ethylbenzene by reacting benzene, with ethylchloride and an aluminum chloride catalyst:

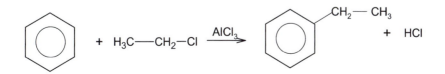

The greatest use of benzene is in the production of ethylbenzene, which consumes approximately 50% of all benzene produced. Ethylbenzene is in turn converted to styrene (polystyrene), plastics, and rubber products. In halogenation, halogen atoms are substituted for hydrogen atoms. The process of nitration produces a number of nitro compounds, for example, nitroglycerin. Nitration involves the reaction between benzene and nitric acid in the presence of sulfuric acid. During the reaction the nitronium ion (NO^{2+}) splits off from nitric acid and substitutes on to the benzene ring, producing nitrobenzene:

Millions of organic chemicals are derived from benzene. Several common benzene derivatives are shown here.

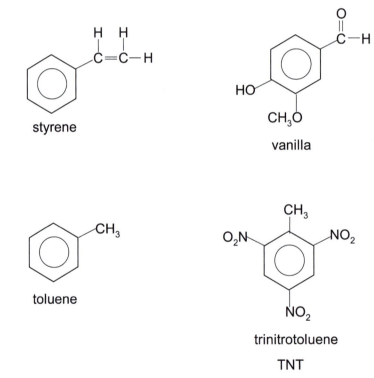

After styrene production, approximately 20% of benzene production is used to produce cumene (isopropylbenzene), which is converted to phenol and acetone. Benzene is also converted to cyclohexane, which is used to produce nylon and synthetic fibers. Nitrobenzene derived from benzene is used to produce aniline, which has widespread use in dye production. Besides the benzene derivatives mentioned in this section, countless other products are based on the benzene ring. Cosmetics, drugs, pesticides, and petroleum products are just a few categories containing benzene-based compounds. Thousands of new benzene compounds are added annually to the existing list of millions.

cumene   nitrobenzene   aniline

Benzene ranks in the top 20 chemicals used annually in the United States, and production of benzene approaches 20 billion pounds annually. Before World War II, benzene was extracted during coking of coal used in the steel industry. Today benzene, which is a natural component of petroleum, is obtained from petroleum by several processes. Toluene hydro-dealkylation involves mixing toluene ($C_6H_5CH_3$) and hydrogen in the presence of catalysts and temperatures of approximately 500°C and pressures of about 50 atmospheres to produce benzene and methane: $C_6H_5CH_3 + H_2 \rightarrow C_6H_6 + CH_4$. Hydrodealkylation strips the methyl group from toluene to produce benzene. Toluene disproportionation involves combining toluene so that the methyl groups bond to one aromatic ring, producing benzene and xylene. Benzene can also be obtained from petroleum reforming in which temperature, pressure, and catalysts are used to convert petroleum components to benzene, which can then be extracted using solvents and distillation processes. Another source of benzene is pyrolysis gasoline or pygas. Pygas is the aromatic rich naphtha steam resulting from ethylene production.

Benzene's toxic effect has received appreciable attention. It is released into the atmosphere through oil and coal combustion from utility plants, gasoline fumes, and automobile exhaust. Inhalation of benzene affects the central nervous system, causing headaches, drowsiness, and dizziness. At high enough levels, benzene irritates the respiratory tract creating irritations, unconsciousness, and in severe cases death. Long-term chronic exposure affects the blood, reducing red cell counts. Benzene is classified as a human carcinogen and leads to an increased rate of leukemia. In addition to exposure from the air, benzene can contaminate water supplies. Health effects associated with water pollution include nervous systems disorders, anemia, and weakening of the immune system.

CHEMICAL NAME = benzoic acid
CAS NUMBER = 65–85–0
MOLECULAR FORMULA = $C_7H_6O_2$
MOLAR MASS = 122.1 g/mol
COMPOSITION = C(68.9%) H(4.9%) O(26.2%)
MELTING POINT = 122°C
BOILING POINT = 249°C
DENSITY = 1.3 g/cm³

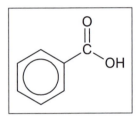

Benzoic acid is a colorless, crystalline solid also known as benzenecarboxylic acid. It is the simplest aromatic carboxylic acid, with a carboxyl group (-COOH) bonded directly to the benzene ring. It is found naturally in the benzoin resin of a number of plants. Benzoin comes from the bark of a number of balsams of the *Styrax* genus, most notably *Styrax benzoin* and *Styrax benzoides*. *S. benzoin* is native to Southeast Asia, and was traded between Indonesia and China as early as the 8th century c.e. Benzoin was used for fragrances, spices, medicines, and incense. Today most benzoin comes from Sumatra (Indonesia) and Laos. Healthy trees do not produce benzoin, but an incision or wound injuring the cambium results in its secretion. Benzoin is harvested after a cut is made in the tree and the exuded benzoin hardens. Benzoic acid was first isolated from the dry distillation of benzoin by Blaise de Vigenère (1523–1596) in the 16th century. Friedrich Wöhler (1800–1882) and Justus von Liebig (1803–1873) prepared benzoic

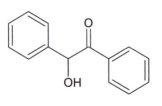

benzoin

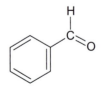

benzaldehyde

acid from oxidizing bitter almond oil (benzaldehyde) in 1832 and determined the formula for each of these compounds. They proposed that bitter almond oil, C_7H_6O, and benzoic acid were derivatives from the benzoyl radical, C_7H_5O; the radical theory was a major early theory in the development of organic chemistry.

Benzoic acid can be synthesized using a number of processes. The industrial method is by the partial oxidation of toluene ($C_6H_5CH_3$) in liquid phase using manganese, cobalt, vanadium-titanium, or other catalysts. The reaction is carried out at temperatures between 150°C and 200°C. It can also be prepared by the oxidation of benzaldehyde, benzyl alcohol ($C_6H_5CH_2OH$), and cinnamic acid ($C_6H_5CHCHO_2$) or by the oxidation of benzene with concentrated sulfuric acid. The hydrolysis of benzonitrile (C_6H_5CN) produces benzoic acid. It is also produced by the carboxylation of a Grignard reagent followed by acidification; typically carbonation occurs by pouring a Grignard ether over dry ice.

Approximately 750,000 tons of benzoic acid is produced globally each year. Benzoic acid's greatest use is as an intermediate in the production of other chemicals. More than 90% of benzoic acid production is converted into phenol (C_6H_5OH, see Phenol) or caprolactam ($C_6H_{11}NO$). Caprolactam is used in the production of nylon and other synthetic fibers.

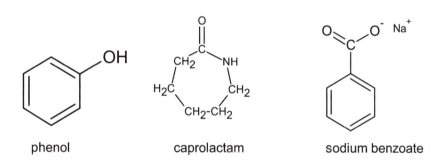

phenol   caprolactam   sodium benzoate

Sodium benzoate is an important benzoic acid derivative produced industrially by neutralization of benzoic acid using sodium hydroxide or sodium bicarbonate solution. Calcium benzoate, potassium benzoate, and other benzoate salts are also produced. Benzoic acid and sodium benzoate (C_6H_5COONa) are used as food preservatives and added to foods, juices, and beverages that are acidic. Although benzoic acid is a better antimicrobial agent than its salts, sodium benzoate is about 200 times more soluble in water, making it the preferable form for preservation. Benzoates require a pH less than 4.5 to be effective and work better as the pH decreases. Benzoate salts also occur naturally in some fruits (prunes, plums, apples, cranberries) and their juices. In the early 1990s, the Food and Drug Administration discovered that benzoates and ascorbic acid in beverages can react under certain conditions to produce benzene, which is a carcinogen. In 2005, several drinks were found to have elevated (> 5 parts per billion) benzene levels. Subsequent surveys of commercial soft drinks indicated that they contain either no detectable benzene or levels below the 5 parts per billion level established for drinking water. Sodium benzoate is also used as a preservative in toothpaste, mouthwashes, cosmetics, and pharmaceuticals. In addition to its use in the food and beverage industry, sodium benzoate is used as an organic corrosion inhibitor. The benzoate ion adsorbs to the metal surface, forming a passive layer that interferes with oxidation by blocking oxidation sites. Sodium benzoate is added to engine coolant antifreezes and industrial boilers to prevent corrosion.

Other derivatives of benzoic acid are diethylene glycol dibenzoate $(C_6H_5CO_2CH_2CH_2)_2O$ and dipropylene glycol dibenzoate $(C_6H_5CO_2C_3H6)_2O$, which are esters used as plasticizers. These are used in polyvinyl chloride and polyvinyl acetates coatings, floor coverings, and roof coatings. They are also used in adhesives, sealants, caulks, resins, and paints. Cosmetics incorporate diethylene glycol dibenzoate and dipropylene glycol dibenzoate as skin softeners; they are also used as wetting agents in deodorants. A small amount of benzoic acid is used to produce benzoyl chloride (C_6H_5COCl) by distilling it with phosphorus pentachloride (PCl_5), which is used as an intermediate to produce cosmetics, dyes, pharmaceuticals, and resins. Esters such as methyl benzoate $(C_6H_5COOCH_3)$ and ethyl benzoate $(C_6H_5COOCH_2CH_3)$ result from condensation reactions of benzoic acid and the corresponding alcohol. These esters have fragrant odors and are used in perfumes, cosmetics, and artificial flavorings.

14. Biphenyl and PCBs

CHEMICAL NAME = biphenyl
CAS NUMBER = 92–52–4
MOLECULAR FORMULA = $C_{12}H_{10}$
MOLAR MASS = 154.2 g/mol
COMPOSITION = C(93.5%) H(6.5%)
MELTING POINT = 68.9°C
BOILING POINT = 254°C
DENSITY = 1.04 g/cm³

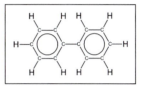

Biphenyl, also called diphenyl, consists of two benzene rings joined by a single bond. It exists as colorless to yellowish crystals, has a distinctive odor, and occurs naturally in oil, natural gas, and coal tar. Biphenyl is used as an antifungal agent to preserve citrus fruit, in citrus wrappers to retard mold growth, in heat transfer fluids, in dye carriers for textiles and copying paper, as a solvent in pharmaceutical production, in optical brighteners, and as an intermediate for the production of a wide range of organic compounds. There are hundreds of biphenyl derivatives.

Biphenyl was once used extensively for the production of polychlorinated biphenyls (PCBs) before their production was banned in the United States in 1979. PCBs are formed by direct substitution of hydrogen atoms in biphenyl with chlorine using chlorine gas under pressure with a ferric chloride ($FeCl_3$) catalyst. There are 209 possible PCB compounds referred to as congeners. PCBs were discovered in 1865 as a by-product of coal tar and first synthesized in 1881. Commercial production of PCBs, originally called chlorinated diphenyls, began in 1929 by the Swann Chemical Company located in Anniston, Alabama. Swann was taken over by Monsanto in 1935.

PCBs are synthetic chemicals that exist as oils or waxy substances; they do not occur naturally. They were once used in many products including hydraulic fluids, pigments, inks, plasticizers, lubricants, and heat transfer fluids, but their primary use was as a dielectric fluid in electrical equipment. Because of their high thermal stability, chemical stability, and electrical insulating properties, PCB fluids were used extensively in transformers, fluorescent light

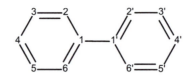

Figure 14.1   Numbering system used for PCBs.

ballasts, capacitors, and other electrical devices. PCBs were produced as mixtures of different congeners, which could contain between 1 and 10 chlorine atoms substituted for the hydrogen atoms in biphenyl. Individual PCB congeners are named using a numbering system where one of the carbon atoms at the single bond in biphenyl is given the number 1, and then other carbon atoms in that ring are sequentially numbered 2–6 (Figure 14.1). The same procedure is used for the other ring, using 1' for the bonded carbon and 2'-6' for the remaining carbons in the ring. The numbers are used to identify the position of the chlorine atoms in the particular congener with the unprimed ring chosen to give the lowest numbered carbon. For example, the following PCB is named 2,3',4'-trichlorobiphenyl.

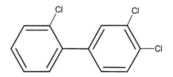

The primary producer of PCBs in North America was Monsanto, which manufactured PCBs under the trade name Arochlor. Arochlor congeners were numbered with a 4-digit number. The first two numbers indicated the number of carbon atoms in the biphenyl, and the last two numbers gave the mass percentage of chlorine in the PCB mixture (Arochlor 1016 was an exception since it had 12 carbons). Different PCB mixtures were used for various applications that expanded over the years from its original applications in electrical devices. Properties of different PCB mixtures varied according to the percentage of chorine in the mixture. As chlorine content increased, so did the PCBs boiling point, persistence, and lipophilicity (ability to dissolve in fat). Water solubility decreased as chlorine content increased.

Production of PCBs increased between 1930 and 1970, with a large increase in production after World War II. It is estimated that 650,000 tons of PCBs were produced in the United States during this 40-year period, with a peak of 42,000 tons in 1970. Global production was approximately twice that of the United States during this same period. Another 400,000 tons have been produced outside the United States after their production was banned in this country in 1979. Health problems started to appear in PCB plant workers soon after their commercialization. The first problem observed among PCB workers was chloracne, a condition resulting in skin lesions associated with overexposure to chlorine compounds.

As PCB production increased, more concerns were raised about the health and environmental effects of PCBs, which entered the environment through leakage, production processes, and improper disposal. The persistence and lipophilicity of PCBs resulted in its biomagnification in the environment (see DDT). Problems associated with PCB contamination in wildlife include deformities, tumors, disruption in the endocrine and reproduction systems, and death. Human exposure to PCBs occurs through environmental and occupational routes. The primary exposure of humans to PCBs is through ingestion of food, especially fish, meat, and dairy products. Acute human health effects of PCBs include skin, eye, and throat irritation; breathing difficulties; nausea; loss of weight; and stomach pain. There is evidence associating long-term increased PCB exposure in occupational settings to an increased incidence of liver and kidney cancer.

Because of public concerns, in October 1976, the U.S. Congress mandated that PCBs be regulated. PCB production in the United States stopped in 1977. The first regulations were put into place in 1978 and dealt with labeling and disposing of PCB materials. In 1979, the manufacture, processing, distribution, and use of PCBs at a concentration of 50 ppm were regulated. The 50 ppm criterion was challenged by environmental groups. Court action resulted in PCBs below 50 ppm being regulated by rules established in the early 1980s, although exemptions were granted depending on the application and form in which the PCBs were used. Although production ceased in the United States, other countries continued to produce PCBs. Russia did not stop production until 1995, and other countries continued to produce PCBs through the early 2000s. The Stockholm Convention on Persistent Organic Pollutants finalized in 2001 prohibits new PCB production after 2005 and calls for eliminating electrical equipment that contains high concentrations of PCBs by 2025.

PCBs will continue to remain in the environment for many years. Because they are still present in buildings, paints, soils, and throughout the environment, PCB remediation and disposal must be considered during remodeling, demolition, or decommissioning activities. Many methods are used to destroy PCBs. Chemical dechlorination is used when PCB concentrations are lower than 12,000 ppm. Chemical dechlorination separates the chlorine molecule from the PCBs, typically using sodium reagents to form inorganic salts. The bioremediation of soils uses select bacteria to break down the chlorinated hydrocarbons in a soil. The degradation reaction is slow and can take several weeks to years. Incineration is the most common method used to destroy PCBs and is required with higher concentrations. Temperatures in the range of 900°C to 1,200°C are used to volatilize and combust (in the presence of oxygen) the PCBs, although the temperature required to degrade PCBs depends on their concentration and residence time in the incinerator. Cement kilns are also used for this purpose. Incinerators are required to destroy 99.9999% of the PCBs. One concern with PCB incineration is the formation of dioxin and dioxin-like compounds such as dibenzo furans, polychlorinated dibenzofurans (PCDFs), and polychlorinated dibenzo-p-dioxins (PCDDs). Dioxin and dioxin-like compounds persist in the environment for decades. They can cause cancer and are toxic to the fetal endocrine system.

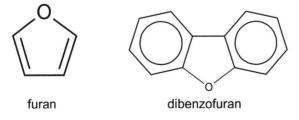

furan   dibenzofuran

The problem of PCB pollution continues into the 21st century. Several legal cases have dominated the media in recent years, such as General Electric's responsibility to remediate the Hudson River and Monsanto's and Solutia's $600 million dollar settlement in 2003 with Alabama over claims concerning Monsanto's Anniston, Alabama plant.

CHEMICAL NAME = butane

MOLECULAR FORMULA = C_4H_{10}

MOLAR MASS = 58.1 g/mol

COMPOSITION = C(82.66%) H(17.34%)

MELTING POINT = –138.3°C for n-butane, –159°C for isobutane

BOILING POINT = –0.50°C for n-butane, –11.6°C for isobutane

DENSITY = 2.6 g/L (vapor density = 2.05, air = 1)

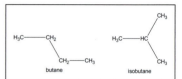

Butane is a flammable, colorless gas that follows propane in the alkane series. Butane is also called n-butane, with the "n" designating it as normal butane, the straight chain isomer. Butane's other isomer is isobutane. The chemical name of isobutane is 2-methylpropane. Isomers are different compounds that have the same molecular formula. Normal butane and isobutane are two different compounds, and the name butane is used collectively to denote both n-butane and isobutane; the names *n-butane* and *isobutane* are used to distinguish properties and chemical characteristics unique to each compound. Butane derives its root word *but* from four-carbon butyric acid, $CH_3CH_2CH_2COOH$. Butyric acid comes from butterfat and the Latin word *butyrum* means butter (see Butyric Acid). Butane, along with propane, is a major component of liquefied petroleum gas (LPG, see Propane). It exists as a liquid under moderate pressure or below 0°C at atmospheric pressure, which makes it ideal for storage and transportation in liquid form. Butane is the common fuel used in cigarette lighters and also as an aerosol propellant, a calibration gas, a refrigerant, a fuel additive, and a chemical feedstock in the petrochemical industry.

Butane is extracted from natural gas and is also obtained during petroleum refining. Butane can be obtained from natural gas by compression, adsorption, or absorption. All three processes were used in the early days of the LPG industry, but compression and adsorption were generally phased out during the 20th century. Most butane now is obtained from absorption and separation from oil. Very little butane is obtained from distillation. Gas stream from cracking units in the refining process contain appreciable amounts of butane, which is separated from the gas mixture by oil absorption, distillation, and various other separation processes. The supply of n-butane is adequate to meet demands for this compound, but generally the demand for isobutane exceeds supply. Normal butane can be converted to isobutane through a process called isomerization. In isomerization, straight chain isomers are converted to branched isomers. Isomerization of n-butane to isobutane takes place through a process in which n-butane is fed to a reactor at approximately 300°C and 15 atmospheres pressure. Hydrogen is added to prevent the formation of olefins (alkenes) and aluminum, platinum, and hydrochloric acids catalysts are used. The product from the reactor contains a mixture of butanes, and isobutane is separated from n-butane in a fractionator.

Isomerization of n-butane to isobutane is important in the oil industry because it provides feedstock for alkylation. Alkylation is a process in which an alkyl group is transferred between molecules. With respect to the refining of gasoline, alkylation refers to the combination of iso-butane with alkenes such as propylene and butylenes to produce branch-chained alkanes. The products of alkylation, called alkylates, are valued because they represent high-octane blend-ing stock that boosts the octane rating of gasoline. Straight-chain alkanes have low octane and produce knocking (premature combustion) in internal combustion engines. Through alkylation reactions, straight-chain alkanes are converted to branched alkanes, and branching increases the octane rating. Isooctane, 2,2,4-trimethylpentane, with an octane rating of 100, is produced through the alkylation reaction of isobutane and isobutylene:

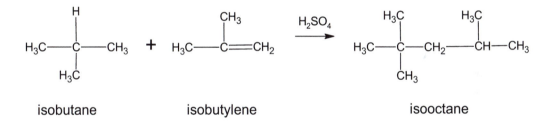

isobutane   isobutylene   isooctane

Butane undergoes typical alkane reactions (see Methane, Ethane, Propane). Pure butane produces a relatively cool flame, and its 0°C boiling point means that it does not vaporize well below this temperature. Butane's use as a camp stove fuel during the summer or in southern locales does not present vaporization problems, but in winter or geographic areas that expe-rience cold temperatures, butane is an inefficient fuel. To overcome vaporization problems, butane-propane mixtures with 20–40% propane are used. An advantage of using blended fuel over a propane fuel is fuel canister weight. Propane has a lower boiling and higher vapor pressure than butane at the same temperature. Pure propane fuel requires a heavier walled container adding weight to stoves used in backpacking situations. Blending of fuels is also used by oil refineries to account for seasonal differences. Butane is added to gasoline in winter to improve performance in cold temperatures.

Butane is used in the petrochemical industry to produce a variety of other compounds. Oxygenated products of n-butane include acetic acid (CH_3COOH), methanol (CH_3OH),

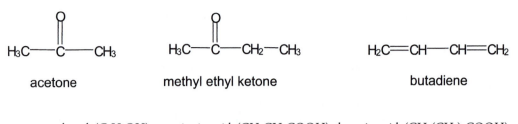

acetone   methyl ethyl ketone   butadiene

ethanol (C_2H_5OH), propionic acid (CH_3CH_2COOH), butyric acid ($CH_3(CH_2)_2COOH$), acetone, and methyl ethyl ketone. Dehydrogenation of butanes produces butylenes (C_4H_8) and butadiene.

16. Butene

CHEMICAL NAMES = see structure diagram
CAS NUMBER = 106–98–9 (1-butene)
 107–01–7 (2-butene)
MOLECULAR FORMULA = C_4H_8
MOLAR MASS = 56.1 g/mol
COMPOSITION = C(85.6%) H(14.4%)

	MELTING POINT	BOILING POINT
1-BUTENE	−185°C	−6.5°C
CIS-2-BUTENE	−139°C	3.7°C
TRANS-2-BUTENE	−106°C	0.9°C
METHYLPROPENE	−141°C	−6.9°C

DENSITY = 2.5 g/L (vapor density = 1.9, air = 1.0)

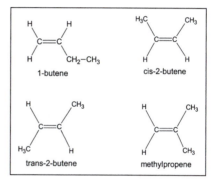

Butenes or butylenes are hydrocarbon alkenes that exist as four different isomers. Each isomer is a flammable gas at normal room temperature and one atmosphere pressure, but their boiling points indicate that butenes can be condensed at low ambient temperatures and/or increase pressure similar to propane and butane. The "2" designation in the names indicates the position of the double bond. The cis and trans labels indicate geometric isomerism. Geometric isomers are molecules that have similar atoms and bonds but different spatial arrangement of atoms. The structures indicate that three of the butenes are normal butenes, n-butenes, but that methylpropene is branched. Methylpropene is also called isobutene or isobutylene. Isobutenes are more reactive than n-butenes, and reaction mechanisms involving isobutenes differ from those of normal butenes.

Most butenes are produced in the cracking process in refineries along with other C-4 fractions such as the butanes. Butenes are separated from other compounds and each other by several methods. Isobutene is separated from normal butanes by absorption in a sulfuric acid solution. Normal butenes can be separated from butanes by fractionation. The close boiling points of butanes and butenes make straight fractional distillation an inadequate separation method, but extractive distillation can be used. Extractive distillation is a vapor-liquid process in which a solvent is used to promote a chemical separation. The less volatile compound has greater solubility in the extractive solvent, increasing separation efficiency. The less volatile compound and the solvent are obtained from the bottom of a distillation column, whereas the more volatile compound is recovered from the top of the column. The less volatile compound is separated and recovered from the solvent, the latter being recycled. Butenes can also be prepared from the dehydrogenation (elimination of hydrogen) of butane.

Butenes are used extensively in gasoline production to produce high-octane gasoline compounds. In alkylation reactions, butenes combine with isobutane to produce branched gasoline-range compounds (see Butane). Isooctane can be produced by dimerization of isobutene in the presence of sulfuric acid. Dimerization is the combination of a molecule with itself to produce a molecule called a dimer. The dimer has exactly twice the number of atoms in the original molecule. Therefore the dimerization of isobutene produces two dimers with the formula C_8H_{16}:

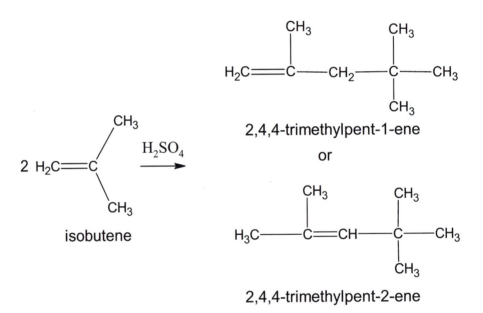

The two trimethylpentenes produced in the dimerization are also called iso-octenes. These can be hydrogenated in the presence of a metal catalyst to produce isooctane (2,2,4-trimethylpentane), which has an octane number of 100.

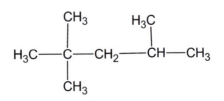

isooctane

Another large use of normal butenes in the petrochemical industry is in the production of 1,3-butadiene ($CH_2 = CH = CH = CH_2$). In the process, a mixture of n-butenes, air, and steam is passed over a catalyst at a temperature of 500°C to 600°C. Butadiene is used extensively to produce synthetic rubbers (see Isoprene) in polymerization reactions. The greatest use of butadiene is for styrene-butadiene rubber, which contains about a 3:1 ratio of butadiene to styrene. Butadiene is also used as a chemical intermediate to produce other synthetic organics such as chloroprene, for adhesives, resins, and a variety of polymers.

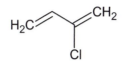

Chloroprene

Isobutylene is more reactive than n-butene and has several industrial uses. It undergoes dimerization and trimerization reactions when heated in the presence of sulfuric acid. Isobutylene dimer and trimers are use for alkylation. Polymerization of isobutene produces polyisobutenes. Polyisobutenes tend to be soft and tacky, and do not set completely when used. This makes polyisobutenes ideal for caulking, sealing, adhesive, and lubricant applications. Butyl rubber is a co-polymer of isobutylene and isoprene containing 98% isobutene and 2% isoprene.

Butene is used in the plastics industry to make both homopolymers and copolymers. Polybutylene (1-polybutene), polymerized from 1-butene, is a plastic with high tensile strength and other mechanical properties that makes it a tough, strong plastic. High-density polyethylenes and linear low-density polyethylenes are produced through co-polymerization by incorporating butene as a comonomer with ethene. Similarly, butene is used with propene to produce different types of polypropylenes.

Another use of 1-butene is in the production of solvents containing four carbons such as secondary butyl alcohol and methyl ethyl ketone (MEK). Secondary butyl alcohol is produced by reacting 1-butene with sulfuric acid and then hydrolysis:

$$H_3C-CH_2-CH=CH_2 \xrightarrow{H_2SO_4} \underset{\underset{OSO_3H}{|}}{H_3C-CH_2-CH_2-CH_3} \xrightarrow{H_2O} \underset{sec\text{-butyl alcohol}}{H_3C-CH_2-\overset{\overset{OH}{|}}{C}=CH_2}$$

Primary alcohol and aldehydes are produced from butene through the Oxo process. The Oxo process involves the addition of carbon monoxide and hydrogen to an alkene under elevated temperature and pressure in the presence of a catalyst.

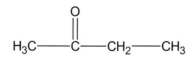

methyl ethyl ketone

17. Butyric and Fatty Acids

CHEMICAL NAME = butanoic acid
CAS NUMBER = 107–92–6
MOLECULAR FORMULA = $C_4H_8O_2$
MOLAR MASS = 88.1 g/mol
COMPOSITION = C(54.5%) H(9.2%) O(36.3%)
MELTING POINT = –7.9°C
BOILING POINT = 163.5°C
DENSITY = 0.96 g/cm³

n-butyric acid   isobutyric acid

Butyric acid is a carboxylic acid also classified as a fatty acid. It exists in two isomeric forms as shown previously, but this entry focuses on n-butyric acid or butanoic acid. It is a colorless, viscous, rancid-smelling liquid that is present as esters in animal fats and plant oils. Butyric acid exists as a glyceride in butter, with a concentration of about 4%; dairy and egg products are a primary source of butyric acid. When butter or other food products go rancid, free butyric acid is liberated by hydrolysis, producing the rancid smell. It also occurs in animal fat and plant oils. Butyric acid gets its name from the Latin *butyrum*, or butter. It was discovered by Adolf Lieben (1836–1914) and Antonio Rossi in 1869.

Butyric acid is one of the simplest fatty acids. Fatty acids, which are the building units of fats and oils, are natural compounds of carbon chains with a carboxyl group (-COOH) at one end. Most natural fatty acids have an unbranched carbon chain and contain an even number of carbon atoms because during biosynthesis they are built in two carbon units from acetyl coenzyme A (CoA). Butyric acid is an unsaturated fatty acid, which means all carbon-carbon bonds are single bonds. Common names for fatty acids stem from their natural sources. In addition to butyric acid, some other common saturated fatty acids include lauric acid, palmitic acid, and stearic acid. Lauric acid was first discovered in Lauraceae (*Laurus nobilis*) seeds, palmitic oil was prepared from palm oil, and stearic acid was discovered in animal fat and gets its name from the Greek word *stear* for tallow.

Monounsaturated fatty acids have a single carbon-carbon double bond present in the chain, and polyunsaturated fatty acids have more than one carbon-carbon double bond. Oleic

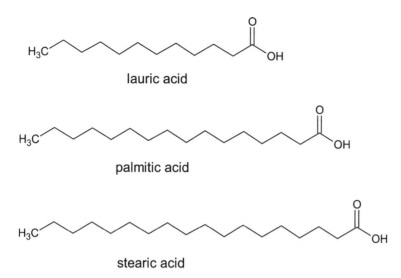

lauric acid

palmitic acid

stearic acid

acid is a monounsaturated fatty acid found in olive oil. In unsaturated fats the arrangement of hydrogen atoms around the double bond can assume a trans or cis configuration. Trans fatty acids have hydrogen atoms on opposite sides of the double bond as opposed to the cis formation where they are on the same side. Research showing that trans fatty acids increase heart disease has led to food labels reporting the amount of trans fat and a decrease of trans fats in processed foods in recent years. Unsaturated fatty acids can also be characterized by the position of the double bond. One system used in physiology numbers carbon atoms starting from the methyl end of the molecule. The methyl carbon is referred to as the omega carbon because omega is the last letter in the Greek alphabet; the first carbon at the carboxyl end of the molecule is the alpha carbon. The double bond can then be located based on its displacement from the omega carbon. For example, an omega-3 fatty acid has its endmost double bond positioned three carbon atoms away from the omega carbon. Omega fatty acids such as omega-3 are used as dietary supplements for a variety of conditions.

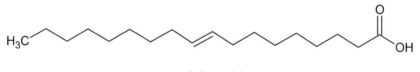

oleic acid

Butyric acid is produced by oxidation of butyraldehyde ($CH_3(CH_2)_2CHO$) or butanol (C_4H_9OH). It can also be formed biologically by the oxidation of sugar and starches using bacteria. It is used in plastics as a raw material for the cellulose acetate butyrate (CAB). CAB is an ester produced by treating fibrous cellulose with butyric acid, butyric anhydride ($(CH_3CH_2CH_2CO)_2O$], acetic acid (CH_3COOH), and acetic anhydride ($(CH_3CO)_2O$) in the presence of sulfuric acid. CAB is a tough plastic that resists weathering and is highly transparent. It is sold in sheets and tubes of various dimensions and is used for signs, goggles, sunglasses, tool handles, pens, and stencils. Reacting butyric acid with alcohols produces esters. For example,

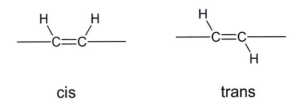

cis   trans

butyric acid and methanol yield methyl butyrate ($C_5H_{10}O_2$). Many butyrate esters have a pleasant smell and are therefore used in food flavorings and perfumes. Other uses of butyric acid are in disinfectants, pharmaceuticals, and feed supplements for plant and animals.

Butyric acid derivatives play an important role in plant and animal physiology. Gamma-aminobutyric acid (GABA)is the main inhibitory neurotransmitter found in humans and is

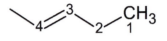

omega-3 fatty acid

present in many other organisms. Indole-3-butyric acid is closely related in structure and function to a natural growth hormone found in plants. Indole-3-butyric acid is used in many nutrient formulations to promote growth and development of roots, flowers, and fruits, and to increase crop yields.

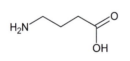

GABA

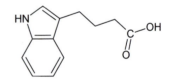

indole-3 butyric acid

18. Caffeine

CHEMICAL NAME = 3,7-dihydro-1,3,7-trimethyl-1H-purine-2,6-dione
CAS NUMBER = 58–08–2
MOLECULAR FORMULA = $C_8H_{10}N_4O_2$
MOLAR MASS = 194.2 g/mol
COMPOSITION = C(49.5%) H(5.2%) N(28.9%) O(16.5%)
MELTING POINT = 237°C
BOILING POINT = sublimes
DENSITY = 1.2 g/cm³

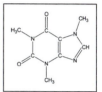

Caffeine is an alkaloid purine belonging to the group of organic compounds called methyl-xanthines. Pure caffeine is a white, crystalline, bitter-tasting compound. Caffeine is found in a number of plants, principally coffee and tea plants, as well as cola and cacao nuts. In plants, caffeine functions as a natural pesticide to deter insects. The consumption of caffeine dates back thousands of years. Tea was consumed in China several thousand years B.C.E. but quite possibly was used in India before that and introduced into China. Coffee consumption is believed to have started in the Kaffa region of Ethiopia around 800 c.e. and over time spread to Arabia, Turkey, and other parts of the Middle East. The port city of Mocha on the Red Sea in Yemen became a principal coffee-growing region and area of export in the Middle East. Rulers in coffee-growing areas imposed strict laws against the exportation of coffee plants to exercise their monopolies over the lucrative product. Coffee plants spread throughout the world from successful smuggling. Tea and coffee were introduced into Europe in the 17th century, well after its use in other parts of the world. Chocolate drinks from cacao beans (cocoa is often used for the drink or powder product made from the cacao beans) were being concocted by native populations in Central and South America several hundred years B.C.E. The Spanish explorer Hernando Cortés (1485–1587) brought cocoa back to Spain around 1528.

Only a small amount of caffeine is found in the cacao plant and chocolate. The principal alkaloid in the cacao plant is theobromine, which is almost identical to caffeine, but differs by having one less methyl group. Theobromine does not contain bromine but derives its name from the genus *Theobroma* of the cacao tree. Theobroma's Greek translation is "food of the gods." Another compound almost identical to caffeine in tea is theophylline. It also contains one fewer methyl group.

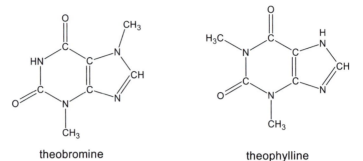

theobromine   theophylline

The discovery of caffeine is attributed to Friedlieb Ferdinand Runge (1795–1867), a German physician and chemist. Runge was working in the laboratory of Johann Wolfgang Döbereiner (1780–1849), when Döbereiner's friend, Johann Wolfgang von Goethe (1749–1832), paid a visit. Runge performed an experiment for Goethe in which he dilated a cat's eye with an extract from a nightshade plant. Goethe awarded Runge with a sample of rare coffee beans and challenged him to determine the compound that gave coffee its stimulating effects. After several months, Runge isolated caffeine from coffee in 1819. Caffeine derives its name from the Kaffa region of Ethiopia. Caffeine comes from the German *kaffeine,* which in turn is derived from the German word for coffee, *kaffee.* In 1827, a compound isolated from tea was named theine, but this was eventually shown to be caffeine.

Caffeine is a stimulant to the central nervous system and cardiac muscle and is a mild diuretic. Caffeine's physiological effects are thought to be the result of caffeine's interference with adenosine in the brain and body. Adenosine moderates nerve transmissions. As adenosine builds up while a person is awake, it produces a self-regulating mechanism to inhibit nerve transmission. As adenosine receptors in the brain acquire more adenosine, the reduction in nerve transmission induces sleep. Caffeine competes with adenosine and interferes with the neural modulation function of adenosine. This is why coffee has a tendency to keep people awake. The effects of coffee and caffeinated drinks vary widely among individuals. Generally, moderate consumption leads to restlessness and tends to energize individuals. Caffeine increases blood pressure and has been indirectly associated with heart and pregnancy problems; it increases stomach acid, which can lead to ulcers. Regular users of caffeinated drinks can experience withdrawal side effects such as anxiety, nervousness, fatigue, and headaches. The effects of caffeine lasts several hours after consumption. It is carried by the blood to all parts of the body and is eliminated primarily through the urine after a half-life from 4 to 10 hours in most adults.

Health experts advise that moderated amounts of caffeine from 100 to 300 mg per day are acceptable. Adult Americans consume approximately 250 mg of caffeine per day. The LD_{50} (the lethal dose that kills 50% of a test population of individuals subjected to a substance) of caffeine for humans is estimated between 150 and 200 mg per kilogram of body weight. The amount of caffeine in some popular food items is given in Table 18.1.

Coffee beans are the primary source of caffeine. These beans are obtained from a variety of plants but can be broadly grouped into two classes: arabica and robusta. Arabica is obtained from the species *Coffea arabica* and robusta from the species *Coffea canephora.* Robusta, as the

Table 18.1 Approximate Caffeine Content
of Selected Food and Medicine Items

	Caffeine in mg
Coffee	30–140 (8-oz cup)
Mountain Dew	54 (12 oz)
Coca-Cola	45 (12 oz)
Pepsi-Cola	37 (12 oz)
Tea (brewed)	20–100 (8 oz)
Red Bull	80 (8.2 oz)
Häagen-Dazs coffee ice cream	58 (1 cup)
Hot chocolate	5 (8 oz)
Excedrin	130 (2 tablets)
NoDoz	200 (1 tablet)
Anacin	64 (2 tablets)

name implies, is more robust than arabica coffee but produces an inferior taste. Arabica plants are grown globally, but robusta plants are grown only in the Eastern Hemisphere. Coffee beans contain 1–2% caffeine, with robusta varieties generally containing twice the content of arabica varieties. Other sources of caffeine contain various caffeine content: kola nut (1–3.5%), tea leaves (1.4–4.5%), and cacao (0.1–0.5%).

Some food processors add caffeine to their products (soft drinks), but others remove caffeine and advertise the product as decaffeinated. Coffee was first decaffeinated in 1906 in Germany through a process founded by the coffee merchant Ludwig Roselius (1874–1943). Roselius's team sought to decaffeinate coffee without destroying the aroma and flavor. Roselius fortuitously worked on beans that had been soaked with seawater during a storm and found a method to remove 97% of the caffeine in coffee beans without destroying the flavor. Roselius then marketed the product under different names in various European countries. In France, the name of the decaffeinated coffee was Sanka, which is derived from *sans* caffeine (*sans kaffee*). Sanka was introduced in the United States in 1923. The traditional method of decaffeinating coffee beans involved steaming the beans and then extracting the caffeine in an organic solvent such as Freon, chloroform, methylene chloride, or ethyl acetate. Because of the environmental problems and costs associated with organic solvents, green decaffeination techniques have been developed in recent years. One popular method is to use supercritical carbon dioxide to extract caffeine (see Carbon Dioxide).

Caffeine has widespread therapeutic use. It is widely used in headache (migraine) remedies such as aspirin and other analgesics. Caffeine is a mild vasoconstrictor and its ability to constrict blood vessels serving the brain explains its use to relieve headache. Individuals who consume caffeine regularly through medications and food are susceptible to what is known as a rebound headache or caffeine rebound. This occurs when regular caffeine intake is suddenly reduced and the vessels dilate. Caffeine is a common substance in medications to treat apnea in premature infants. Apparently, the area of the brain controlling respiration in premature infants is not fully developed and caffeine helps to stimulate this portion of the brain. The combination of caffeine and ephedrine is used in dietary and athletic supplements, and their role as appetite suppressant and energy boosters has been extensively studied. Some individuals claim that a modest dose (200 mg) of caffeine can enhance athletic performance, but its exact effect is unclear. Caffeine is used for the treatment of attention deficit disorder/ attention-deficit hyperactivity disorder, but health experts do not recommend its use for this condition.

19. Calcium Carbonate

CHEMICAL NAME = calcium carbonate
CAS NUMBER = 471–34–1
MOLECULAR FORMULA = $CaCO_3$
MOLAR MASS = 100.1 g/mol
COMPOSITION = Ca(40.0%) C(12.0%) O(48.0%)
MELTING POINT = 825°C
BOILING POINT = decomposes
DENSITY = 2.7 g/cm³

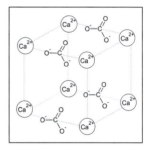

Calcium carbonate is a naturally occurring compound found in organisms and throughout the earth's crust. After quartz, calcium carbonate, primarily in the form of calcite, is the most common mineral found in the crust. Geologically, calcium carbonate exists in several mineral forms: calcite, aragonite, and vaterite. Calcite is the most common calcium carbonate mineral, whereas vaterite is a very rare form. The different mineral forms of calcium carbonate are based on their crystalline structure. The form of calcium carbonate depends on the conditions at its formation such as temperature and pressure. The structure of calcium carbonate in minerals can be viewed as a triangle, with calcium in the center and a carbonate group at each of the vertices. In calcite, the carbonates lie in the same plane, whereas in aragonite the carbonates lie in two different planes. Aragonite is more closely packed than calcite and therefore has a greater density. The density of calcite is about 2.7 g/cm³; that of aragonite is about 2.9 g/cm³. Calcium carbonate is found in many rocks, but it is most commonly associated with chalk, limestone, and marble, which is metamorphic limestone.

Numerous organisms extract and synthesize calcium carbonate from seawater, which is used in skeletal structures. Marine organisms including mollusks, sponges, foraminiferans, coccolithophores, coralline algae, and corals use dissolved calcium and carbon dioxide to make calcium carbonate. Corals secrete calcium carbonate in the form of aragonite skeletons, which are formed from the reactions of calcium and bicarbonate: $Ca^{2+}_{(aq)} + 2HCO_{3\ (aq)}^{-} \leftrightarrow CaCO_{3(s)} + CO_{2(g)} + H_2O_{(l)}$. Pearls are formed by oysters secreting calcium carbonate to encase foreign objects lodged inside their shells. The remains of marine organisms accumulated over millions of years form sedimentary deposits of chalk and limestone. Humans primarily use calcium carbonate as a primary source of calcium to combat osteoporosis.

The abundance of limestone deposits throughout the world has resulted in the use of calcium carbonate as a primary building material since antiquity. The ancient pyramids and the Sphinx in Egypt were made almost exclusively with limestone 5,000 years ago. Most limestone is used today as construction material. One estimate is that the United States used a billion tons of crushed limestone for roads, dams, fill, buildings, and various other construction uses in 2005.

In addition to its use as a construction material, calcium carbonate is also used in numerous industrial processes. Two forms commonly used are ground calcium carbonate (gcc) and precipitated calcium carbonate (pcc). Ground calcium carbonate is pulverized limestone that has been reduced in particle size, with diameters of a fraction of a micron to several microns. Precipitated calcium carbonate is made by subjecting calcium carbonate to processes in order to produce a product with specific characteristics with respects to form (calcite or aragonite), size, and properties. Precipitated calcium carbonate is made by heating (calcining) limestone to calcium oxide (lime, CaO) at temperatures between 600°C and 900°C: $CaCO_{3(s)} \rightarrow CaO_{(s)} + CO_{2(g)}$. Calcium oxide is then slacked with water to produce calcium hydroxide $(Ca(OH)_2)$: $CaO_{(s)} + H_2O_{(l)} \rightarrow Ca(OH)_{2(aq)}$. The calcium hydroxide is then combined with carbon dioxide, which is produced when the limestone is calcined, to give a suspension of calcium carbonate and water: $Ca(OH)_{2(aq)} + CO_{2(g)} \rightarrow CaCO_{3(s)} + H_2O_{(l)}$. The calcium carbonate produced in the last step is separated by filtration and various other separation methods to give precipitated calcium carbonate. By modifying the process of making pcc using different temperatures and limestone sources, numerous forms of calcium carbonate that vary in structure, size, and properties can be made for various applications.

Calcium carbonate is used widely in papermaking as filler and coating pigment to whiten paper. Papermaking plants often contain a satellite plant devoted to the production of pcc. Calcium carbonate is used in place of more expensive optical brightening agents in paper and as a fill to replace more expensive wood pulp fiber; it also helps control the pH in an alkaline range. The second most common industrial use of calcium carbonate (after papermaking) representing the largest use of gcc is in the production of plastics. It is used in the production of polyvinyl chloride (PVC), thermoset polyesters, and polyolefins. Calcium carbonate can be used to replace resins that are more expensive. Similar to its use in the paper industry, it is used as an optical brightener and whitening agent. It also is used to increase strength and absorb heat during exothermic processes. Calcium carbonate is also used in the production of polyethylene and polypropylene. It is an additive to paints and coatings for several purposes including particle size distribution, opacity control, weather resistance, pH control, and anticorrosion.

Calcium carbonate is used to buffer acidic soils. In soils that contain sulfuric acid calcium carbonate, it will react with the acid to produce calcium sulfate ($CaSO_4$), carbon dioxide, and water: $H_2SO_{4(aq)} + CaCO_{3(s)} \rightarrow CaSO_{4(s)} + CO_{2(g)} + H_2O_{(l)}$. The ability of various limes to neutralize acid in a soil is given in terms of calcium carbonate equivalents. In this system, limestone has a calcium carbonate equivalent of 100. If a slaked lime (calcium hydroxide) has a calcium carbonate equivalent of 150, then only two-thirds as much of slaked lime would be needed to achieve the same neutralizing effect. Calcium carbonate has also been used to mitigate the effects of acid precipitation on water bodies. Another environmental application of calcium carbonate is for gas desulfurization in scrubbers used to reduce sulfur emissions from air pollution sources. In the most popular of these methods, called wet scrubbing, process calcium carbonate reacts with sulfur dioxide to produce calcium sulfite ($CaSO_3$), which can be further oxidized to gypsum (see Calcium Sulfate).

20. Calcium Oxide (Lime)

CHEMICAL NAME = calcium oxide
CAS NUMBER = 1305–78–8
MOLECULAR FORMULA = CaO
MOLAR MASS = 56.0 g/mol
COMPOSITION = Ca(71.4%) O(28.6%)
MELTING POINT = 2,572°C
BOILING POINT = 2,850°C
DENSITY = 3.3 g/cm³

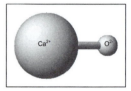

Calcium oxide is a white caustic crystalline alkali substance that goes by the common name lime. The term *lime* is used both generically for several calcium compounds and with adjectives to qualify different forms of lime. This entry equates lime, also called quicklime or burnt lime, with the compound calcium oxide. Hydrated lime, made by combining lime with water, is calcium hydroxide and is often referred to as slaked lime $(Ca(OH)_2)$. Dolomite limes contain magnesium as well as calcium. Limestone is the compound calcium carbonate. The term *lime* comes from the Old English word *lïm* for a sticky substance and denotes lime's traditional use to produce mortar. *Calx* was the Latin word for lime and was used to name the element calcium.

Calcium oxide dates from prehistoric times. It is produced by heating limestone to drive off carbon dioxide in a process called calcination: $CaCO_{3(s)} \xrightarrow{\Delta} CaO_{(s)} + CO_{2(g)}$. At temperatures of several hundred degrees Celsius, the reaction is reversible and calcium oxide will react with atmospheric carbon dioxide to produce calcium carbonate. Efficient calcium oxide production is favored at temperatures in excess of 1,000°C. In prehistoric times limestone was heated in open fires to produce lime. Over time, lined pits and kilns were used to produce lime. Brick lime kilns were extensively built starting in the 17th century and the technology to produce lime has remained relatively constant since then.

Modern lime kilns operate at approximately 1,200°C to 1,300°C. Limestone, which has been crushed and screened into pieces with diameters of several inches, is fed into the top of the kiln. Air fed into the kiln's bottom fluidizes the limestone, allowing for greater reaction efficiency. Rotating horizontal kilns several meters in diameter and as long as 100 meters are also used to produce lime. In these kilns, limestone is fed into one end that is elevated and moves down the rotating kiln by gravity as it is heated and converted into lime. Much of the lime produced is hydrated with water in a process called slaking to produce slaked lime or calcium hydroxide ($Ca(OH)_2$): $CaO_{(s)} + H_2O_{(l)} \rightarrow Ca(OH)_{2(s)}$. The hydration of calcium oxide is highly exothermic and one reason for converting lime to slaked lime is for safety.

The oldest use of lime is as a mortar. There is evidence that Egyptians used lime mortars as early as 4000 B.C.E. Lime mortars consist of a mixture of lime, water, and sand. Mortars are based on the reaction of slaked lime reacting with atmospheric carbon dioxide to produce calcium carbonate and water: $Ca(OH)_2 + CO_{2(g)} \rightarrow CaCO_{3(s)} + H_2O_{(l)}$. The conversion of slaked lime to limestone is a slow process and lime mortars can take several years to cure. Modern mortars contain a mixture of substances such that curing times can be controlled. The most common mortar is cement, which uses lime as a major ingredient in cement. Portland cement is made by heating limestone, clay, and sand in rotating kilns at a temperature of 2,700°C. The limestone is converted to lime and the clay and sand are sources of silicates, iron, and aluminum. Portland cement is 85% lime and silicates by mass. The "Portland" name comes from an English bricklayer named Joseph Aspdin (1799–1855) who patented the cement in 1824 and noted that it resembled limestone quarried from the Isle of Portland in the English Channel.

Lime has also been used since ancient times in the production of glass. Early glasses were made by producing a melt of silica (SiO_2), sodium carbonate (soda ash, Na_2CO_3), and lime and by letting the melt cool into an amorphous solid. Other additives were included in the melt to impart color or various other properties to the glass; for example, copper produced a blue glass and calcium strengthened it. Most modern glass produced is called soda-lime glass and consists of approximately 70% silica, 15% soda (Na_2O), and 5% lime.

Lime is traditionally one of the top ten chemicals produced annually. Approximately 15 million tons of calcium oxide and 22 million tons of all compounds called limes (burnt, slaked, dolomite) are produced annually in the United States. Global production is approximately 135 million tons. The major uses of lime are metallurgy, flue gas desulfurization, construction, mining, papermaking, and water treatment. About one third of calcium oxide production in the United States is used for metallurgical processes, principally in the iron and steel industry. Calcium oxide is used to remove impurities during the refining of iron ore. Calcium oxide combines with compounds such as silicates, phosphates, and sulfates contained in iron ores to form slag. Slag is immiscible with molten iron, allowing the slag and iron to be separated. Lime is also used for purification in other metal refining and to control pH in mining processes such as leaching and precipitation. The calcium oxide is also used in remediation of mine wastes to recover cyanides and to neutralize acid mine drainage.

Both lime and slaked limes are use to reduce sulfur emissions, which contribute to acid precipitation, from power plants, particularly coal-fired plants. By using lime, more than 95% of the sulfur can be eliminated from the emissions. Calcium oxide reacts with sulfur dioxide to produce calcium sulfite: $CaO_{(s)} + SO_{2(g)} \rightarrow CaSO_{3(s)}$. Sulfur dioxide is also removed by spraying limewater in the flue gas. Limewater, also called milk of lime, is a fine suspension of calcium hydroxide in water. Other pollutants removed with lime include sulfur trioxide, hydrofluoric acid, and hydrochloric acid.

Lime is used in drinking water treatment to control pH, soften water, and control turbidity. Lime, in combination with sodium carbonate, is used to precipitate the major bivalent cations Ca^{2+} and Mg^{2+} that cause hardness. The bivalent ions Fe^{2+}, Sr^{2+}, Mn^{2+}, and Zn^{2+} are removed as well. Calcium oxide is also used to treat wastewater. It is used to keep pH between 6.0 and 9.0 and aids in the precipitation of solids, especially nitrogen and phosphorus compounds. Phosphates from wastewater discharges can lead to algae blooms and eutrophication in receiving waters. Calcium oxide precipitates out the phosphate from the discharge as calcium phosphate: $3CaO_{(s)} + 3H_2O_{(l)} + 2PO_4^{3-}{}_{(aq)} \rightarrow Ca_3(PO_4)_{2(s)} + 6OH^-{}_{(aq)}$. Lime is used to treat sludge and biosolids to decrease odor and control pH.

Lime is used in the paper industry to produce the bleaching agent calcium hypochlorite $(Ca(OCl)_2)$. Slaked lime is used to recover sodium hydroxide from soda ash after the pulping process: $Ca(OH)_{2(aq)} + Na_2CO_{3(aq)} \rightarrow 2NaOH_{(aq)} + CaCO_{3(s)}$. The recovered calcium carbonate can be used to regenerate lime. Lime is also used to treat process water in the paper industry.

Lime is used in the chemical industry as a feedstock. It is heated with carbon in the form of coke to make calcium carbide (CaC_2): $2CaO_{(s)} + 5C_{(s)} \xrightarrow{\Delta} 2CaC_{2(s)} + CO_{2(g)}$. Calcium carbide is used to produce acetylene. Some of the other chemicals made with lime include calcium hypochlorite, citric acid, and sodium alkalis. Lime is used to produce precipitated calcium carbonate (PCC), which is a fine-grained form of calcium carbonate. To produce PCC, lime is hydrated to produce slaked lime and the slaked lime is combined with water to produce limewater. Carbon dioxide is added to the limewater, causing calcium carbonate to precipitate as PCC. PCC is used widely in plastics production, papermaking, pharmaceuticals, and the petrochemical industry.

Calcium oxide is sometimes used in the building industry to treat soils at construction sites. Treating the soil serves several purposes. Calcium oxide serves as a drying agent by combining with water in the soil to form calcium hydroxide. The heat generated in this process aids in the drying. The addition of CaO changes the soil chemistry and combines with clays to stabilize the soil. This provides a stronger building foundation and reduces soil plasticity.

Lime has limited use as a liming agent. Liming refers to the process of adding substances that neutralize acidity by increasing the pH of the soil. This is referred to as sweetening the soil. Lime, as both CaO and $Ca(OH)_2$, can be used as liming agents, but these are not the preferred forms for agricultural purposes. Lime and slaked lime have high neutralizing capacity, but lime is difficult to store and is caustic. Clumping on the soil surface of these forms is also a problem. The lime used in agriculture and gardening is typically crushed limestone (calcitic lime) or dolomitic limestone (a calcium-magnesium carbonate). Calcitic and dolomitic limes are what is meant by agricultural lime.

21. Calcium Sulfate (Gypsum)

CHEMICAL NAME = calcium sulfate   calcium sulfate dihydrate

CAS NUMBER = 7778–18–9   10101–41–4 (gypsum)

MOLECULAR FORMULA = $CaSO_4$ $CaSO_4{\cdot}2H_2O$

MOLAR MASS = 136.1 g/mol   172.1 g/mol

COMPOSITION = Ca(29.4%) O(47%)   Ca(23.2%) O(55.8%)

                S(23.6%)   S(18.6%) H(2.3%)

MELTING POINT = decomposes at 1,450°C becomes hemihydrate at 128°C

BOILING POINT = decomposes at 1,450°C becomes anhydrous at 160°C

DENSITY = 2.96 g/cm³   2.32 g/cm³

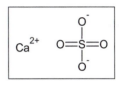

Calcium sulfate and its hydrates are important industrial compounds that have been used throughout history. Calcium sulfate dihydrate, $CaSO_4{\cdot}2H_2O$, is commonly called gypsum and calcium sulfate hemihydrate, $CaSO_4{\cdot}\frac{1}{2}H_2O$, is known as plaster of Paris. Calcium sulfate is obtained naturally from mined gypsum rock, but it also exists in mineral form. Gypsum forms in beds as sedimentary rock when calcium sulfate, which is a natural component of seawater, is deposited as shallow marine water bodies evaporate. Gypsum deposits are found widely throughout the world. Gypsum is a transparent, soft, white mineral and is the most common sulfate mineral. Alabaster is a dense translucent form of gypsum used for sculpture, engravings, and carvings. Selenite is a form that exists in crystallized transparent sheets.

Gypsum's use dates back to prehistoric times. Archeological evidence shows that it was mined from caves and used to paint ancient gravestones. The earliest evidence of its use as a building material dates from 6000 B.C.E. in the southwest Asian areas of ancient Anatolia and Syria. Egyptians used gypsum and plaster in their buildings and monuments, with both found in the Great Pyramids built around 3700 B.C.E. Calcium sulfate has been used for more than 2,000 years in China to produce tofu. The word gypsum comes from the Greek word for chalk, *gypsos.* The Greek natural philosopher Theophrastus (371–286 B.C.E.) referred to gypsum in his writings. Gypsum was extensively used in Roman times and throughout the Middle Ages.

Calcination forms the hemihydrate type of calcium sulfate by the reaction: $CaSO_4{\cdot}2H_2O \xrightarrow{\Delta} CaSO_4{\cdot}\frac{1}{2}H_2O + \frac{3}{2}H_2O$. Various calcining methods have been used throughout history to convert gypsum into plaster. The calcined or hemihydrate form is commonly known as plaster of Paris. In the 1700s, Paris was a leading plaster center, with most of its buildings made using plaster. After the fire of London in 1666 destroyed 80% of the city, the king of France ordered all wooden houses in France to be covered with plaster as protection against fire. The hemihydrate form got the name plaster of Paris from the extensive gypsum deposits quarried in the Montmartre district of Paris. In 1765 and 1766, Antoine Lavoisier (1743–1794) presented papers to the French Academy of Science on gypsum that explained the setting of plaster. Lavoisier determined that gypsum is a hydrated salt and that set plaster occurs when the hemihydrate form rehydrates back to gypsum. The hardening of plaster results from hemihydrate calcium sulfate's greater solubility in water than gypsum. When the hemihydrate plaster form of calcium sulfate is mixed with water, it absorbs water to reform into gypsum, which precipitates out as solid fibrous crystals.

Calcination of $CaSO_4 \cdot 2H_2O$ to produce the hemihydrate calcium sulfate, plaster of Paris, takes place in the temperature range of approximately 150°C to 200°C. Further heating beyond 200°C up to 300°C results in complete dehydration to produce an anhydrous form of calcium sulfate, called soluble anhydrite. Soluble anhydrites are used as drying agents; Drierite, a popular drying agent, is a soluble anhydrite that changes from blue to pink when it absorbs water. When heated above 300°C, calcium sulfate loses its ability to absorb water and becomes insoluble calcium sulfate, also known as insoluble anhydrite, dead burnt calcium sulfate, or dead burnt gypsum.

The United States is the leading producer of gypsum in the world, accounting for approximately 15% of the world's production of 120 million tons of mined gypsum. The leading gypsum-producing states are Oklahoma, Texas, Nevada, Iowa, and California. Gypsum is mined throughout the world, with gypsum mines in 90 countries. In addition to mined gypsum, gypsum is produced synthetically as a by-product of chemical processes. Synthetic gypsum accounts for 25% of gypsum production in the United States. Most synthetic gypsum comes from flue gas desulfurization used to control sulfur emissions from electric power generation, especially coal-burning power plants. In this process calcium carbonate reacts with sulfur dioxide to produce calcium sulfite ($CaSO_3$): $CaCO_{3(s)} + SO_{2(g)} \rightarrow CaSO_{3(s)} + CO_{2(g)}$. Calcium sulfite is then oxidized to gypsum: $2CaSO_{3(s)} + O_{2(g)} + 4H_2O_{(l)} \rightarrow 2CaSO_4 \cdot 2H_2O_{(s)}$. The second major source of synthetic gypsum is acid neutralization. The sulfate production of titanium dioxide, TiO_2, used as a whitener in many commercial products, yields gypsum as a by-product in the process of neutralizing acidic waste. Gypsum is also generated from sugar and citric acid production. In the latter, dilute citric acid ($C_6H_8O_7$) is precipitated with calcium hydroxide as calcium citrate ($Ca_3(C_6H_5O_7)_2$). Calcium citrate is then treated with sulfuric acid to produce citric acid, with gypsum as a by-product.

The largest use of calcium sulfate is in the construction industry, which accounts for more than 90% of its production. The hemihydrate or calcined form, as noted, is mixed with water to make plaster. The plaster slurry is poured between sheets of paper to make common drywall, also know as wallboard or sheetrock (Sheetrock is a brand name of USG corporation, the parent corporation of United States Gypsum Company). Drywall is highly fire resistant because of water's high specific heat and its presence in gypsum's hydrated form. The other large construction use of calcium sulfate is in its dihydrate uncalcined form as an ingredient in Portland cement. Portland cement contains 3.5% gypsum by weight. The gypsum retards hydration and prevents the cement from hardening too rapidly.

Gypsum has hundreds of other applications outside the construction industry. It can be used agriculturally for several purposes: to supply calcium and sulfur to soils, to balance pH, and to condition soil. Food grade calcium sulfate is used as a calcium supplement in enriched foods such as flour, cereals, and baked goods. It is used as a gelling and firming agent with canned vegetables. Calcium sulfate is the most common tofu coagulant. The positively charged calcium ion in calcium sulfate attracts the negatively charged groups in protein molecules, causing thermally denatured proteins to coagulate. Anhydrous calcium sulfate is used as a filler to whiten food and consumer products such as frostings, ice creams, paper, paints, and toothpaste. Powdered gypsum can be used as a chalk for marking athletic fields and other large areas.

CHEMICAL NAME = carbon dioxide
CAS NUMBER = 124–38–9
MOLECULAR FORMULA = CO_2
MOLAR MASS = 44.01 g/mol
COMPOSITION = C(27.3%) O(72.7%)
MELTING POINT = –56.6°C
BOILING POINT = –78.5°C (sublimation point)
DENSITY = 2.0 g/L (vapor density = 1.53, air = 1)

$$O=C=O$$

Carbon dioxide is a colorless, odorless gas present throughout the atmosphere and is an essential compound for life on Earth. It is found on other planets in the solar system. Mars's icecaps are primarily frozen carbon dioxide and Venus's atmosphere is mostly carbon dioxide. The discovery of carbon dioxide, credited to Joseph Black (1728–1799), played a critical role in supplanting the phlogiston theory and advancing the development of modern chemistry. Black, in his medical studies, was searching for a substance to dissolve kidney stones, but he switched his subject to a study of stomach acidity. Black was working with the carbonates magnesia alba (magnesium carbonate) and calcium carbonate (limestone) and observed that when magnesia alba was heated or reacted with acids, it produced a gas and a salt. Black, who published his work in 1756, called the gas "fixed air" and noted that it had properties similar to those described by Jan Baptista van Helmont (1577–1644) for *spiritus sylvestrius. Spiritus sylvestrius* was the gas produced during combustion processes, and van Helmont realized that this was the same gas produced during fermentation and when acids reacted with seashells.

Carbon dioxide accounts for 0.037% by volume of the atmosphere. Its low concentration means that most commercial supplies of carbon dioxide are acquired as by-products of industrial chemical reactions. Several methods can be used to produce large volumes of CO_2. The combustion of coke or other carbonaceous substances produces results in CO_2: $C_{(coke)} + O_2 \rightarrow CO_{2(g)}$. In combustion processes, CO_2 is concentrated by separating it from other gases using scrubbing and absorption techniques. Another source of CO_2 involves the calcination (slow heating) of carbonates such as limestone, $CaCO_3$: $CaCO_{3(s)} \xrightarrow{\Delta} CaO + CO_{2(g)}$.

This process takes place in a lime kiln in the production of precipitated calcium carbonate at temperatures of from 500°C to 900°C. Carbon dioxide is also produced as a by-product in fermentation reactions to produce alcohols. An example is the fermentation of glucose, $C_6H_{12}O_6$ to ethanol (C_2H_5OH): $C_6H_{12}O_{6(aq)} \xrightarrow{\text{yeast}} 2C_2H_5OH_{(aq)} + 2CO_{2(g)}$. Carbon dioxide is produced as a by-product in a number of syntheses, such as the Haber process, to produce ammonia.

Carbon dioxide has several major uses. Solid carbon dioxide, dry ice, is used as a refrigerant. Dry ice was first prepared in France by Charles Thilorier (1797–1852) in 1834. He observed dry ice when carbon dioxide expanded from pressurized containers. Thilorier reported his findings in 1835, but dry ice was not used commercially until the 1920s. The term *dry ice* was the original trademark of Prest Air Devices of Long Island in 1924. The company built a commercial plant to produce dry ice and changed its name to DryIce. Dry ice has been adopted generically as the name for solid CO_2. Dry ice is an appropriate name because at atmospheric pressure, carbon dioxide exists as a solid or gas depending on the temperature. The phase change from solid to gas without passing through the liquid stage is called sublimation. The reverse process from gas to solid is called deposition. Sublimation and deposition of carbon dioxide occur at −78.5°C.

Another major use of carbon dioxide is in the soda industry. Soda is sodium carbonate monohydrate $(Na_2CO_3 \cdot H_2O)$. Other forms of soda include washing soda, which is sodium carbonate decahydrate $(Na_2CO_3 \cdot 10H_2O)$, and baking soda, which is sodium bicarbonate $(NaHCO_3)$. Sodium carbonate is also referred to as soda ash because it was prepared by leaching the ashes of burnt wood. The Solvay process for producing soda ash is a series of reactions that uses ammonia, carbon dioxide, and water. Baking soda is produced from sodium carbonate and carbon dioxide according to the reaction: $Na_2CO_{3(s)} + H_2O_{(l)} + CO_{2(g)} \rightarrow 2NaHCO_{3(s)}$.

Carbon dioxide is used to produce carbonated beverages by bubbling CO_2 into them. The carbonation of beverages was discovered by the English chemist and Presbyterian minister Joseph Priestley (1733–1804). Priestley's first investigations in chemistry dealt with carbon dioxide produced in a brewery near his parish in Leeds. Priestley discovered that he could dissolve the gas in water to produce a pleasant-tasting beverage. Priestley had actually produced soda water, the equivalent of sparkling water. Priestley's process for producing carbonated water was made commercially feasible in the 1790s by Jacob Schweppe (1740–1821), the founder of the company that still bears his name.

Carbon dioxide is used as a gas in fire extinguishers, as an inflation gas for flotation devices, and as a propellant (for example in air guns). In recent years, the use of carbon dioxide as a supercritical fluid in green chemistry applications has increased. A supercritical fluid is a fluid with a temperature and pressure above its critical point. For CO_2, the critical temperature is 31.1°C and the critical pressure is 73 atmospheres. Beyond the critical point, the properties of liquid and gas merge. The phase diagram for CO_2 shown in Figure 22.1 illustrates sublimation. Below 5.1 atmospheres, which defines the pressure at the triple point where all three phases exist, CO_2 passes directly from solid to gas. Supercritical CO_2 has the penetrating power of a gas but the solubility properties of a liquid. Supercritical CO_2, abbreviated scCO$_2$, has a number of advantages when used to replace traditional organic solvents. These include its relative inertness, nonflammability, low toxicity, abundance, and low expense. Supercritical CO_2 has replaced the Freon substance methylene chloride, CH_2Cl_2, to extract caffeine from coffee and tea to produce decaffeinated products. Supercritical CO_2 is also used in place of other organic solvents to extract compounds used for pharmaceuticals, spices, flavorings, and industry. It can be used as a cleaner and degreaser. It is also finding increasing use in the dry cleaning industry, where it has started to replace perchloroethylene, C_2Cl_4.

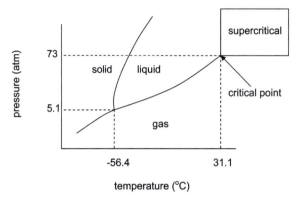

Figure 22.1   Phase diagram for carbon dioxide.

Plants require carbon dioxide for photosynthesis, and during respiration carbon dioxide is produced. Photosynthesis involves a series of biochemical reactions in which plants (and some bacteria) take inorganic carbon dioxide and water and use energy supplied from the sun to build carbohydrates. The carbohydrates are oxidized for energy during cellular respiration and are also used to build other compounds such as lipids, proteins, and nucleic acids. The general reaction for photosynthesis can be represented as: $nCO_2 + nH_2O \xrightarrow{\text{light}} (CH_2O)_n + nO_2$. The production of the carbohydrate glucose, $C_6H_{12}O_6$, is therefore represented by the equation: $6CO_2 + 6H_2O \xrightarrow{\text{light}} C_6H_{12}O_6 + 6O_2$

This reaction representing photosynthesis is highly simplified; the actual process involves numerous biochemical reactions, which take place in two sets called the light-dependent and light-independent (sometimes called light and dark) reactions. Oxygen is produced in the light-dependent reaction when water is split, and carbon dioxide is converted to carbohydrates in the light-independent reactions. The conversion of carbon dioxide into carbohydrates is termed carbon fixation.

Carbon dioxide readily dissolves in water to produce the weak acid carbonic acid, H_2CO_3. The presence of CO_2 in aqueous solution results in equilibria involving carbon dioxide, carbonic acid, bicarbonate, and carbonate: $CO_{2(aq)} + H_2O_{(l)} \rightleftharpoons H_2CO_{3(aq)} \rightleftharpoons H^+ + HCO_{3-} \rightleftharpoons H^+ + CO_3^{2-}$. The dissolution of CO_2 in water explains why rainwater is naturally acidic. Pure rainwater has a pH of about 5.6, owing to the presence of carbonic acid. The solubility of carbon dioxide in water has a significant impact on the earth's carbon cycle. Dissolved CO_2 can be incorporated into carbonate sediments in oceans and lakes. The solubility of CO_2 in blood as carbonic acid is one of the most important buffering systems in the human body. The pH of blood is about 7.4. If the pH of blood falls below 6.8 or above 7.8, critical problems and even death can occur. Three primary buffer systems control the pH of blood: carbonate, phosphate, and proteins. The primary buffer system in the blood involves carbonic acid and bicarbonate. The carbonic acid neutralizes excess base in the blood and the bicarbonate ion neutralizes excess acid. The excessive amount of bicarbonate in the blood means that blood has a much greater capacity to neutralize acids. Many acids accumulate in the blood during strenuous activity such as lactic acid. Excretion of bicarbonate through the kidneys and the removal of carbon dioxide through respiration also regulate the carbonic acid/bicarbonate blood buffer.

Carbon dioxide is a greenhouse gas, which makes it a major concern in the global warming and climate change controversy. Carbon dioxide is the most important anthropogenic greenhouse gas and is thought to be responsible for roughly 60% of recent global warming. The atmosphere's CO_2 concentration has risen from about 280 ppmv (parts per million by volume) to 370 ppmv in the last 160 years. This increase is due primarily to the burning of fossil fuels. Land clearing has also contributed to increased CO_2 levels. The greenhouse effect continues to be a controversial topic in the early 21st century. The controversy has shifted from uncertainty concerning whether humans affect the earth's climate to how much humans are affecting the climate and what action, if any, should be taken. A major difficulty surrounding the greenhouse effect is separating human influences from the climate's natural variability. Although there are still skeptics, the consensus seems to be that humans are affecting the climate, but exactly how, and to what extent, continues to be debated. The issue is complicated by numerous factors. The oceans play a major role in the carbon cycle because they have a great capacity to absorb carbon dioxide. Vegetation also affects global carbon concentrations through photosynthesis. Some studies have indicated that primary productivity increases with elevated CO_2, but others show no increase.

23. Carbon Monoxide

CHEMICAL NAME = carbon monoxide
CAS NUMBER = 630–08–0
MOLECULAR FORMULA = CO
MOLAR MASS = 28.0 g/mol
COMPOSITION = C(42.9%) O(57.1)
MELTING POINT = –205°C
BOILING POINT = –192°C
DENSITY = 1.26 g/L (vapor density = 0.97, air = 1)

Carbon monoxide is a colorless, odorless, tasteless, flammable, toxic gas. It was first identified by the Spanish alchemist Arnold of Villanova (1235–1313), who noted the production of a poisonous gas when wood was burned. The formal discovery of carbon monoxide is credited to the French chemist Joseph Marie François de Lassone (1717–1788) and the British chemist Joseph Priestley (1733–1804). The former prepared carbon monoxide by heating carbon in the presence of zinc, and for a time the compound was incorrectly identified as hydrogen. William Cumberland Cruikshank (1745–1800) correctly determined that carbon monoxide was an oxide of carbon in 1800.

Carbon monoxide is produced when carbon and carbon compounds undergo incomplete combustion. The inefficient combustion of carbon fuels for heating results in the production of carbon monoxide, which may result in high CO concentrations in indoor environments. The use of carbon fuel heaters without adequate ventilation can result in deadly conditions. Each year several hundred people in the United States die from CO poisoning, and 10,000 patients are treated in hospitals for CO exposure. Most of these cases result from faulty heating systems, but barbeques, water heaters, and camping equipment (stoves, lanterns) are also sources of CO.

Cars and other forms of transportation are a major source of carbon monoxide pollution in cities. Carbon monoxide concentrations are generally highest in the winter, especially when meteorological conditions create inversions trapping pollutants near the ground. To reduce these pollution episodes, governments use strategies to reduce CO emissions. The use of catalytic converters on vehicles became standard in the 1970s. Catalytic converters promote the complete combustion of emissions from engines by oxidizing carbon monoxide to carbon dioxide using a platinum catalyst.

Elevated carbon monoxide concentrations can lead to various health problems depending on exposure levels and duration of exposure. Health problems are a consequence of the blood hemoglobin's high affinity for carbon monoxide. Hemoglobin's affinity for CO is more than 200 times that of its affinity for oxygen. Carbon monoxide bonds to the iron in hemoglobin to form carboxyhemoglobin, which interferes with oxygen's ability to bind to hemoglobin to form oxyhemoglobin. Thus carbon monoxide is a chemical asphyxiant, which prevents oxygen from reaching body tissues.

The health effects of carbon monoxide make it a primary air pollutant. The federal government has established national standards for CO to protect the general population from this toxic gas. The national standard is 9 ppm (parts per million) averaged over 8 hours or 35 ppm averaged over 1 hour. Table 23.1 summarizes the health effects associated with different CO concentrations.

Table 23.1   Health Effects of Carbon Monoxide

CO level (ppm)	Effect
0–3	Normal levels found in ambient air
30–60	Shortness of breath during physical exertion and exercise
61–150	Shortness of breath, headache
151–300	Headache throbs, dizziness, nausea, blurred vision
301–600	Severe headache, nausea, vomiting, disorientation, possible loss of consciousness
601–1000	Convulsions, coma
1001–2000	Heart and lung damage
> 2000	Unconsciousness and possible death

Carbon monoxide is an important industrial chemical. It is produced, along with hydrogen, by steam reforming. In this process, methane is heated in the presence of a metal catalyst, typically nickel, to a temperature of between 700°C and 1100°C: $CH_{4(g)} + H_2O(g) \rightarrow CO_{(g)} + 3H_{2(g)}$. A mixture of hydrogen and carbon monoxide is called synthesis gas or syngas. Methyl alcohol is produced from syngas. The synthesis is conducted at high pressures, from 50 to 100 atmospheres, and in the presence of catalysts consisting of copper and oxides of zinc, manganese, and aluminum. The reaction is: $CO_{(g)} + 2H_{2(g)} \rightarrow CH_3OH_{(l)}$.

The sustained elevated price of crude oil seen in 2005 has led to increased interest in synthetic fuels. Synthetic fuels have been produced for more than 80 years through processes known as Fischer-Tropsch chemistry. Carbon monoxide is a basic feedstock in these processes. Franz Fischer (1852–1932) and Hans Tropsch (1889–1935) produced liquid hydrocarbons in the 1920s by reacting carbon monoxide (produced from natural gas) with hydrogen using metal catalysts such as iron and cobalt. Germany and Japan produced synthetic fuels during World War II. Low crude oil prices dictated little interest in synthetic fuels after the war, but as petroleum prices have soared, there is much greater interest in synthetic fuels. Several companies have built plants to produce diesel and convert natural gas to liquid fuel, a process known as gas-to-liquid technology.

Carbon monoxide is also useful as a reducing agent. It is used in metallurgy to obtain metals from their oxides. For example, during iron and steel production coke in a blast furnace is converted to carbon monoxide. The carbon monoxide reduces the Fe^{3+} in the iron (III) oxide contained in the iron ore to produce elemental iron according to the reaction: $Fe_2O_{3(s)} + 3CO_{(g)} \rightarrow 2Fe_{(l)} + 3CO_{2(g)}$.

24. Chloroform

CHEMICAL NAME = Trichloromethane
CAS NUMBER = 67–66–3
MOLECULAR FORMULA = $CHCl_3$
MOLAR MASS = 119.38 g/mol
COMPOSITION = C(10.1%) H(0.8%) Cl(89.1%)
MELTING POINT = –63.5°C
BOILING POINT = 61.7°C
DENSITY = 1.48 g/cm³

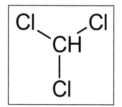

Chloroform is a clear, colorless liquid with a pleasant odor and sweet burning taste. It is used to make hydrochloroflurocarbons (HCFCs), as a solvent for organic chemicals, and in chemical synthesis. Its use in many commercial products has been eliminated in recent decades because of its toxic and carcinogenic properties. It was once used extensively as an anesthetic, in medicines, in dry cleaning, and in refrigerants. Several individuals discovered chloroform independently in 1831: Samuel Guthrie (1782–1848) in the United States, Eugéne Souberian (1797–1859) in France, and Justus von Liebig (1803–1873) in Germany. The French physiologist Marie Jean Pierre Flourens (1794–1867) reported on the anesthetic effect of chloroform on animals in 1847, but it was the Scottish physician James Young Simpson (1811–1870) who introduced its use in humans. Simpson administered chloroform as a substitute for ether, which was first used as an anesthetic in 1846, in 1847 to relieve pain during childbirth. After Simpson's demonstrated chloroform's efficacy in relieving labor pains, it was commonly administered during childbirth and as a general anesthetic during surgery and dentistry until the 1920s. Queen Victoria's use of chloroform for childbirth in 1853 popularized its use in Europe, whereas ether was a more widely used anesthetic in North America. Chloroform's medical use was controversial, as it was first administered to humans. Death associated with its use was not uncommon; the first death occurred in 1848 within a year of its first use. Chloroform was subsequently replaced by anesthetics and analgesics that had fewer detrimental side effects, which included cardiac arrhythmia, liver and kidney damage, and nausea.

Chloroform was first synthesized by treating acetone or ethanol with calcium hypochlorite or sodium hypochlorite bleaching powder. Chlorination of ethanol produces acetaldehyde and then trichloroacetaldehyde. Acetaldehyde yields chloroform and the formate ion by action of hydroxide ion. Acetone is chlorinated to trichloroacetone, which then splits into chloroform and the acetate ion. The modern industrial preparation of chloroform involves the chlorination of methane or methyl chloride, CH_3Cl, using heat to substitute the chlorine atoms for hydrogen (Figure 24.1). The reaction is carried out at approximately 500°C. Hydrochlorination by reacting methanol and hydrogen chloride can also be used to produce chloroform.

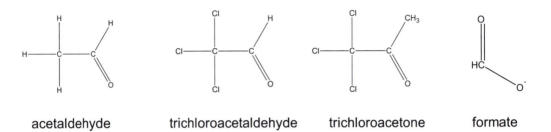

acetaldehyde   trichloroacetaldehyde   trichloroacetone   formate

Large volumes of chloroform were once used for the production of chlorofluorocarbons (CFCs), but the Montreal Protocol enacted in 1989 to eliminate CFCs as a result of their role in ozone destruction has decreased their use for this purpose. Chloroform is used to produce HCFCs, and hydrofluorocarbon (HFCs), which have been substituted for CFCs in recent years. HCFCs are due for phase out in the next decade. HCFC-22 (CHF_2Cl) is the primary HCFC produced (see Freon for information on the numbering scheme) and accounts for about 80% of chloroform's use. In its production, chloroform reacts with anhydrous hydrogen fluoride to produce HCFC-21 and HCFC-22: $CHCl_3 + HF \rightarrow CHFCl_2 + CHF_2Cl$. In addition to its major use as a feedstock for HCFCs and HFCs, chloroform is used as an organic solvent in a variety of applications including pharmaceuticals, resins, lacquers, rubbers, dyes, and pesticides.

Chloroform is produced naturally through the reaction of chlorine and organic compounds, most notably when chlorine used for disinfecting water reacts with organic compounds found

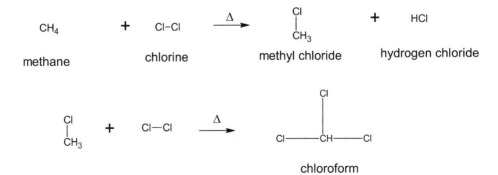

Figure 24.1 Modern production of chloroform.

in water bodies receiving treated wastewater to produce chloroform. In particular, hypochlorous acid (HOCl) formed when chlorine is added to water reacts with humic acids under certain conditions to form chloroform and other compounds known as trihalomethanes (THMs). THMs have the general formula CHX_3, where X represents chlorine or bromine atoms or a combination of the two. Chloroform is listed as a probable human carcinogen as a result of evidence suggesting that it causes liver and kidney cancers in animals. Because of health concerns, the Environmental Protection Agency has established a drinking water standard of 80 parts per billion for THMs. Some states have separate standards specifically for chloroform that may be as low as several parts per billion. The World Health Organization's water standard is 200 parts per billion.

CHEMICAL NAME = chlorophyll a

CAS NUMBER = 42617–16–3

MOLECULAR FORMULA =

$C_{55}H_{72}MgN_4O_5$

MOLAR MASS = 893.5 g/mol

COMPOSITION = C(73.9%) H(8.1%)

Mg(2.7%) O(9.0%) N(6.3%)

MELTING POINT = 117–120°C

BOILING POINT = decomposes

DENSITY = unknown

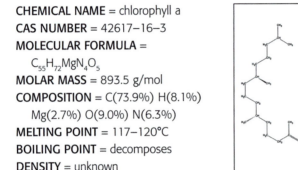
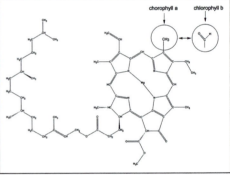

Chlorophyll is the principal green pigment found in plants and acts as the light-absorbing molecule responsible for photosynthesis. Several forms of chlorophyll exist that differ slightly in their molecular structure. The most common form of chlorophyll is chlorophyll a and the second most common form is chlorophyll b. The basic chlorophyll structure is characterized by a porphyrin ring system surrounding a single magnesium atom. The porphyrin ring is made of four pyrrole subunits. A long hydrocarbon chain attaches to the porphyrin ring. The difference between the two main forms of chlorophyll is in the side chain attached to one of the pyrrole group. In chlorophyll a, a methyl group is attached to the pyrrole, and in chlorophyll b the side chain consists of CHO. This is depicted in the structure's diagram by showing the two different side chains in the circles.

Pyrrole

Knowledge on chlorophyll paralleled advances in deciphering the photosynthetic process and the birth of modern chemistry. Joseph Priestley (1733–1804) discovered, during the 1770s, that plants replenished oxygen when placed in a container filled with fixed air (carbon dioxide). Building on the work of Priestley, Jean Senebier (1742–1809) discovered that oxygen was replenished while carbon dioxide was consumed by plants, and Jan Ingenhousz (1730–1799) determined that the green part of plants was responsible for replenishing oxygen and that plants required light to do this. At the beginning of the 19th century, Nicholas Theodore de Saussure (1767–1845) discovered that water and carbon dioxide were the source of hydrogen and carbon in plants, respectively. In 1818, Pierre Joseph Pelletier (1788–1842) and Joseph Bienaimé Caventou (1795–1877) isolated chlorophyll and gave it its name. The name was derived from the Greek words *chloros* meaning yellow-green and *phyllon* meaning leaf; therefore chlorophyll can be interpreted as green leaf. René-Joachim-Henri Dutrochet (1776–1847) was the first to recognize that chlorophyll was necessary for photosynthesis in 1837, and in 1845 Julius Robert von Mayer (1814–1878) proposed that plants convert light to chemical energy. Different chlorophylls were separated by chromatography in a process developed by the Russian Mikhail Semenovich Tsvett (1872–1919) at the beginning of the 20th century. Richard Martin Willstätter (1872–1942) used chromatography to isolate plant pigments and found that the structure of chlorophyll was similar to that of hemoglobin. He also isolated the two main types of chlorophyll: the blue-green compound known as chlorophyll a and the yellow variety know as chlorophyll b. Willstätter received the 1915 Nobel Prize in chemistry primarily for his work on chlorophyll. He determined the basic structure of chlorophyll. The complete structure of chlorophyll was determined by Hans Fischer (1881–1945), who received the 1930 Nobel Prize in chemistry. Fischer showed the relationship between chlorophyll and hemin and developed a synthesis for the latter.

The main chlorophylls found in green plants are chlorophyll a and chlorophyll b, with the former being dominate. All plants, green algae, and cyanobacteria that photosynthesize contain chlorophyll a; chlorophyll b occurs in plants and green algae. In addition to chlorophyll a and b, chlorophylls c, d, and e exist in various plants. Chlorophyll, located in the thylakoid membranes of chloroplasts, serves as the light-harvesting antennae in plants, gathering the energy that drives a series of biochemical reactions that ultimately convert radiant energy to chemical energy. Although numerous reactions take place in photosynthesis, the overall reaction is represented by: $6CO_2 + 6H_2O \rightarrow 6O_2 + C_6H_{12}O_6$.

The structure of chlorophyll is key to its role in energy transfer. The conjugated system of alternating single and double bonds produce delocalized electrons that can be excited into higher molecular orbitals by light. The release of energy when an excited electron returns to a lower molecular orbital produces visible light when its wavelength falls between 400 nm and 700 nm (light's visible range). When a photon of light strikes chlorophyll, an electron can absorb this energy and then transfers it to a neighboring molecule. The photon's energy can be transferred through the system of chlorophyll molecules until it arrives at a location in the chlorophyll called the reaction center. At the reaction center, an electron is transferred to an electron acceptor. The light-gathering chlorophyll antennae, the electron transfer chlorophyll, and the reaction center make up a photosystem. Two photosystems are associated with green plants: Photosystem I and Photosystem II, referred to as P700 and P680, respectively. The numbers 700 and 680 designate the wavelength of light in nanometers at which these systems are most efficient.

In Photosystem II, chlorophyll absorbs a photon of light, with maximum absorption occurring at 680 nm. The photon excites an electron in the chlorophyll, and this excited electron moves through the chlorophyll to chlorophyll's reaction center. Here the photon's energy is used by electron-transfer proteins to pump protons (hydrogen ions, H^+) into the thylakoid.

This establishes a proton gradient in which protons diffuse out of the thylakoid through ATP synthase, synthesizing adenosine triphosphate (ATP) from adenosing diphosphate (ADP) and P_i (inorganic phosphorus). At this point, the excited electron's energy has been partially spent and the electron moves to the Photosystem I reaction center. The electron absorbs additional light energy, with maximum absorption occurring at 700 nm, and the excited electron is used to produce NADPH (nicotinamide adenine dinucleotide phosphate hydrogen) by the reduction of NADP⁺. The hydrogen ions required by Photosystem II, as well as the electron balance, are maintained by the oxidation of water: $2H_2O \rightarrow 4H^+ + O_2 + 4e^-$. The reactions taking place in Photosystems I and II make up the light reactions in photosynthesis. The light reactions require light to produce the primary products of ATP and NADPH. The dark reactions in photosynthesis use ATP and NADPH to reduce carbon dioxide for the synthesis of carbohydrates. Chlorophyll a and b absorb strongly in the red and blue-green regions of the visible spectrum as shown by its absorption spectrum in Figure 25.1.

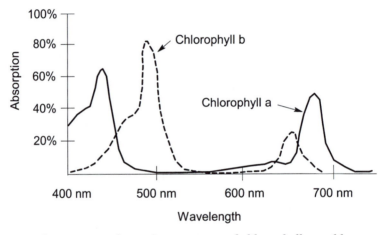

Figure 25.1   Absorption spectrum of chlorophyll a and b.

Because the blues and red hues are strongly absorbed, and the green wavelengths are transmitted and reflected, chlorophyll plant tissues such as leaves and stems appear green. Chlorophyll a is the primary pigment in plants, but plants contain accessory pigments including other chlorophylls as well as carotenes, anthocyanins, and xanthophylls. The range of pigments in chlorophyll b enables plants to capture light over a broader spectrum than is available for chlorophyll a. During the summer the abundance of chlorophyll a masks the color of accessory pigments. In autumn, changes in the photoperiod and cooler temperatures signal the end of summer and chlorophyll production decreases and eventually ceases; concurrently the production of other pigments may be stimulated. The loss of chlorophyll allows the display of other pigments, which produces the fall colors.

26. Cholesterol

CHEMICAL NAME = Cholest-5-en-3ß-ol
CAS NUMBER = 57–88–5
MOLECULAR FORMULA = $C_{27}H_{46}O$
MOLAR MASS = 386.7 g/mol
COMPOSITION = C(83.9%) H(12.0%)
   O(4.1%)
MELTING POINT = 148°C
BOILING POINT = 360°C
DENSITY = 1.05 g/cm³

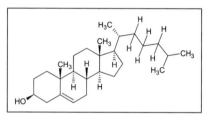

Cholesterol is a soft waxy substance that is a steroidal alcohol or sterol. It is the most abundant steroid in the human body and is a component of every cell. Cholesterol is essential to life and most animals and many plants contain this compound. Cholesterol biosynthesis occurs primarily in the liver, but it may be produced in other organs. A number of other substances are synthesized from cholesterol including vitamin D, steroid hormones (including the sex hormones), and bile salts. Cholesterol resides mainly in cell membranes.

Cholesterol was discovered in 1769 by Poulletier de la Salle (1719–1787), who isolated the compound from bile and gallstones. It was rediscovered by Michel Eugène Chevreul (1786–1889) in 1815 and named cholesterine. The name comes from the Greek words *khole* meaning bile and *steros* meaning solid or stiff. The "ine" ending was later changed to "ol" to designate it as an alcohol.

Humans produce about 1 gram of cholesterol daily in the liver. Dietary cholesterol is consumed through food. High cholesterol foods are associated with saturated fats and trans-fatty acids (commonly called trans fats). Dietary cholesterol comes from animal products (plants contain minute amounts of cholesterol) such as meats and dairy products. Table 26.1 shows the amount of cholesterol in common foods.

Cholesterol is commonly associated with cardiovascular disease and its routine measurement is used to measure its potential health risk. High blood serum cholesterol levels are often correlated with excessive plaque deposits in the arteries, a condition known as atherosclerosis

Table 26.1   Cholesterol in Common Foods

Item	Quantity	Cholesterol in mg.
Butter	1 tablespoon	30
Mozzarella cheese	1 oz.	22
Cheddar cheese	1oz.	30
Egg	1	200
Chicken	4 oz.	70
Liver	4 oz.	340
Ham	4 oz	80
Skim milk	1 cup	5
Whole milk	1 cup	35
Ice cream	½ cup	30
Low-fat ice cream	½ cup	10

or hardening of the arteries. Although high total blood cholesterol levels are associated with heart disease, it is important to distinguish between types of cholesterol when interpreting cholesterol levels. Cholesterol has been labeled as "good" and "bad" depending on its physiological role. Forms of cholesterol depend on the lipoproteins that are associated with it. Low-density lipoprotein cholesterol (LDL cholesterol) is often referred to as "bad" cholesterol and high-density lipoprotein (HDL) is identified as "good" cholesterol. An understanding of the difference between LDL and HDL cholesterol requires an understanding of substances associated with cholesterol in the body. Cholesterol is a lipid so it has very low solubility in water and blood. For the cholesterol synthesized in the liver to be delivered by the bloodstream to the rest of the body, the liver manufactures lipoproteins that can be viewed as carriers for cholesterol (and triglycerides). Lipoproteins, as the name implies, are biochemical assemblages of fat and protein molecules. Several different types of lipoproteins are found in the human blood. A lipoprotein can be viewed as a globular structure with an outer shell of protein, phospholipid, and cholesterol surrounding a mass of triglycerides and cholesterol esters. The proteins in lipoproteins are called apolipoproteins, with different apolipoproteins associated with different lipoproteins.

Cholesterol leaves the liver in the form of very-low-density lipoprotein (VLDL). VLDL has a high percentage (50–65%) of triglycerides and relatively low protein composition of 10% or less. The percentage of fat and protein in different forms of lipoproteins dictates their density; a greater proportion of protein gives a higher density. As the VLDL moves through the bloodstream, it encounters an enzyme called lipoprotein lipase in the body organs' capillaries, which causes the triglycerides to be delivered to cells. Triglycerides are used for energy or stored as fat. As the triglycerides are depleted from the lipoprotein, it becomes intermediate density lipoprotein (IDL). As IDL circulates in the blood, cell structures called LDL receptors bind to the apolipoprotein called Apo B-100 with its enclosed cholesterol and in the process converts IDL to LDL. This delivers cholesterol to the cell. Apo B-100 allows LDL cholesterol to be delivered to the tissues, but it has a tendency to attach to blood vessel walls. The accumulation impedes blood flow and can build up as plaque and lead to atherosclerosis. Atherosclerosis is a type of arteriosclerosis (the latter being a more general term to include normal aging processes) that occurs when excess cholesterol combines with a number of other substances including other fats, lignin, and calcium to form a hard deposit on the inner lining of the blood vessels. Because of the problems associated with Apo B and LDL, LDL cholesterol is labeled as "bad" cholesterol. HDL is produced in the liver, intestines, and other tissues. It has a low level of triglycerides, but a high protein content of approximately 50%. HDL cholesterol or "good" cholesterol transports cholesterol in the bloodstream back to the liver where it is broken down and excreted. Although HDL are labeled as good, it has not been demonstrated definitively that high HDL reduces heart disease, but there is an inverse relation between HDL and heart disease.

A proper balance of cholesterol in the bloodstream requires having an adequate balance of receptors to process the amount of cholesterol in the blood. Receptors are continually regenerated, produced, and disappear in the cell in response to blood biochemistry. The liver contains the greatest concentration of receptors. Too few receptors or excess dietary cholesterol intake can lead to elevated blood cholesterol. A genetic disorder called familial hypercholesterolemia results when a person inherits a defective gene from one parent resulting in the inability to produce sufficient receptors. A diet with too much cholesterol represses the production of LDL receptors and leads to high blood cholesterol and Apo B.

Standard lipid screening to obtain a cholesterol profile for the risk of cardiovascular disease routinely reports total cholesterol, LDL cholesterol, HDL cholesterol, and triglycerides. Cholesterol values are reported in milligrams per deciliter of blood (mg/dL). Different organizations have made recommendations for normal cholesterol levels, but these must be interpreted carefully, as they are contingent on other risk conditions. For example, the recommendations for smokers or those with a family history of heart disease will be lower for someone without these conditions. The National Center for Cholesterol Education (NCEP) endorsed by the American Heart Association believes that LDL is the primary cholesterol component to determine therapy. LDL cholesterol accounts for 60–70% of blood serum cholesterol. An LDL less than 160 mg/dL is recommended for individuals with no more than one risk factor and less than 100 mg/dL for individuals with coronary heart disease. NCEP classifies HDL, which comprises between 20% and 30% of blood cholesterol, below 40 mg/dL as low. Triglycerides are an indirect measure of VLDL cholesterol. The NCEP considers a normal triglyceride level as less than 150 mg/dL.

Heart disease is the leading cause of death in adults over 35 years old in the United States, with more than 1 million deaths annually. Cholesterol's role in contributing to heart disease has led to several broad strategies to lower blood cholesterol. Major treatment strategies to control cholesterol include changes in diet, lifestyle changes, and drug therapy. Dietary changes for lowering cholesterol primarily involve reducing fat intake, especially saturated fats and trans-fats. Trans-fats are made when liquid oils are hydrogenated (or more likely partially hydrogenated) to solidify them. During this process the hydrogens bonded to carbons are reconfigured from being on the same side of the double bond (cis position) to a cross or trans position. Saturated and trans-fats raise the LDL cholesterol, but trans-fats also lower HDL. Starting on January 1, 2006, the Food and Drug Administration (FDA) required trans-fat content to be included on nutritional labels of foods sold in the United States. Another dietary strategy is to eat foods high in soluble fiber such as oatmeal, oat bran, citrus fruits, and strawberries. Soluble fiber binds to cholesterol and eliminates it in the feces.

Eating foods and taking supplements containing omega-3 fatty acids are another strategy for lowering cholesterol. In addition to changes in the diet, other lifestyle changes include exercise, smoking cessation, and losing weight.

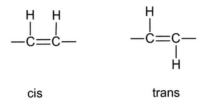

cis   trans

In the last 20 years the use of statin drugs has revolutionized the treatment of heart disease. Statins work by inhibiting the enzyme HMG-CoA reductase, which is required to produce cholesterol in the liver. During cholesterol biosynthesis, HMG-CoA (3-hydroxy-3-methyl-glutaryl-CoA) is converted to mevalonate. All statin drugs contain a structure similar to mevalonate. Generic names of statins are sold under specific brand names. For example, lovastatin, which was the first FDA-approved statin in 1987, was marketed as Mevacor by Merck. Atorvastatin is sold as Lipitor by Pfizer and Merck sells simvastatin as Zocor. Several of the top-selling drugs worldwide are statins. Lipitor has been the top-selling drug for several years, with annual sales in 2005 approximately $13 billion and Zocor binging in over $5 billion.

27. Citric Acid

CHEMICAL NAME = 2-hydroxypropane-1,2,3-tricar-
boxylic acid
CAS NUMBER = 77–92–9
MOLECULAR FORMULA = $C_6H_8O_7$
MOLAR MASS = 192.1 g/mol
COMPOSITION = C(37.5%) H(4.2%) O(58.3%)
MELTING POINT = 153°C
BOILING POINT = decomposes at 175°C
DENSITY = 1.66 g/cm³

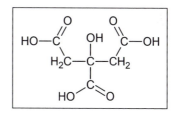

Citric acid is a white, crystalline, weak organic acid present in most plants and many animals as an intermediate in cellular respiration. Citric acid contains three carboxyl groups making it a carboxylic, more specifically a tricarboxylic, acid. The name citrus originates from the Greek *kedromelon* meaning apple of melon for the fruit citron. Greek works mention *kitron, kitrion,* or *kitreos* for citron fruit, which is an oblong fruit several inches long from the scrublike tree *Citrus medica.* Lemons and limes have high citric acid content, which may account for up to 8% of the fruit's dry weight. The discovery of citric acid is credited to Jabir ibn Hayyan (Latin name Geber, 721–815). Citric acid was first isolated in 1784 by the Swedish chemist Carl Wilhelm Scheele (1742–1786), who crystallized it from lemon juice.

Citric acid is a weak acid and loses hydrogen ions from its three carboxyl groups (COOH) in solution. The loss of a hydrogen ion from each group in the molecule results in the citrate ion, $C_3H_5O(COO)_3^{3-}$. A citric acid molecule also forms intermediate ions when one or two hydrogen atoms in the carboxyl groups ionize. The citrate ion combines with metals to form salts, the most common of which is calcium citrate. Citric acid forms esters to produce various citrates, for example trimethyl citrate and triethyl citrate.

Industrial citric acid production began in 1860 and for the next 60 years was dominated by Italian producers. The original production method was based on extraction from the juice of citrus fruits by adding calcium oxide (CaO) to form calcium citrate, $Ca_3(C_6H_5O_7)_2$, as an insoluble precipitate that can then be collected by filtration. Citric acid can be recovered from its calcium salt by adding sulfuric acid. Citric acid production using sugar fermentation was first reported in 1893 by Carl Wehmer. Wehmer fermented sugar using the fungi he named *Citromyces*. Wehmer was also studying oxalic acid production and attributed its production to the fungus *Aspergillus niger*. Wehmer published numerous articles on the production of acids from fungi and other researchers sought to commercialize the process. In 1917, James N. Currie (1883–?), who worked as a food chemist for the U.S. Department of Agriculture, published an article on the production of citric acid using *Aspergillus niger*. Currie's article presented the conditions that increased the yield of citric acid over oxalic acid during fermentation. Currie found that by controlling the cultures, culture medium, temperature, acidity, etc., he could obtain a high yield of citric acid. That same year Currie was hired by Pfizer where he worked on the industrial fermentation process to produce citric acid.

Citric acid was Pfizer's leading product. Pfizer had sold citric acid since the 1880s and was interested in acquiring an alternative to its Italian sources for raw materials (concentrated lemon and lime juice), which was prone to disruption because of wars, weather, and political instability. Pfizer started to mass produce citric acid in 1919 using methods developed under Currie's leadership. During the 1920s, Pfizer refined and increased its citric acid production methods, and this led to the collapse of the Italian citric acid industry. In 1922, Italy controlled approximately 90% of the world citric acid market and by 1927 most exports had ceased. By 1929, Pfizer was producing all of its own citric acid. Pfizer's development of citric acid production continued through the 1930s. In the original method of citric acid production, fermentation was carried out in trays on the surface of the sugar solution. Pfizer developed a deep tank method in which fermentation could take place in a submerged environment using aeration inside a tank. The deep tank method resulted in greater productivity and efficiency, further resulting in lower cost citric acid. In 1920, a pound of citric acid cost $1.25 and 20 years later it was selling for about $0.20 cents per pound.

Both the surface tray and deep tank methods are used to produce citric acid today. In the surface process, sterilized air circulates over a layer of the medium consisting of sugar (typically dextrose or molasses), salts, and nutrients. *Aspergillus niger* spores introduced on the surface ferment the sugar over 6 to 10 days. This method is not used in the United States but is more common in less industrialized nations. The fermentation cycle runs 5 to 14 days. In the deep tank (submerged) process, fermentation takes place over 5 to 10 days in stirred stainless steel tanks or aerated towers. The deep tank process is preferred for large-volume production in the industrial world. It requires less labor and less space per volume of citric acid produced, is easier to maintain sterile conditions, and results in higher production capacity. One disadvantage of the deep tank process is the higher energy costs. Citric acid yield from submerged culture fermentation processes can range between 80% and 95% per weight of sugar. After fermentation, the citric acid is separated from the broth by treating the broth with calcium hydroxide, $Ca(OH)_2$ to precipitate calcium citrate. Citric acid is regenerated from the calcium citrate by treating it with sulfuric acid.

Citric acid and its citrate compounds are widely used in hundreds of applications. Global production of citric acid in 2005 was 1.6 million tons, with China producing approximately 40% of the world supply. In the United States, approximately 65% of citric acid use is in the food and beverage industry. Citric acid is used as an acidulant to impart tartness, to control pH, as a preservative and antioxidant, as a metal chelator, and to stabilize color and taste. Citrate salts can be used as mineral and metal dietary supplement; for example, calcium citrate is used as a calcium supplement. The second greatest use of citric acid is in detergents and cleaning products. Sodium citrate is used as a builder. Citric acid's ability to chelate metals makes it useful as a water-softening agent, which can also assist in cleaning. Approximately 10% of citric acid production is used in the pharmaceutical industry. Citric acid's largest use in pharmaceuticals is as an effervescent when combined with carbonates or bicarbonates such as in Alka-Seltzer. As an effervescent, it improves tastes, buffers, and improves solubility of ingredients. It is also used in pharmaceuticals to impart tartness to mask unpleasant medicinal flavors, maintain stability, and as a buffering agent.

Citric acid is formed during cellular respiration in most organisms' mitochondria through a series of chemical reactions called the citric acid or Krebs cycle. It is called the citric acid cycle because citric acid is the first intermediate produced in the process. The key pathways in this cycle were determined by Hans Adolf Krebs (1900–1981) in 1937 for which he received the Nobel Prize in physiology or medicine in 1953. The cycle starts when acetyl coenzyme A, which is synthesized from digested food, combines with oxaloacetate to produce citryl coenzyme A, which then hydrolyzes to citrate. The citrate then goes through a set of reactions to change it to isocitrate. Further oxidation and reduction reactions result in overall energy production. The cycle ends with the regeneration of oxaloacetate, which can combine with another acetyl coenzyme A molecule and the process starts over. Each cycle of the process reduces three molecules of NAD^+ to NADH (nicotinamide adenine dinucleotide), whereas a molecule of FAD (flavin adenine dinucleotide) is converted to its reduced form, $FADH_2$. NADH and $FADH_2$ move to the electron transport chain, where they lose hydrogen and electrons in another series of reactions. The electron transport chain in turn releases energy synthesizing ATP through oxidative phosphorylation (see Adenosine Triphosphate).

CHEMICAL NAME = 8-Azabicyclo[3.2.1]
octane-22carboxylic acid, 3-(benzoyloxy)-
8-methyl-, methyl ester
CAS NUMBER = 50–36–2
MOLECULAR FORMULA = $C_{17}H_{21}NO_4$
MOLAR MASS = 303.4 g/mol
COMPOSITION = C(67.3%) H(7.0%)
N(4.6%) O(21.1%)
MELTING POINT = 98°C
BOILING POINT = 187°C
DENSITY = 1.22 g/cm³ (calculated)

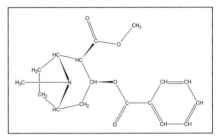

Cocaine is best known as an illegal drug that produces a euphoric "high" in individuals who use it. Cocaine is an alkaloid obtained from the leaves of the coca plant, *Erythroxylum coca,* which is native to northwestern South America and Central America. Native Indians in the Andes have chewed coca leaves for thousands of years, and early Spanish explorers noted the stimulating effect that chewing the leaves had on these people. Coca was traditionally reserved for royalty and religious ceremonies in many of these cultures. It was the most sacred plant for the Incas, and its use was reserved for priests and nobility. For many indigenous populations coca was an important food that provided nourishment and essential nutrients and was therefore widely cultivated. Natives carried pouches of coca leaves called *chuspas* and distances and time would be measured by the length of a chew. The time of a chew was called a *cocada.* Its traditional use among pre-Columbian cultures varied, but the subsequent conquest of these groups helped establish its general use among common people. This was related to the apparent ability of coca to provide stamina as well as induce insomnia to users.

Spanish conquistadors introduced coca to Europe and the original missionaries unsuccessfully attempted to ban it use. The Catholic Church viewed its use as an act of paganism and a remnant of native religious ceremonies. Despite this position and the Church's destruction of cultured coca crops, coca's widespread use among many isolated cultures prevented significant elimination. As conquered native populations were enslaved, Europeans saw the utility of coca as a stimulant to induce greater work out of people. Furthermore, control of coca plants provided early European settlers a valuable economic commodity to obtain goods and labors from natives. Subsequently, King Philip II of Spain (1527–1598) lifted any ban on coca, gave land grants to establish coca plantations, and imposed a tax on it. Discovery of Andean silver further stimulated coca use, as Europeans used coca leaves to boost slave labor.

Coca was touted in Europe as a great elixir and its use increased between the 16th and 19th centuries. In 1855, the German chemist Friedrich Gaedcke (1828–1890) succeeded in isolating the active ingredient in coca leaves and called it erythroxyline. An improved process for isolating cocaine was discovered by Albert Niemann (1834–1861) during his dissertation work in 1860. Soon after Niemann's success, an explosion of cocaine in numerous therapeutic products ensued. It was widely used as a topical anesthetic in dentistry (Figure 28.1) and in ophthalmology; it also found use as an appetite suppressant, a drug used to treat morphine addiction, a stimulant, and a general elixir. Its popularity to treat depression was originally advanced by Sigmund Freud (1856–1939). In 1900, it was among the top five medicinal products in the United States.

Coca extracts were also added to common food items. A popular wine called Vin Mariani was concocted by the chemist Angelo Mariani (1832–1914) in 1863. Mariani's Bordeaux mixture used coca leaves. Cocaine from the coca leaves was extracted by the ethanol in the

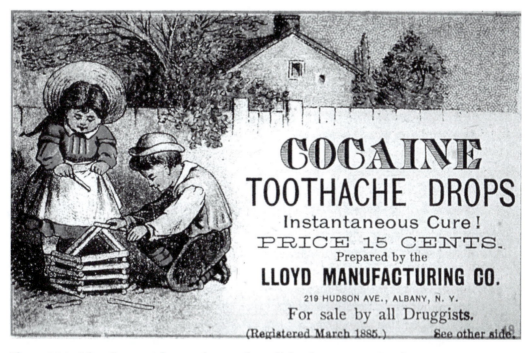

Figure 28.1. Advertisement for cocaine tooth medicine from 1885. Cocaine was extensively used in the latter part of the 19th century in medicines, as a stimulant, as an anesthetic, and to treat morphine addiction. *Source*: National Library of Medicine of the National Institute of Health.

wine. After Mariani's success a number of other vintners used coca leaves to infuse cocaine into their wines. One of these was a druggist from Atlanta named John Stith Pemberton (1831–1888). Pemberton produced his own version of a coca wine, but he also added the cola nut, which was also believed to have therapeutic properties. Reacting to the temperance movement, Pemberton sought to produce a nonalcoholic version of his beverage. Pemberton obsessively worked to find a new formula using the coca leaf and cola nut ingredients. On May 8, 1886, the first coke was served at Jacobs Pharmacy in Atlanta. In establishing his new drink, which was to be marketed as an invigorating tonic, Pemberton sought a unique name. Frank Robinson was one of Pemberton's partners and his bookkeeper. Robinson, who was a keen marketer, joined the names of the two ingredients Coca and Cola together and designed the unique script of Coca-Cola. Coca-Cola was not an instant success. Pemberton, who suffered from morphine addiction, sold his company interests and Coca-Cola's formula during his last months and never realized the eventual success of the company. Although the cocaine extract of coca was eliminated from Coca-Cola in 1906 because of passage of the Pure Food and Drug Act, the formula still calls for other coca extracts for flavoring.

At the end of the 19th century, people started to become aware of the addictive and medical problems associated with cocaine. Cocaine is a stimulant and its use produces a euphoric high accompanied by increased motivation, energy, and libido. Concurrently, it has physiological effects that include increased pulse rate, breathing, and blood pressure (it is a vasoconstrictor); muscle tension; loss of appetite; and insomnia. Unfortunately, after the initial euphoric high, which may last several minutes to several hours, the user experiences a letdown. This state of depression leaves the user craving another dose and the vicious cycle of drug addiction has begun. One theory for cocaine's effect is related to its role in disrupting the neurotransmitter dopamine. Cocaine occupies receptor areas on nerve cells blocking dopamine from the cell. The dopamine in the cell discharges its signal in the synapse and leads to a prolonged and extended buildup of dopamine. Increased dopamine affects the pleasure center of the brain and the elevated dopamine produces the high. The neurons respond to cocaine use by reducing the number of dopamine receptors; therefore when the brain returns to normal conditions, the lack of receptors and decrease in dopamine results in depression.

The effects of cocaine vary according to how it is consumed, individual differences, dose, and frequency of use. The most common form is the crystalline salt cocaine hydrochloride. Cocaine in this form is water-soluble and can be pulverized into a fine powder and can be "snorted" or inhaled through the nose. Here it is adsorbed onto the mucous membranes and then absorbed into the bloodstream through mucous membranes. Cocaine can also be prepared as an aqueous solution and directly injected into the bloodstream. This method delivers cocaine quickly to the brain and the user can experience a high in a matter of minutes. Free base cocaine is cocaine in which the hydrochloride has been removed to produce a more pure product that can be used for smoking. Cocaine hydrochloride is not suitable for smoking because it vaporizes at too high a temperature. Free base is prepared by making an aqueous solution of cocaine hydrochloride with baking soda (sodium bicarbonate) or ammonia and then boiling the solution down to give the free base. Cracking sounds during the process led to the name "crack" for free-base cocaine.

Colombia is the world's leading producer of cocaine, with about 75% of the world's production. Coca is grown locally and is also imported from Peru and Bolivia. The processing of coca involves mashing the leaves with a base, kerosene, and sulfuric acid to produce a paste containing between 40% and 70% cocaine. It can then be exported where it is dried and purified into cocaine hydrochloride. Estimates of global consumption of cocaine vary, but a reasonable approximation is roughly 750 tons. Of this amount, approximately one-third is imported into the United States, which is the leading consumer of cocaine.

Cocaine is used medicinally for local anesthesia and vasoconstriction, especially in surgery involving the ear, nose, and throat. It is the only naturally occurring anesthetic. Although it is still used in limited quantities for surgery, many surgeons and anesthesiologists have turned to safer alternatives such as lidocaine and benzocaine. Also, the use of alternatives eliminates the storage of a well-known addictive drug in clinics and hospital pharmacies.

29. Cytosine, Thymine, and Uracil

Cytosine

CHEMICAL NAME = 4-Aminopyrimidin-2(1*H*)-one
CAS NUMBER = 71–30–7
MOLECULAR FORMULA = $C_4H_5N_3O$
MOLAR MASS = 111.1 g/mol
COMPOSITION = C(43.2%) H(4.5%) N(37.8%) O(14.4%)
MELTING POINT = 100°C
BOILING POINT = Decomposes at 320°C
DENSITY = 1.55 g/cm^3 (calculated)

Thymine

CHEMICAL NAME = 5-Methylpyrimidine-2,4(1*H*,3*H*)-dione
CAS NUMBER = 65–71–4
MOLECULAR FORMULA = $C_6H_6N_2O_2$
MOLAR MASS = 126.1 g/mol
COMPOSITION = C(47.6%) H(4.8%) N(22.2%) O(25.4%)
MELTING POINT = 316°C
BOILING POINT = Decomposes at 335°C
DENSITY = 1.23 g/cm^3 (calculated)

Uracil

CHEMICAL NAME = Pyrimidine-2,4(1*H*,3*H*)-dione
CAS NUMBER = 66–22–8
MOLECULAR FORMULA = $C_4H_4N_2O_2$
FORMULA WEIGHT = 112.1 g/mol
COMPOSITION = C(42.9%) H(3.6%) N(25.0%) O(28.6%)
MELTING POINT = 335°C

BOILING POINT = Decomposes
DENSITY = 1.32 g/cm³

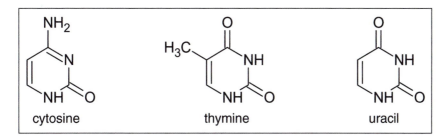

cytosine   thymine   uracil

Cytosine, thymine, and uracil are pyrimidines; along with adenine and guanine they account for the five nucleic acid bases. Pyrimidines are heterocyclic single-ringed compounds based on the structure of pyrimidine. Cytosine, thymine, and uracil, like adenine and guanine, form nucleosides and nucleotides in RNA and DNA. When the bases combine with ribose, a ribonucleoside forms; and when it attaches to deoxyribose, a deoxyribosenucleoside is formed. Names of the nucleoside are summarized in Table 29.1. These in turn combine with phosphoryl groups, in a process called phosphorylation, to form their respective nucleotides that form nucleic acids. The nucleotides can be tri, di, and mono phosphate nucleotides similar to the way in which adenine forms ATP, ADP, and AMP.

In nucleic acid base pairing, adenine (A) groups bond to thymine (T) groups and guanine (G) groups bond to cytosine (C) in DNA. In RNA, uracil (U) replaces thymine so that the base pairing is adenine to uracil in RNA. Cytosine, thymine, and uracil have similar structures and human genetics involves reactions where one base can be converted to another. The deamination (removal of NH_2 group) of cytosine produces uracil. This helps explains why uracil is not normally present in DNA. If U were present rather than thymine, a G-U pairing could come from the cytosine in the C-G pairing converting to uracil. This would produce a U-G mismatch. This mismatch could also occur if the adenosine in an A-U pairing, which would normally occur if uracil replaced thymine, mutated to guanine. In short, U would not form

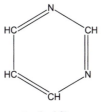

Pyrimidine

Table 29.1   Nucleosides

	Ribonucleoside	**Deoxyribonucleoside**
Cytosine	Cytidine	Deoxycytidine
Thymine	Thymidine	Deoxythymidine
Uracil	Uridine	Deoxyuridine

a distinct base pair, making it possible to distinguish normal and mutagenic conditions. The presence of thymine means that any uracil found in DNA is a general indication of mutation. Biochemical processes occur to remove the uracil in DNA and replace it with cytosine. Thus the methylation of uracil to thymine in DNA synthesis protects the DNA molecule.

Cytosine and thymine were first isolated by hydrolysis of calf thymus tissue by Albrecht Kossel (1853–1927) and A. Neumann during 1893–1894. Thymine's structure was published in 1900 and confirmed over the next several years when it was synthesized by several investigators. In 1903, cytosine was synthesized by Henry Lord Wheeler (1867–1914) and Treat B. Johnson, confirming its structure. Uracil was first isolated in 1900 from yeast nucleic acid found in bovine thymus and herring sperm. The methylation of uracil produces thymine; thymine is also called 5-methyluracil because methylation takes place at the fifth carbon in uracil to produce thymine.

The amine bases in DNA are adenine, thymine, cytosine, and guanine, symbolized by A,T, C, and G, respectively. RNA contains adenine, cytosine, and guanine, but thymine is replaced by the base uracil. The primary structure of nucleic acids is given by the sequence of the amine side chains starting from the phosphate end of the nucleotide. For example, a DNA sequence may be T-A-A-G-C-T. Genes, residing in the chromosomes, are segments of the DNA molecule. The sequence of nucleotides, represented by their letters, corresponding to a specific gene may be hundreds or even thousands of letters long. Humans have between 50,000 and 100,000 genes contained in their 46 chromosomes, and the genetic code in humans consists of roughly 5 billion base pairs.

The process by which the information in DNA is used to synthesize proteins is called transcription. Transcription involves turning the genetic information contained in DNA into RNA. The process starts just like DNA replication with the unraveling of a section of the two strands of DNA (see Adenine). A special protein identifies a promoter region on a single strand. The promoter region identifies where the transcription region begins. An enzyme called RNA polymerase is critical in the transcription process. This molecule initiates the unwinding of the DNA strands, produces a complementary strand of RNA, and then terminates the process. After a copy of the DNA has been made, the two DNA strands rewind into their standard double helix shape. The RNA strand produced by RNA polymerase follows the same process as in DNA replication except that uracil replaces thymine when an adenosine is encountered on the DNA strand. Therefore, if a DNA sequence consisted of the nucleotides: C-G-T-A-A, the RNA sequence produced would be G-C-A-U-U. The transcription process occurs in the cell's nucleus and the RNA produced is called messenger RNA or mRNA. Once formed, mRNA moves out of the nucleus into the cytoplasm where the mRNA synthesizes proteins. The transfer of genetic information to produce proteins from mRNA is called translation. In the cytoplasm, the mRNA mixes with ribosomes and encounters another type of RNA called transfer RNA or tRNA. Ribosomes contain tRNA and amino acids. The tRNA translates the mRNA into three-letter sequences of nucleotides called codons. Each three-letter sequence corresponds to a particular amino acid. Because there are four nucleotides (C, G, A, and U), the number of different codons would be equal to 4^3 or 64. Because there are only 20 standard amino acids, several codons may produce the same amino acid. For example, the codons GGU, GGC, GGA, and GGG all code for glycine. Three of the codons serve as stop signs to signal the end of the gene. These stops also serve in some case to initiate the start of a gene sequence. The sequential translation of mRNA by tRNA builds the amino acids into the approximately 100,000 proteins in the human body.

CHEMICAL NAME = 1-chloro-2-[2,2,2-trichloro-1-
(4-chlorophenyl)ethyl]benzene
CAS NUMBER = 50–29–3
MOLECULAR FORMULA = $C_{14}H_9Cl_5$
MOLAR MASS = 354.5 g/mol
COMPOSITION = C(47.4%) H(2.6%) Cl(50.0%)
MELTING POINT = 108.5°C
BOILING POINT = 260°C
DENSITY = 1.5 g/cm³

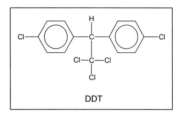

DDT is a polychlorinated persistent chemical that exists as a solid under normal conditions. In 1939, the Swiss chemist Paul Müller (1899–1965), working for the Geigy chemical company, discovered that the compound dichlorodiphenyltrichloroethane (DDT) was an effective insecticide. DDT was first synthesized in 1873 by an Austrian student, but it was Müller who discovered its efficacy as an insecticide. DDT was initially marketed in 1941 and found its first widespread use during World War II. During World War I several million deaths, including 150,000 soldiers, were attributed to typhus. There are several forms of typhus, but the most common form is due to bacteria carried by lice. During World War II, fearing a repeat of World War I typhus outbreaks, the Allied forces used DDT to combat typhus in addition to malaria, yellow fever, and other diseases carried by insects. Soldiers liberally applied talcum powder containing 10% DDT to clothes and bedding to kill lice. America and its European allies were relatively free from typhus and other diseases, whereas the Germans, who did not use DDT, had many more noncombat deaths resulting from infectious diseases. DDT solutions were sprayed in areas of the Pacific Theater to prevent malaria and yellow fever. In addition to its use in the war, DDT was used by civilians in tropical areas as a generic insecticide to prevent infectious diseases, especially malaria. Once the war ended, the use of DDT to advance public health in tropical developing countries was expanded for use in agriculture in developed countries. Paul Müller was awarded the Nobel Prize in physiology or medicine in 1948 for his discovery of the insecticide potential of DDT. By 1950, DDT and several related compounds were viewed as miracle insecticides that were inexpensive and that could be used indiscriminately.

Even though DDT seemed to be a cheap and effective pesticide, enough was known in its early development to raise concerns. DDT is a persistent chemical that lasts a long time in the environment. DDT is fat-soluble and not readily metabolized by higher organisms. This meant that DDT accumulated in the fat tissues of higher organisms. Organisms with longer life spans residing higher on the food chain continually fed on organisms lower on the food chain, accumulating DDT in their tissues. For example, the concentration of DDT in a lake might be measured in parts per trillion, plankton in the lake may contain DDT in parts per billion, fish a few parts per million, and bird feeding on fish from the lake several hundred parts per million. The accumulation of a chemical moving up the food chain is a process known as biological magnification (Figure 30.1). Another concern was that certain pests seemed to develop immunity to DDT and the application rate had to be increased to combat insects. This immunity occurred because natural selection favored insects that had the genetic characteristics to survive DDT and passed this ability on to their offspring. Direct deaths of bird and fish populations had also been observed in areas with heavy DDT use. Problems associated with DDT and other post World War II organic pesticides became a national concern with Rachael Carson's (1906–1964) publication of *Silent Spring* in 1962. Carson's book alerted the public to the hazards of insecticides, and although the book did not call for a ban, Carson challenged the chemical and agricultural industry to curtail its widespread use of chemical pesticides. Most developed countries started to ban the use of DDT and related compounds in the late 1960s. DDT was banned in the United States in 1973. Although it has been banned in developed countries, its use to improve public health in developing countries continues. The World Health Organization estimates that DDT has saved 25 million lives from malaria and hundreds of millions of other lives from other diseases.

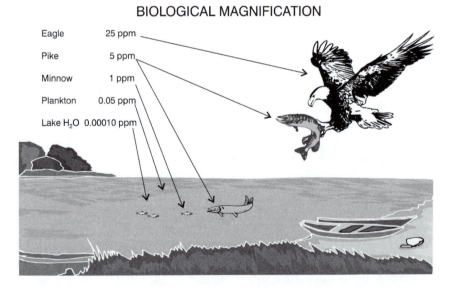

BIOLOGICAL MAGNIFICATION

Eagle   25 ppm
Pike   5 ppm
Minnow   1 ppm
Plankton   0.05 ppm
Lake H$_2$O 0.00010 ppm

Figure 30.1 Pollutants can be concentrated as they move up the food chain.
Drawing by Rae Déjur.

The United Nations' Stockholm Treaty on persistent organic pollutants calls for the phase out of DDT but recognizes its efficacy as a deterrent to vector-borne diseases such as malaria and typhus. According to the treaty, the continued use of DDT is discouraged, but until effective economical alternatives are found, DDT use will be continued in countries with high rates of vector diseases. A number of developing countries still use DDT. It is applied primarily in the interior of homes to prevent malaria. Currently DDT is produced only in India and China, and current production volumes are unknown.

DDT belongs to a group of chemical insecticides know as organochlorides. These contain hydrogen, carbon, and chlorine and kill by interfering with nerve transmission, making them neurotoxins. Organochlorides were the dominant type of chemical insecticide used from 1940 to 1970. Some common organochlorides besides DDT are chlordane, heptachlor, aldrin, and dieldrin. Because of their problems and subsequent ban in many regions, numerous other classes of insecticides have been synthesized to replace organochlorides.

CHEMICAL NAME = N,N-diethyl-3-methylbenzamide
CAS NUMBER = 134–62–3
MOLECULAR FORMULA = $C_{12}H_{17}NO$
MOLAR MASS = 191.3 g/mol
COMPOSITION = C(75.4%) H(9.0%) N(7.3%) O(8.4%)
MELTING POINT = –45°C
BOILING POINT = 285°C
DENSITY = 1.0 g/cm³

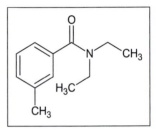

DEET has been used for more than 50 years as the active ingredient in many insect repellent formulations. It is used to repel biting pests such as mosquitoes, flies, midges, gnats, and ticks. Approximately one-third of the U.S. population and 200 million people worldwide use DEET in some form each year; it is also used on dogs, cats, horses, and other animals. DEET is available in various liquids, lotions, sprays, and impregnated materials such as wrist bands. Formulations registered for direct human application contain from 4% to 100% DEET. DEET was developed as a joint effort by the Department of Defense and U.S. Department of Agriculture (USDA). After examining hundreds of compounds for their repellent capabilities in the 1940s, DEET was selected and patented by the U.S. Army in 1946. The USDA did not announce DEET's discovery until 1954, and it was registered for public use in 1957. DEET is prepared from m-toluoyl chloride and diethylamine in benzene or ether.

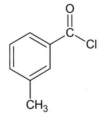

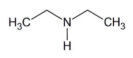

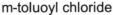
m-toluoyl chloride   diethylamine

Insects are attracted to people by visual, thermal, and olfactory stimuli. Visual stimuli are more important for attraction at greater distance, with thermal and olfactory being more important at closer range. Humans emit several hundred volatile compounds directly from the body and breath during metabolism. Two compounds thought to attract mosquitoes are carbon dioxide and lactic acid. Carbon dioxide is exhaled with each human breath and lactic acid is a component of sweat. Chemoreceptors on mosquitoes' antennae are stimulated by lactic acid. Although the exact mode of action for DEET is unknown, it is believed that DEET works by disrupting the olfactory senses of the target insect. DEET also is thought to affect insect lactic acid chemoreceptors, interfering with its ability to locate the host.

The amount of insect protection from DEET depends on its concentration, the host (for example, whether the host is male or female), sweat production, and environmental factors such as wind, temperature, and rain. DEET is sold in formulations that range from a few percent up to 100%. Higher concentrations of DEET are no more effective than lower concentrations but provide longer lasting protection. A 5% DEET repellent lasts for about an hour, a 15% solution lasts about 3 hours, a 25% DEET formulation should provide about 5 hours of protection, and a 100% DEET should protect for about 12 hours. Health concerns have been raised about the use of DEET, especially when used frequently and in high concentrations. Studies on rats subjected to high prolonged exposure indicated destruction of brain neurons that control muscle movement, learning, memory, and concentration; rats treated with doses similar to that used by humans showed much less ability to complete motor control and strength tasks than control rats. DEET absorption through the skin depends on its concentration and other chemicals used in combination with DEET formulations; absorption ranges between a few a percent up to 20%. DEET distributes into the skin and fat tissue (owing to its lipophilicity), where in the lower skin layers it is absorbed into the body. It is metabolized and excreted in the urine within 12 hours after removal from the skin.

An important aspect raised about skin absorption of DEET is its health effects when used in combination with sunscreens. Animal and *in vitro* studies indicate that sunscreen or certain sunscreen ingredient such as oxybenzone increase DEET's absorption into the skin. Researchers performing the studies have recommended further research to determine the potential health effects of DEET acting in combination with sunscreens. The medical community and government health agencies such as the Centers for Disease Control and Prevention recommend that repellents and sunscreens should be applied as two separate formulations rather than combined in a single product. The reason for this is that sunscreens are often applied repeatedly, which could result in much more DEET being used than is needed for protection. Because physicians recommend that only enough repellent be applied to provide the necessary protection, using individual products for sun and insect protection helps ensure against excessive DEET application.

DEET use on children under the age of two months is not recommended. The American Academy of Pediatrics recommends that DEET repellents should not be applied more than once per day on children and that the maximum DEET concentration used be no more than 30%. Spraying on hands, under clothing, and in the vicinity of food is not recommended. The Academy also recommends against using combined sunscreen-repellent formulations.

The threat of West Nile virus has led government agencies to recommend the use of DEET repellents. West Nile virus is a mosquito-borne infectious disease that is common in Africa, west Asia, and the Middle East; it was first detected in North America in 1999.

People infected usually experience only mild flulike symptoms, but West Nile virus can result in life-threatening encephalitis or meningitis. DEET is the most popular pesticide used as a repellent for humans, with several hundred products containing DEET available in the United States. Because it is a pesticide, products containing DEET are required to be labeled with information concerning the method of application, directions for medical attention, list special precautions for children, and the percent DEET in the product. DEET is still available at 100% strength in the United States, but Canada bans formulation that are more than 30% and various groups have recommended that this standard be used in the United States.

32. Dichlorodifluoromethane, CFC-12

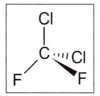

CHEMICAL NAME = dichlorodifluoromethane
CAS NUMBER = 75–71–8
MOLECULAR FORMULA = CCl_2F_2
MOLAR MASS = 120.9 g/mol
COMPOSITION = C(9.9%) Cl(58.6%) F(31.4%)
MELTING POINT = −157.7°C
BOILING POINT = −29.8°C
DENSITY = 5.5 g/L (vapor density = 4.2, air = 1)

Dichlorodifluoromethane is known as CFC-12, also called R-12, or Freon-12. R-12 is a general name for Refrigerant-12. Freon is a trade name for DuPont. CFC stands for chlorofluorocarbons, which are nontoxic, nonflammable, synthetic chemicals containing atoms of carbon, chlorine, and fluorine. CFCs were developed in the 1930s as coolants for refrigerator, freezer, and motor vehicle air conditioners. They subsequently found use as metal cleaners, degreasers, propellants, solvents, and blowing agents in the production of foams. CFCs have received widespread environmental attention because of their potential to deplete stratospheric ozone.

The number "12" in CFC-12 is based on the unique numbering system for CFC compounds. According to this system the CFC number is added to 90, and the result indicates the compound's number of carbon, hydrogen, and fluorine atoms, respectively. For example, in CFC-12 the sum is 102, so there is one carbon atom, 0 hydrogen atoms, and 2 fluorine atoms. The number of chlorine atoms in the compound can be inferred from the structure; because carbon bonds to four atoms in CFC-12, there must be two chlorine atoms.

The basic chemistry on producing fluorinated organic compounds was discovered at the end of the 19th century. The Belgian chemist Frédéric Swarts (1866–1940) had produced CFC compounds in the 1890s. Swarts discovered that pentavalent antimony catalyzed the fluorination of chlorinated organic compounds. The discovery of CFCs was led by Thomas Midgley (1889–1944) working in a cooperative effort between General Motors, DuPont, and Frigidaire (Frigidaire was owned by General Motors from 1919 to 1979) to find a safe refrigerant. Midgley was employed by the General Motors Chemical Company and had developed tetraethyl lead as an octane booster (see Octane). Household refrigerators of the late 19th century and early 20th century used ammonia, methyl chloride, and sulfur dioxide as refrigerants, and deaths had occurred as a result of leakage of these refrigerants. The refrigerant CFC-12 was discovered in 1928 by Thomas Midgley with Albert Leon Henne and Robert Reed McNary. Midgley's team prepared CFC-12 based on Swart's process by reacting carbon tetrachloride with antimony trifluoride using antimony (V) chloride (SbCl$_5$) as a fluorine exchange catalyst: $3CCl_4 + 2SbF_3 \rightarrow 3CCl_2F_2 + 2SbCl_3$. Midgley, Henne, and McNary filed the patent on the used of chlorofluorocarbons as refrigerants in February 1930, and it was registered in November 1931 (U.S. Patent Number 1833847). It should be noted that Midgley's team did not discover CFCs, but their patent was based on the application of fluorine compounds as refrigerants. Midgley also applied for the use of CFC as a fire suppressant in 1930, which was granted in 1933 (U.S. Patent Number 1926395). In 1930, General Motors and Du Pont formed the Kinetic Chemical Company to produce CFC-12 and to develop other refrigerants. The synthesis of fluorocarbon refrigerants was announced in April 1930. In that same year, Midgley demonstrated the safety of CFC-12 at a meeting of the American Chemical Society by inhaling it and then blowing out a candle. The Freon name was filed for in 1931 by DuPont and registered in 1932. Closely related compounds were introduced over the next several years: CFC-11 (1932), CFC- 114 (1933), and CFC-113 (1934). By the mid-1930s, several million refrigerators that used Freon-12 had been made in the United States.

CFC use climbed steadily worldwide as it was incorporated in refrigeration and air conditioning, as well as being used as propellants, blowing agents, and solvents. CFC production peaked in the late 1980s, with a worldwide annual production of just over 1.2 million tons (Figure 32.1).

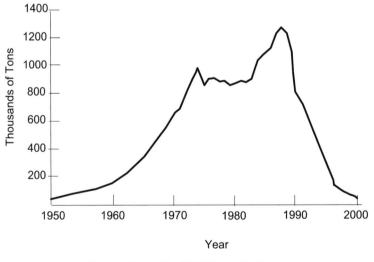

Figure 32.1   World CFC production.

CFCs possess several ideal characteristics: they are chemically inert, nonflammable, and noncorrosive. In 1974, F. Sherwood Rowland (1927–) and Mario Molina (1943–) predicted the destruction of stratospheric ozone. Rowland and Molina theorized that inert CFCs could drift into the stratosphere where they would be broken down by ultraviolet radiation. Once in the stratosphere, the CFCs would become a source of ozone-depleting chlorine. The destruction of ozone by CFCs can be represented as occurring by the following series of reactions:

$$CF_2Cl_2 \xrightarrow{\ uv\ } Cl + CF_2Cl$$
$$Cl + O_3 \rightarrow ClO + O_2$$
$$ClO + O \rightarrow Cl + O_2$$
$$\text{net reaction: } CF_2Cl_2 + O_3 + O \rightarrow Cl + CF_2Cl + 2O_2$$

This set of reactions shows that ultraviolet radiation strikes a CFC-12 molecule, removing a chlorine atom. The chlorine atom in the form of a free radical collides with an ozone molecule and bonds with one of ozone's oxygen atoms. The result is the formation of chlorine monoxide, ClO, and molecular oxygen, O_2. Chlorine monoxide is a free radical and reacts with atomic oxygen, producing another atom of chlorine that is released and available to destroy more ozone. In this manner, a single chlorine atom cycles through the atmosphere many times, causing ozone destruction. The actual destruction of ozone in the stratosphere actually involves hundreds of different reactions involving not only CFC compounds but numerous other chemicals.

The pioneering work of Rowland and Molina raised questions concerning human activity at least a decade before serious depletion of the ozone layer was confirmed. These scientists shared the 1995 Nobel Prize in chemistry along with Paul Crutzen (1933–), another researcher who pioneered work on ozone destruction. Ozone measurements taken in the mid-1980s demonstrated that humans were doing serious damage to the ozone layer. This prompted international action and adoption of the Montréal Protocol in 1987. The original agreement called for cutting in half the use of CFCs by the year 2000. Subsequent strengthening of the agreement banned the production of CFCs and halons in developed countries after 1995. Halons are compounds similar to CFCs that contain bromine such as CF_2BrCl. Production of other ozone-depleting chemicals such as methyl bromide, CH_3Br, is also banned.

The phasing out of CFCs has prompted the use of new compounds to take their place. Hydrochlorofluorocarbons (HCFCs) are being used as a short-term replacement for hard CFCs. Hard CFCs do not contain hydrogen, but soft CFCs do contain hydrogen. The presence of hydrogen atoms in HCFCs allows these compounds to react with hydroxyl radicals, •OH, in the troposphere. Some HCFCs can eventually be converted in the troposphere to compounds such as HF, CO_2, and H_2O. Other HCFCs can be converted into water-soluble compounds that can be washed out by rain in the lower atmosphere. HCFCs provide a near-term solution to reversing the destruction of stratospheric ozone. Their use signaled the first step in repairing the damage that has occurred during the last 50 years. Signs are positive that the ozone layer can repair itself, but it may take another 30 years to bring ozone levels back to mid-20th century levels.

The Montreal Protocol and its subsequent amendments outlined the phase-out of CFCs. This agreement established numerous deadlines for eliminating the production and use of CFC for specific purposes. Developed countries such as the United States were banned from production of CFCs by 2002. Use was severely restricted in developed countries, but developing countries were allowed production, more liberal use, and longer deadlines. A "total" phase-out of CFC for all parties is set for January 1, 2010, but even this deadline has exemptions for essential uses. Essential uses include CFC propellants used in certain inhalers for asthmatics and patients with pulmonary disease, applications related to the space program, and use in laboratory research.

33. *Dopamine, L-Dopa*

CHEMICAL NAME = 4-(2-aminoethyl) benzene-1,2-diol	3-(3,4-Dihydroxyphenyl)-L-alanine (L-Dopa)
CAS NUMBER = 51–61–6	59–92–7 (L-Dopa)
MOLECULAR FORMULA = $C_8H_{11}NO_2$	C9H11NO4 (L-Dopa)
MOLAR MASS = 153.2 g/mol	197.2 (L-Dopa)
COMPOSITION = C(62.7%) O(20.9%) H(7.2%) N(9.1%)	C(54.8%) O(32.5%) H(5.6%) N(7.1%) (L-Dopa)
MELTING POINT = 128°C	295°C (L-Dopa)
BOILING POINT = decomposes	
DENSITY (CALCULATED) = 1.25 g/cm³	1.47 g/cm³ (L-Dopa)

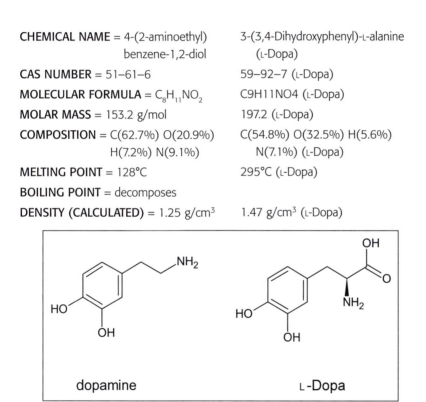

dopamine   L-Dopa

Dopamine, abbreviated DA, is a biosynthetic compound and neurotransmitter produced in the body from the amino acid tyrosine by several pathways. It is synthesized in the adrenal gland where it is a precursor to other hormones (see Epinephrine) and in several portions of the brain, principally the substantia nigra and hypothalamus. Dopamine is stored in vesicles in the brain's presynaptic nerve terminals. It is closely associated with its immediate precursor, L-Dopa (levodopa). Casmir Funk (1884–1967) first synthesized Dopa in racemic form in 1911 and considered Dopa a vitamin. In 1913, Marcus Guggenheim, a biochemist from Hoffman-LaRoche, isolated L-Dopa from seedlings of *Vicia faba,* the Windsor bean plant native to northern Africa and southwest Asia. Guggenheim used beans from the garden of Felix Hoffman (1868–1946), the discoverer of aspirin. Guggenheim ingested a 2.5-gram dose of L-Dopa, resulting in nausea and vomiting; he also administered small dosages to animals and did not observe any significant effects. This led him to believe that L-Dopa was biologically inactive. Studies commencing in 1927 reported that Dopa played a role in glucose metabolism and affected arterial blood pressure. Interest in dopamine accelerated in 1938 when the German physician and pharmacologist Peter Holtz (1902–1970) and co-workers discovered the enzyme L-Dopa decarboxylase and that it converted L-Dopa into dopamine in humans and animals. Research over the next two decades focused on L-Dopa's role as a precursor to other catecholamine hormones, its vascular effects, and its role in brain chemistry.

One large area of dopamine research involves its role in Parkinson's disease. Parkinson's disease is a progressive neurological disorder resulting from the degeneration of neurons in regions of the brain that control movement. Symptoms include tremors in the limbs, slow movement, shuffling gait, rigidity in the limbs, and stooped posture. Parkinson's disease is associated with a shortage of dopamine leading to impaired coordination of movement. Parkinson's disease was first described in *An Essay on the Shaking Palsy,* published in 1817 by a London physician named James Parkinson (1755–1824). Parkinson's disease has probably existed for thousands of years; its symptoms and suggested treatments appear in ancient medical texts. A significant medical breakthrough involving dopamine and L-Dopa occurred around 1960 when it was demonstrated that Dopa could be used to treat Parkinson's disease. In the previous decade, Arvid Carlsson (1923–) demonstrated that dopamine played a direct role in brain chemistry as a neurotransmitter. In a 1957 article, he reported on how he was able to reverse the effects Parkinson-like effects in reserpinized animals by administering L-Dopa. Reserpine is an alkaloid compound that causes depletion of neurotransmitters such as dopamine when administered to animals. Carlsson received the 2005 Nobel Prize in physiology or medicine for his work on dopamine's role as a neurotransmitter. Another study in 1960 involving autopsies on individuals with Parkinson's disease showed that the subjects had extremely low dopamine levels in the brain. Studies in the 1960s established the use of dopamine in treating Parkinson's patients. One of the problems with using dopamine was that it did not penetrate the blood-brain barrier. To overcome this problem, the precursor Dopa, which did pass the blood-brain barrier, could be used, although injected Dopa was toxic. Clinical trials produced progress, but it was not until 1967 that George C. Cotzias (1918–1977) showed that, by starting with small oral doses that were progressively increased over time, remission of Parkinson's symptoms occurred. Cotzias also found that L-Dopa was effective in treatment of Parkinson's disease, whereas D-Dopa provided no therapeutic result, but contributed to Dopa's toxicity.

L-Dopa is still the standard treatment for relieving symptoms of Parkinson's disease. Because its prolonged use can lead to complications and unpleasant side effects, it is often used in combination with other drugs. L-Dopa was highlighted in Oliver Sachs's (1933–) book, *Awakenings,* which was made into a movie in the 1970s. Sachs's book tells the story of how a group of individuals suffering from encephalitis lethargica, a disease which in some cases causes its victims to exist in an unconscious sleeplike state, were "awakened." The victims were individuals who had contacted the disease during an epidemic that swept the world in the 1920s. Unfortunately, L-Dopa relieved the symptoms for only a brief period.

Dopamine is used as a drug to treat several conditions. It can be injected as a solution of dopamine hydrochloride, such as in the drug Intropin. It is used as a stimulant to the heart muscle to treat heart conditions; it also constricts the blood vessels, increasing systolic blood pressure and improving blood flow through the body. Dopamine is used in renal medications to improve kidney function and urination. Dopamine dilates blood vessels in the kidneys, increasing the blood supply and promoting the flushing of wastes from the body. Dopamine is used to treat psychological disorders such as schizophrenia and paranoia.

Dopamine levels in the brain affect centers of reward and pleasure and is therefore associated with the action of drugs like alcohol, cocaine, heroin, and nicotine. Addiction is associated with increased dopamine levels in the reward and pleasure centers of the brain. Different mechanisms affect how psychoactive agents affect dopamine in the brain. Some agents excite the dopamine-containing neurons in the brain, increasing the production and release of dopamine from vesicles. Tobacco binds to dopamine receptors and postsynaptic neurons. Over time it decreases the number of dopamine receptors, which leads to desensitization to the drug. Increase nicotine use is required to derive comparable pleasure and thus promotes addiction. Amphetamines increase release of dopamine from vesicles. Cocaine decreases the reuptake of dopamine at pre-synaptic sites, which increase the amount of dopamine available at the postsynaptic receptor. Specific therapies to combat addiction are based on modifying dopamine brain chemistry; for example, drugs to decrease dopamine production or block dopamine receptors can be used to reduce the craving for an agent. Dopamine is the active ingredient in Zyban, a drug designed to help smokers quit.

L-Dopa was produced industrially by Hoffmann-LaRoche, using a modification of the Erlenmeyer synthesis for amino acids. In the 1960s, research at Monsanto focused on increasing the L-Dopa form rather than producing the racemic mixture. A team led by William S. Knowles (1917–) was successful in producing a rhodium-diphosphine catalyst called DiPamp that resulted in a 97.5% yield of L-Dopa when used in the Hoffmann-LaRoche process. Knowles's work produced the first industrial asymmetric synthesis of a compound. Knowles was awarded the 2001 Nobel Prize in chemistry for his work. Work in the last decade has led to green chemistry synthesis processes of L-Dopa using benzene and catechol.

34. Epinephrine (Adrenaline)

CHEMICAL NAME = 4-(1-hydroxy-2-(methylamino)ethyl)
    benzene-1,2-diol
CAS NUMBER = 51–43–4
MOLECULAR FORMULA = $C_9H_{13}NO_3$
MOLAR MASS = 183.2 g/mol
COMPOSITION = C(59%) N(7.6%) H(7.2%) O(26.2%)
MELTING POINT = 211°C–212°C
BOILING POINT = decomposes at 215°C
DENSITY = 1.3 g/cm³ (calculated)

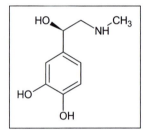

Epinephrine, also known as adrenaline, is a hormone continually secreted by the medulla of the adrenal gland, which is located on the top of each kidney. Epinephrine comes from the Greek *epi nephros* meaning "on kidneys"; adrenaline is the English equivalent of epinephrine. Both epinephrine and adrenaline were named by original researchers without the "e" at the end and this "e" was added over time. Epinephrine is also secreted at nerve endings as a neurotransmitter. It was isolated by Jokichi Takamine (1854–1922) in 1900 and was the first hormone to be isolated in pure form. Takamine's success marked several years of efforts in attempting to obtain the compound from adrenal gland secretions of animals. English researchers George Oliver (1841–1915) and Edward Albert Sharpley-Schaffer (1850–1935) had injected adrenal secretions into animals in the mid-1890s, producing a rise in blood pressure; researchers believe adrenal compounds held promise for medical applications. In 1897, John Jacob Abel (1857–1938) and Albert C. Crawford (1869–1921), working at Johns Hopkins Medical School, isolated a compound they named epinephrin, but it turned out to be the monobenzoyl derivative of epinephrine. Takamine, who worked for the Parke, Davis & Company drug producer, visited Abel's laboratory in 1900. Takamine's assistant, Keizo Uenaka, successfully crystallized pure epinephrine in 1900. Takamine applied for a patent on a "Glandular Extractive Product" on November 5, which he called adrenalin; on April 16, 1901, Takamine was granted a trademark for Adrenalin. Takamine presented and published the first articles on epinephrine in 1901. Concurrently, another Parke-Davis chemist, Thomas Bell Aldrich

(1861–1938), also produced epinephrine and determined its correct formula. Parke, Davis began promoting Adrenalin soon after the discoveries of Takamine and Aldrich. It was promoted as a treatment for heart disease, goiter, deafness, and Addison's disease.

Epinephrine is synthesized in the body from the nonessential amino acid tyrosine. Tyrosine undergoes hydroxylation to produce DOPA (3,4-dihydroxyphenylalanine). DOPA decarboxylation produces dopamine, which is hydroxylated to norepinephrine. Norepinephrine, which is closely related to epinephrine, performs a number of similar functions in the body. The prefix "nor" associated with a compound is used to denote an alkylated nitrogen in the compound that has lost an alkyl group. It comes from the German *N-ohne-radical,* which means Nitrogen without the radical. Therefore norepinephrine is epinephrine minus the methyl, CH_3, radical on the nitrogen. The methylation of norepinephrine gives epinephrine. The synthesis is summarized in Figure 34.1.

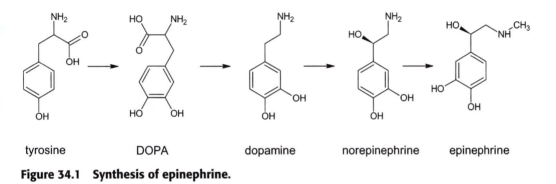

| tyrosine | DOPA | dopamine | norepinephrine | epinephrine |

Figure 34.1   Synthesis of epinephrine.

Epinephrine has several important physiological functions in the body. Its effect is produced when it binds to receptors associated with different organs. Receptors are highly specialized, and the effect of a hormone such as epinephrine depends on the type of receptor to which it binds. For this reason, epinephrine can produce different effects in different organs, so it is important to realize that physiological effects produced by epinephrine are not absolute. The physiological effects of epinephrine are the same whether it is produced in the adrenal gland or at the nerve endings, but because the adrenal source delivers the hormone to organs via the bloodstream, its effect lasts considerably longer. In general, epinephrine in the blood produces an effect lasting several minutes, which is several times as long as when it is produced at nerve endings.

Epinephrine is vital for normal physiological function and maintaining homeostasis, but it is secreted in large quantities during times of stress (norepinephrine is also secreted and many of its effects are similar to those of epinephrine). The stress response, sometimes called the "fight or flight" response, highlights the effects of epinephrine on the body. Epinephrine increases heart rate and stroke volume, resulting in an increase of blood flow to muscles. It produces vasoconstriction in peripheral arteries and veins, but vasodilation in other organs such as muscles, liver, and the heart. Epinephrine's effect on blood vessels depends on the type of receptor it acts upon. When it acts on alpha receptors, it results in vasoconstriction; with beta receptors it produces vasodilation. There is evidence showing that vasoconstriction dominates at high epinephrine concentrations and vasodilation at low concentrations.

High epinephrine results in an increase in blood pressure because of vasoconstriction. High epinephrine increases lipid metabolism and the conversion of glycogen to glucose providing increased energy input to cells. During times of stress, epinephrine inhibits nonessential functions such as gastric secretions and insulin production.

Epinephrine belongs to a class of hormones called catecholamines, which are derived from tyrosine and have a structure related to catechol. It is used in drugs and medications, often in the salt form as epinephrine hydrochloride. It is best known for treating allergic reactions, a condition called anaphylaxis. Anaphylaxis is caused by insect bites, foods, medications, latex, and other causes. A common device familiar to many is the epi-pen, which is an autoinjector that delivers a single dose of epinephrine. EpiPen, the most popular pen, is a registered trademark of Dey Laboratories. Adult pens are designed to deliver 0.3 mg of epinephrine, and child pens deliver a 0.15 mg. dose. Injection of epinephrine almost immediately improves breathing, stimulates the heart, and reverses swelling to the face and lips. Epinephrine is also used for heart conditions, bronchitis, bronchial asthma, emphysema, and glaucoma. It is a heart stimulant. The use of epinephrine has recently been adopted in hair transplant surgeries to reduce bleeding.

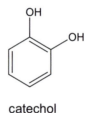

catechol

CHEMICAL NAME = ethane
CAS NUMBER = 74–84–0
MOLECULAR FORMULA = C_2H_6
MOLAR MASS = 30.1 g/mol
COMPOSITION = C(79.9%) H(20.1%)
BOILING POINT = –88.6°C
MELTING POINT = –182.8°C
DENSITY = 1.37 g/L (vapor density = 1.05, air = 1)

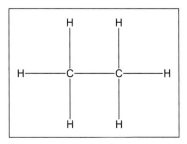

Ethane is a colorless, odorless, flammable hydrocarbon gas that follows methane as the second simplest alkane. The root of ethane, "et," is derived from the Greek word *aithein,* which means to burn; it was first applied to the compound ether ($CH_3CH_2OCH_2CH_3$). Ether is a highly flammable compound that was first prepared from the two-carbon alcohol ethanol (C_2H_5OH), and ethane is the two-carbon alkane. Ethane is the second most abundant component of natural gas, with sources typically containing 1–5% by volume, but some sources may contain up to 30% ethane. Ethane was first synthesized in 1834 by Michael Faraday (1791–1867) through the electrolysis of acetate solutions, although Faraday believed the compound was methane. Twenty years later Adolph Wilhelm Hermann Kolbe (1818–1884) incorrectly identified ethane as the methyl radical in his research, and Edward Frankland (1825–1899) prepared ethane by treating ethyl iodine (C_2H_5I) with metals.

The synthesis of ethane takes place through a process called Kolbe synthesis. In this process acetic acid (CH_3COOH) undergoes electrolysis to oxidize acetate ions at the anode of an electrochemical cell to produce acetate radicals: $CH_3COO^- \rightarrow CH_3COO\bullet$. Two acetate radicals then combine to give ethane and carbon dioxide: $CH_3COO\bullet + CH_3COO\bullet \rightarrow C_2H_6 + 2CO_2$.

Alkanes such as ethane are relatively unreactive and reactions involving alkanes require high-energy atoms or free radicals. Three general types of reactions involving alkanes are combustion, halogenation, and pyrolysis. The most common reactions of alkanes involve combustion. Combustion of alkanes has been the primary source of heat for human civilizations throughout modern history. The combustion of ethane is given by the equation: $2C_2H_6 + 7O_2 \rightarrow 4CO_2 + 6H_2O$. Ethane can also be halogenated with hydrogen substituting for hydrogen. The reactivity of ethane decreases from fluorine through iodine. Because fluorine is explosively reactive, it is hard to control and iodine is generally unreactive, so most practical halogenation reactions involve chlorine and bromine. For example, ethane reacts with chlorine in an endothermic reaction to produce chloroethane (C_2H_5Cl): $C_2H_6 + Cl_2 +$ energy $\rightarrow C_2H_5Cl + HCl$. Pyrolysis of ethane yields a host of compounds used in the petrochemical industry such as ethene (C_2H_4) and ethyne (C_2H_2). Pyrolysis is the decomposition of a compound using heat. Pyrolysis comes from the Greek root word for fire *pyr* and *lysis*, meaning to loosen. Ethane's primary use is as a source of ethylene. Ethylene is one of the most important compounds in the petrochemical industry owing to its high reactivity because of its double bond. Ethylene is produced from ethane by steam cracking. In this process ethane is mixed with steam and heated to 750°C–900°C converting ethane to ethylene: $C_2H_6 \rightarrow C_2H_4 + H_2$. The high temperature provides the energy needed to cause the decomposition of ethane. Steam cracking of ethane produces a number of other products besides ethylene. Different procedures and catalyst may be used to increase the yield of ethylene.

36. Ethene (Ethylene)

CHEMICAL NAME = ethene
CAS NUMBER = 74–85–1
MOLECULAR FORMULA = C_2H_4
MOLAR MASS = 28.1 g/mol
COMPOSITION = C(85.6%) H(14.4%)
BOILING POINT = –103.7°C
MELTING POINT = –169.4°C
DENSITY = 1.26 g/L (vapor density = 0.98, air = 1)

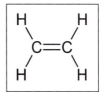

Ethylene is a colorless, odorless gas that is the simplest alkene hydrocarbon. It is a natural plant hormone and is produced synthetically from natural gas and petroleum. The double bond in ethylene makes this compound highly reactive, and the volume of ethylene used in the chemical industry is greater than any other organic compound. The name ethylene goes back to the mid-19th century. At that time the ending "ene," which comes from ancient Greek and means "daughter of," was added to names to indicate one fewer hydrogen atom that the substance from which it was derived. Thus ethylene was the daughter of ethyl, C_2H_5.

Ethylene is primarily obtained from the ethane and propane components of natural gas and from the naphtha, kerosene, and gas oil components of crude oil. It can also be synthesized through the dehydration of ethanol (C_2H_5OH). The production of ethylene from hydrocarbon feedstocks involves mixing with steam and then subjecting the hydrocarbons to thermal or catalytic cracking. Cracking is a process in which organic molecules are broken down into smaller molecules. Thermal cracking involves the use of heat and pressure. Catalytic cracking uses various catalysts to reduce the amount of heat and pressure required in the process. Thermal cracking of hydrocarbons to ethylene occurs between approximately 650°C and 800°C (1200°F and 1500°F). After hydrocarbons are cracked, a mixture containing ethylene and other gases such as methane, ethane, and propane is obtained. Ethylene is separated from these through physical processes such as fractional distillation, refrigeration, absorption, or adsorption.

Ethylene is highly reactive and is one of the most important compounds for the chemical industry. The highest use of ethylene is in polymerization reactions. The singular term *polyethylene* implies a single repeating polymer based on the ethylene monomer, but it actually refers to thousands of different compounds with molar masses ranging from several hundred to several millions. Polyethylene polymers are linear, but they contain side branchings of methyl groups. Among these are several groups defined by their density produced under different pressure regimens. High-pressure polymerization was the first process developed, starting in 1935 when 8 grams of polyethylene were accidentally produced. The production of the first polyethylene involved serendipity as researchers were investigating reactions under high pressure at Imperial Chemical Industries in London. In one experiment a white, waxy substance was obtained rather than the desired product. Initial attempts to duplicate the experiment failed, but then it was realized that oxygen must have been present when the first polyethylene was produced. It was determined that in the initial experiment, a small leak in the reaction chamber required recharging the chamber with additional ethylene, which contained just the right quantity of air providing oxygen needed to produce polyethylene.

The high-pressure process, which occurs at approximately 200°C and 2,000 atmospheres, produces polyethylene with numerous areas of side branching. Regions along the molecule where side branching is abundant produces an amorphous structure, whereas straight-chain regions along the polymer are described as crystalline (Figure 36.1). The side branching amorphous structure means that polymers cannot pack as closely together as polymers with less branching. This results in low-density polyethylene (LDPE). In addition to lower density, they also melt at lower temperature, are softer, and have less tensile strength. LDPEs have densities between 0.910 and 0.940 g/cm^3 and can be identified by the recycling symbol:

Crystalline HDPE Structure   Amorphous LDPE Structure

Figure 36.1   Crystalline HDPE is characterized by a straight-chain arrangement; LDPE has an amorphous structure.

LDPEs are flexible and include numerous plastic items such as plastic wraps (cellophane), grocery and trash bags, and diaper lining.

Low-pressure polymerization began 20 years after the high-pressure process when chemical processes and catalysts were discovered that polymerized ethylene. Low-pressure polymerization takes place between 20°C and 150°C and from several atmospheres pressure up to 50 atmospheres. The first catalysts used were organometallic compounds developed by Karl Ziegler (1898–1973) in Germany and referred to as Ziegler catalysts. Ziegler received the 1963 Nobel Prize in chemistry for his work on developing organometallic catalysts. The polyethylenes produced using the low-pressure process resulted in a high degree of straight chain polymers. Therefore they showed a predominantly crystalline structure. Crystalline straight molecules could pack more tightly than in LDPE, resulting in high-density polyethylene (HDPE). HDPEs have densities between 0.941 and 0.970 g/cm^3. The tight packing of HDPEs produces a hard, rigid structure and is used in applications where flexibility is not critical. Containers for juice, motor oil, bleach, and CDs are made from HDPE and contain the recycling symbol. HDPE is also used in furniture, cabinets for electronic items such as televisions and computers, and numerous other items.

The second highest use of ethylene involves oxidation to ethylene oxide and its derivative ethylene glycol (HO-CH_2-CH_2-OH). Ethylene glycol is used mainly as antifreeze and in the production of polyesters. Other important compounds produced through oxidation of ethylene are acetaldehyde (H_3C-CH = O) and vinyl acetate (CH_2 = CH-O-CO-CH_3). Ethylene may also be halogenated to produce a number of other compounds. The most important of these are ethylene dichloride (1,2-dichloroethane, ClCH_2CH_2Cl), which is then used in the production of polyvinyl chloride. Polyvinyl chloride (PVC) is the polymer of the monomer vinyl chloride (H_2C = CHCl). Another important industrial use of ethylene is in the alkylation reaction to produce ethylbenzene, which is used in the production of polystyrene. Two widely used compounds derived from substituting halogens for hydrogens in ethylene are tetrachloroethylene (C_2Cl_4) and tetrafluoroethylene (C_2F_4). Tetrachloroethylene, also called perchloroethylene or perc, is an industrial solvent that is the primary chemical used for dry cleaning. Tetrafluoroethylene is the monomer used to form polytetrafluoroethylene (PTFE). PTFE is marketed by DuPont under the trade name Teflon. In addition to the specific polymer know as Teflon, there are numerous other Teflons formed from other polymerization reactions (see Tetrafluorethene).

ethylene oxide

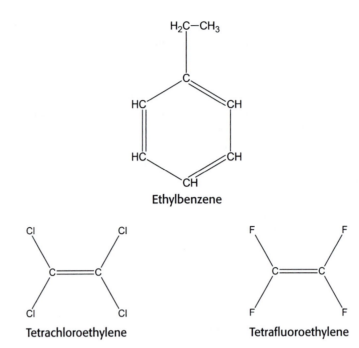

Ethylbenzene

Tetrachloroethylene

Tetrafluoroethylene

CHEMICAL NAME = ethoxyethane
CAS NUMBER = 60–29–7
MOLECULAR FORMULA = $C_2H_5OC_2H_5$
MOLAR MASS = 74.1g/mol
COMPOSITION = C(64.8%) H(13.6%) O(21.6%)
MELTING POINT = –116.3°C
BOILING POINT = 34.5°C
DENSITY = 0.71 g/cm³

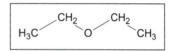

Ether is a volatile, flammable, colorless liquid with a distinctive odor. Ether is the common name for diethyl ether, which belongs to the large group of organic compounds called ethers. The names ether, diethyl ether, and ethyl ether are all used for the ether given by the formula $C_2H_5OC_2H_5$. In this entry, ether is considered the compound $C_2H_5OC_2H_5$. Ethers are characterized by an oxygen atom singly bonded to two carbon atoms; they have the general formula R-O-R'. Ethers are commonly named by naming the groups attached to the oxygen followed by the word ether. Diethyl ether is the most common ether and was historically used as an anesthetic. Petroleum ether is not an ether but a mixture of hydrocarbons, which are typically alkanes such as pentane and hexane.

Ether was supposedly discovered by Raymundus Lullus (1232–1315) around 1275, although there is no extant evidence of this in his writings. The discoverer of ether is often credited to the German physician and botanist Valerius Cordus (1515–1554), who gave the first description of the preparation of ether in the mid-16th century. Cordus called the substance *oleum vitrioli dulce,* which is translated as sweet oil of vitriol. Cordus used sulfuric acid (oil of vitriol) to catalyze the conversion of alcohol to ether. At approximately the same time Paracelsus (1493–1541), a Swiss physician who is also cited as a discoverer of ether, observed that chickens were safely put to sleep by breathing vapors from sweet oil of vitriol. In 1730, August Siegmund Frobenius changed the name of sweet vitriol to ether.

Ether was applied topically, inhaled, and consumed for medical purposes well before it was used as an anesthetic. As early as the late 18th century and during the 19th century, ether was used recreationally to induce drunklike, stupor-state conditions in those who inhaled it. Ether (and nitrous oxide) was inhaled during parties called ether frolics in which the partygoers were entertained by the behavior of those under the influence of ether. Crawford Williamson Long (1815–1878) was the first physician known to use ether as an anesthetic in medicine. Long removed a tumor from a patient anaesthetized with ether on March 20, 1842; he subsequently used ether for other surgeries and in childbirth. Long did not publish an article on his use of ether until 1848, several years after William Thomas Green Morton (1819–1868) had publicly demonstrated its use in dental surgery and received credit for its discovery as an anesthetic.

Morton's demonstration on the use of ether started a 20-year feud between Morton, Long, and two colleagues: Charles T. Jackson (1805–1880), Morton's mentor at Harvard medical school, and Horace Wells (1815–1848), a dentist who used nitrous oxide as an anesthetic in his dental practice. Morton was a former partner of Wells, and Wells had taught Morton dental techniques. Jackson had consulted Wells about his use of nitrous oxide on patients and conducted his own experiments with ether on animals. Morton successfully used ether on a patient in his office in September of 1846 and then gave a public demonstration the next month at Massachusetts General Hospital. Morton applied for a patent for an ether-based substance called Letheon and included Jackson's name on the application. Letheon contained aromatic oils and opium in addition to ether. The patent was granted in November 1846. Meanwhile, Wells claimed that he should share in any financial rewards from the patent. Wells, who used anesthetics recreationally, committed suicide at age 33 while in prison in 1868. There is some indication that Long had filed an earlier claim on ether. Morton fought for priority and sole possession of its use, and Jackson sought credit and any subsequent monetary rewards from the patent. As it became known that ether was the active ingredient in Letheon, it was widely adopted as a general anesthetic. Furthermore, ether had already been used for centuries in medical applications, making a patent priority impossible to enforce. Morton continued his battle for monetary reimbursement for its use, appealing to Congress for compensation, which he never received. He died in poverty at age 49. Jackson died in an asylum at age 75.

Ether and chloroform (see Chloroform) transformed surgery in the middle of the 19th century. Before these substances were available, surgery was a dreaded last alternative. To mitigate the pain during these procedures patients were sedated with opium, alcohol, and herbal mixtures; in some cases hypnotism was used. Ether and chloroform increased surgical procedures and alleviated much of the associated suffering. Ether became the principal anesthetic used in the United States until the middle of the 20th century, although it had side effects such as nausea and prolonged recovery, and it was highly flammable. As new anesthetics began to be developed starting in the 1930s, use of ether decreased. Today ether is not used as an anesthetic in developed countries, but it is still used in developing countries.

Ether is only slightly soluble in water (6.9%), but it is a good solvent for nonpolar organic compounds. Approximately 65% of ether production is used as a solvent for waxes, fats, oils, gums, resins, nitrocellulose, natural rubber, and other organics. As a solvent, it is used as an extracting agent for plant and animal compounds in the production of pharmaceuticals and cosmetics. Another 25% of total ether production is used in chemical synthesis. It is an intermediate used in the production of monoethanolamine (MEA, C_2H_7NO). Ether is used in the production of Grignard reagents. A Grignard reagent has the general form RMgX, where R is an alkyl or aryl group and X is a halogen. Grignard reagents are widely used in industrial organic synthesis. A Grignard reagent is typically made by reacting a haloalkane with magnesium in an ether solution, for example, $CH_3I + Mg \xrightarrow{ether} CH_3MgI$. Ether is a common starting fluid, especially for diesel engines.

Ether is produced by the dehydration of ethanol using sulfuric acid: $2CH_3CH_2OH + 2H_2SO_4 \xrightarrow{\Delta} (CH_3CH_2)_2O + H_2SO_4 + H_2O$. The temperature of the reaction is carried out at about 140°C to control for unwanted products. The volatile ether is distilled from the mixture. Ether can also be prepared by Williamson synthesis. In this reaction, ethanol reacts with sodium to form sodium ethanolate ($Na^+C_2H_5O^-$). Sodium ethanolate then reacts with chloroethane to form ether and sodium chloride: $Na^+C_2H_5O^- + C_2H_5Cl \rightarrow C_2H_5OC_2H_5 + NaCl$. Ether is also produced as a by-product in the production of ethanol.

38. Ethyl Alcohol (Ethanol)

CHEMICAL NAME = ethanol
CAS NUMBER = 64–17–5
MOLECULAR FORMULA = C_2H_5OH
MOLAR MASS = 46.1 g/mol
COMPOSITION = C(52.1%) H(13.1%)
O(34.7%)
MELTING POINT = −114.1°C
BOILING POINT = 78.5°C
DENSITY = 0.79 g/cm³

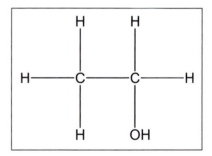

Ethyl alcohol, also called ethanol, absolute alcohol, or grain alcohol, is a clear, colorless, flammable liquid with a pleasant odor. It is associated primarily with alcoholic beverages, but it has numerous uses in the chemical industry. The word alcohol is derived from the Arabic word *al kuhul,* which was a fine powder of the element antimony used as a cosmetic. In Medieval times, the word *al kuhul* came to be associated with the distilled products known as alcohols. The hydroxyl group, -OH, bonded to a carbon, characterizes alcohols. Ethyl is derived from the root of the two-carbon hydrocarbon ethane.

The earliest use of alcohol is unknown. Alcohol is produced naturally from the fermentation of sugars, and it is assumed that prehistoric humans consumed alcohol when eating fermented fruits. The earliest direct evidence of alcohol consumption dates from the Neolithic period 10,000 years ago and consists of stone jugs used for holding alcoholic beverages. Ancient records and art from Egypt, Babylon, Mesopotamia, and other early civilizations indicate the use of alcohol as a beverage, medicine, and ceremonial drink. Records also show that the intoxicating effects of alcohol were known for thousands of years B.C.E. Alcoholic drinks were stored in Egyptian burial tombs, and deities devoted to alcoholic beverages were worshiped by different civilizations. As the human population expanded, alcoholic drinks assumed a prominent role in different cultures; for example, numerous references are made to wine in the Bible. Ancient Islamic alchemists advanced the practice of alcohol production by using distillation techniques. Distilled alcohols began to appear in the Middle Ages and was used in many remedies and medicines. A common practice by alchemists in different regions was the preparation of special liquors and brews with healing power. *Aqua vitae* (water of life) could refer to brandy, gin, whiskey, wine, or another form of alcoholic depending on the geographic area.

Ethyl alcohol is prepared by the fermentation of numerous plants including grains, fruits, and vegetables. Some of the most common products used to produce ethyl alcohol are corn, sugar cane molasses, rice, potatoes, and sugar beets. Natural fermentation is used to produce alcoholic beverages and was also used for most of the industrial ethyl alcohol produced until the middle of the 20th century. Fermentation is the conversion of sugars to carbon dioxide and alcohol by the action of enzymes. The enzymes typically come from yeasts, bacteria, or molds. In the production of ethyl alcohol, the fungus *Saccharomyces cerevisiae* produces the enzyme zymase. The common name for *Saccharomyces cerevisiae* is baker's or brewer's yeast. Fermentation of ethyl alcohol can be represented by the action of zymase on glucose: $C_6H_{12}O_{6(s)} \xrightarrow{\text{zymase}} + 2C_2H_5OH_{(l)}$. Fermentation is a process that organisms use to supply their energy needs under anaerobic conditions. Different strains of *Saccharomyces* promote either carbon dioxide or alcohol production. The former is more prevalent with baker's yeasts and the latter is favored in brewer's yeast.

Alcohol's physiological effect on humans is well known. Alcohol is soluble in water and is distributed throughout the body through the blood and body fluids. Once consumed, alcohol passes to the stomach and is absorbed primarily in the small intestine. Alcohol is toxic and is processed in the liver where it is converted to acetaldehyde (C_2H_4O) and subsequently metabolized to acetic acid, carbon dioxide, and water. Alcohol penetrates the blood-brain barrier where it impairs the central nervous system. Alcohol is a depressant that suppresses brain functions as the concentration of alcohol in the blood increases. Increasing alcohol levels in the blood produces the classical conditions associated with intoxication. These include slurred speech, loss of motor skills, and slower response time to stimuli. Alcohol acts as an anesthesia at moderately high concentrations, inducing drowsiness and sleep. Intoxication is quantified by measuring the amount of blood alcohol. Although effects vary among individuals, most individuals do not exhibit abnormal behavior until the blood alcohol level is about 0.03%. Between 0.03% and 0.10%, people are euphoric, experience loss of judgment and control, and start to lose motor skills. These conditions are intensified as the blood alcohol level climbs to approximately 0.20%. At approximately 0.20% speech becomes increasingly slurred, the person is disoriented and has difficulty walking, visual perception is blurred, and there is a general state of confusion. At levels greater than 0.20%, the person may vomit, lose motor functions, and lose consciousness. Severe problems ensue as the blood alcohol concentration approaches 0.40%, impairing the respiratory and circulatory system, and can ultimately induce a coma or lead to death. The legal limit to determine drunk driving in the majority of states is 0.08%;in the remainder it is 0.10%. During the last decade, many states lowered the legal limit from 0.10% to 0.08%. Most states also have zero tolerance laws for drivers under the age of 21, and a few states specify for under the age of 18. Zero tolerance is an attempt to curtail underage drinking and driving. The most common blood alcohol limit for zero tolerance is 0.02%, but zero tolerance limits can range from 0.00% all the way to 0.08%. Penalties for violating zero tolerance include the loss of driving privileges, imprisonment, and fines.

Industrial ethyl alcohol was primarily produced by fermentation during the first half of the 20th century. The fermentation process involved production of a mash consisting of a sugar source such as molasses, potatoes, or corn. The mash is loaded into a large steel fermentor, which can hold several hundred thousands gallons of mash. The mash is inoculated with yeast and other chemicals are added to provide nutrients and control pH. Carbon dioxide is recovered from the process and purified for use as a by-product. The ethyl alcohol produced from the fermentor has an alcohol content of 6–10% and is called a beer (a general term used for the product from the fermentor). The concentration of alcohol is limited because, as the alcohol content increases above 14%, zymase loses its activity. The beer is distilled to a concentration of approximately 95%. Distillation cannot produce ethyl alcohol higher than 95.6% because alcohol and water form an azeotrope (a constant boiling point mixture in which the vapor has the same composition as the mixture). To produce pure ethyl alcohol, called absolute alcohol or anhydrous alcohol, extractive distillation or a molecular filtering process can be used to remove the remaining 5% water. The concentration of alcoholic beverages is defined by its proof. Pure ethyl alcohol is 200 proof. The proof of an alcohol is twice the percent alcohol content. Typical proofs are beer, 7–12; wine,14–30; distilled liquors, 80–190.

The making of alcoholic drinks, such as beers, wines, and liquors, by fermentation involves various fermentation processes. Starch-rich grains used in the production of beer (and industrial alcohol) must first be converted to sugars through a process called malting. Grinding the grain and soaking in water creates a malt extract. Barley is the grain used to produce beer. The mixture produced after the malting process is called a wort. Brewing the wort in a large kettle, which is essentially a pressure cooker, converts the starches into sugars. During brewing the wort is concentrated and sterilized. Additives such as hops are added during the brewing process. After brewing, yeasts are added to the cooled wort, and it is fermented and aged to the final product. The production of wine is much simpler than beer. Essentially, grapes are passed through a mechanical device to crush them and remove the stems. The grapes are then pressed and the juice fermented and aged in wooden barrels. Numerous variations of this process are used, giving rise to thousands of different varieties of wine. Distilled liquors start out as beers produced from various substances and are processed according to the final product desired. These liquors can be named for where they are produced: Champagne, Irish Whiskey, and Scotch. Liquors are also dependent on the starting grain or additives. Bourbon must be produced from at least 51% corn, rum is made from sugar molasses, and gin is grain liquor incorporating juniper berries in some form.

Industrial ethyl alcohol is produced from petrochemicals. The traditional process involved the hydration of ethylene with sulfuric acid to ethyl sulfate followed by hydrolysis to ethyl alcohol:

$$C_2H_4 + H_2SO_4 \rightarrow CH_3CH_2OSO_3H$$

$$CH_3CH_2OSO_3H + H_2O \rightarrow C_2H_5OH$$

The current synthesis of ethyl alcohol eliminates sulfuric acid and uses phosphoric acid suspended on zeolite substrates. Zeolites are porous aluminosilicate crystalline minerals. The use of phosphoric acid as a catalyst allows the direct hydrolysis of ethylene into ethyl alcohol: $C_2H_4 + H_2O \rightarrow C_2H_5OH$. Industrial alcohol is rendered inconsumable by adding a small amount of a poisonous substance such as methanol or acetone to it. Alcohol unfit for consumption because of a poisonous additive is termed denatured alcohol.

Ethyl alcohol's most familiar use is in alcoholic beverages, but it has hundreds of industrial uses. It is widely used as an organic solvent in the production of numerous products including cosmetics, pharmaceuticals, paints, resins, lacquers, antiseptics, antifreezes, fuel additives, and inks. Some common household items that typically contain ethyl alcohol are mouthwashes, cough syrups, vanilla extract, almond extract, and antibacterial soaps. Ethyl alcohol can substitute for petroleum in the synthesis of traditional petrochemicals. One of the largest uses of ethyl alcohol is in the production of acetic acid (see Acetic Acid), which is further processed into other products. Acetaldehyde and acetic anhydride ($C_4H_6O_3$) can be made from ethyl alcohol.

One of the most prominent uses of ethyl alcohol is as a fuel additive and increasingly as a fuel itself. Ethyl alcohol is added to gasoline to increase its oxygen content and octane number. In the United States, the Environmental Protection Agency has mandated that oxygenated fuels be used in certain geographic areas to help meet air quality standards for carbon monoxide, especially in winter. A gasoline blended for this purpose may contain a few percent ethyl alcohol. Gasoline blended with ethyl alcohol is called gasohol. A typical gasohol may contain 90% gasoline and 10% ethanol. Gasohol reduces several common air pollutants including carbon monoxide, carbon dioxide, hydrocarbons, and benzene. Conversely, nitrogen oxides increase with gasohol.

A fuel consisting of 85% ethanol and 15% gasoline is called E85. Promoters of E85 claim several advantages of this fuel: it burns cleaner than gasoline, is a renewable fuel, has an octane rating of more than 100, and promotes energy independence. The reduction of pollutants must be balanced by the reduced energy content of E85, which is only about 70% that of regular gasoline. Therefore a vehicle will get about 70% of the miles per gallon using E85. Vehicles that can burn either gasoline or E85 are required to take advantage of E85 blends. Vehicles with the ability to burn either gasoline or ethanol are called flexible fuel vehicles. Many of these models are available only for fleet purchases. As might be expected, corn-producing states (Illinois, Kansas, and Iowa) heavily promote the use of these vehicles. The use of ethanol as a fuel is expected to increase in the future if the United States seeks to become energy independent. It is already used heavily in several other countries, such as Brazil. The United States uses about four billion gallons of ethyl alcohol per year and can meet most of this demand domestically but must import ethyl alcohol from Brazil, Saudi Arabia, and the Caribbean. The use of ethanol as a fuel is a hot political issues with environmental, economic, and industrial implications.

39. Ethylenediaminetetraacetic Acid (EDTA)

CHEMICAL NAME = ethylenediaminetetraacetic
  acid
CAS NUMBER = 60–00–4
MOLECULAR FORMULA = $C_{10}H_{16}N_2O_8$
MOLAR MASS = 292.3 g/mol
COMPOSITION = C(41.1%) H(5.5%) N(9.6%)
  O(43.8%)
MELTING POINT = decomposes at 240°C
BOILING POINT = decomposes at 240°C
DENSITY = 0.86 g/cm³

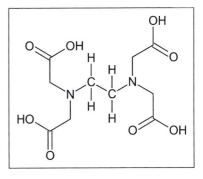

EDTA, also known as editic acid, is a colorless crystalline substance widely used to chelate metal ions. Before discussing EDTA, a brief overview of chelation is presented. The term *chelation* comes from the Greek word *chele or khele,* which means claw; it is used to describe the ability of some substances to grab atoms or ions to form complexes. Chelation occurs when a ligand binds to a metal ion. Ligand is derived from the Latin word *ligare* meaning to bind. A ligand is a chemical species (atom, ion, or molecule) that acts as a Lewis base. Lewis bases are electron pair donors. Water often acts as a Lewis base when salts dissolve in water, and water molecules donate a pair of electrons to form a coordinate covalent bond with the metal ion from the salt. For example, when a Cu^{2+} compound dissolves in water, four water molecules will each donate a pair of electrons binding to copper to form the complex ion $Cu(H_2O)_4{}^{2+}$. In this example, a single oxygen atom from each water molecule attaches to the metal ion, making water a monodentate ligand. Monodentate means it has "one tooth" with which to bite into the metal ion. Polydentate ligands have more than one electron pair donor atoms with which to bind (bite) into a metal.

EDTA is a common polydentate ligand. In EDTA, the hydrogen atoms are easily removed in solution to produce anionic $EDTA^{4-}$. In its anionic form EDTA has six binding atoms, two nitrogen and four oxygen as depicted in Figure 39.1.

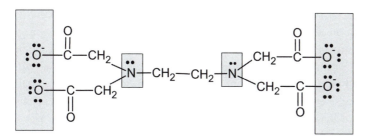

Figure 39.1   Potential binding sites in EDTA.

EDTA binds to a metal ion at the six binding sites, wrapping itself around the metal ion, forming a very stable complex. The strong grasp of EDTA on the metal ion is analogous to a crab or lobster clamping down on an object with its claw, hence the name chelation. EDTA is such an effective chelating agent because it can deactivate a metal at up to six sites (Figure 39.2).

EDTA was first synthesized in the early 1930s by the German chemist Ferdinand Münz working for I. G. Farben. Münz, who was looking for a substitute for citric acid to use with dye solutions in the textile industry, was the first to patent a process for EDTA synthesis in Germany in 1935. Münz subsequently applied for United States patents in 1936 and 1937 (U.S. Patent Number 2130505); his method involved reacting monochloroacetic acid ($C_2H_3ClO_2$) and ethylene diamine ($C_2H_8N_2$). Concurrent with Münz's work, Frederick C. Bersworth in the United States synthesized EDTA using different methods that gave greater yields and made EDTA's commercial production economically viable. Bersworth syntheses involved reacting formaldehyde, amines, and hydrogen cyanide. Bersworth and Münz obtained patents for EDTA production in the 1940s (U.S. Patent Numbers 2407645 and 2461519).

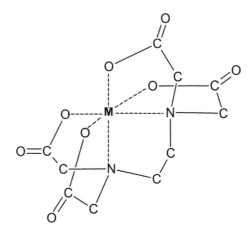

Figure 39.2   EDTA^{-4} chelating a metal, M.

EDTA is marketed in its salt forms such as sodium EDTA or calcium EDTA. EDTA has industrial and medical uses as a chelating agent. Much of its utility is related to the fact that metals and metal compounds are important catalysts in numerous reactions. By chelating metals, EDTA prevents the metal from catalyzing reactions, thereby limiting degradation, oxidation, and other undesirable reactions. The major industries using EDTA and other chelating agents are paper and pulp, cleaning products, chemicals, agriculture, and water treatment. The paper and pulp industry is the major user of EDTA, where it is used to stabilize bleaches by sequestering metals that catalyze the degradation of bleaches. EDTA's ability to stabilize bleaches also makes them useful in laundry detergents and various other cleaning products. In addition to improving bleaching efficiency, EDTA use in detergents and cleansers also softens hard water by tying up divalent metal ions responsible for water hardness, primarily Ca^{2+} and Mg^{2+}. Its softening ability helps EDTA reduce scale formation and improves foaming properties in cleaning formulations. EDTA is applied in general water treatment to soften water, helping to prevent scale and corrosion. EDTA has low toxicity and is used in the food and beverage industry. Foods naturally contain small traces of metals and small quantities are added during food processing. EDTA is used with foods to preserve color and preserve flavor, prevent odors, maintain nutrient content, and extend shelf life. When used in beverages, EDTA preserves color and stabilizes other ingredients such as citric acid and benzoates. In the chemical industry, EDTA is used to control metal catalytic processes during reactions. EDTA salts are used in agriculture to provide metal micronutrients in fertilizers.

EDTA has been used in the medicine since the 1950s in chelation therapy. Chelation therapy involves administering a calcium EDTA salt solution intravenously to patients to remove metal toxins from the bloodstream. The Food and Drug Administration has approved chelation therapy using EDTA for the treatment of heavy metal (lead, mercury) poisoning for more than 40 years. Hundreds of thousands of patients are treated annually using chelation therapy. EDTA is administered to patients over a variable number of sessions, with sessions lasting from an hour to several hours. Because EDTA chelation therapy removes nutrients and vitamins from the body with the target toxic substances, chelation solutions contain supplements in addition to EDTA.

Chelation therapy has been used for numerous treatments other than heavy metal poisoning, which has created controversy among the medical community. It is most often used as an alternative treatment for heart disease, atherosclerosis, and cancer, but it has also been advocated for numerous other diseases. Chelation therapy's use for clearing arteries is based on EDTA's ability to remove calcium and other substances that contribute to plaque in blood vessels. Proponents of chelation therapy claim that it is an alternative to standard coronary procedures such as bypass surgery or angioplasty. National health organizations such as the American Heart Association and American Cancer Society do not recommend it as an alternative treatment owing to the lack of scientific support for its efficacy. A five-year study launched in 2003 by the National Institute of Health to examine EDTA chelation therapy for individuals with coronary artery disease is currently in progress.

40. Fluoxetine (Prozac)

CHEMICAL NAME = N-methyl-3-phenyl-3-[4-(trifluoromethyl) phenoxy]propan-1-amine

CAS NUMBER = 54910–89–3

MOLECULAR FORMULA = $C_{17}H_{18}F_3NO$

MOLAR MASS = 309.3 g/mol

COMPOSITION = C(66.0%) H(5.9%) F(18.4%) N(4.5%) O(5.2%)

MELTING POINT = 179°C (158°C for hydrochloride salt)

BOILING POINT = decomposes

DENSITY = 1.16 g/cm³ (calculated)

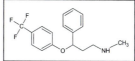

Fluoxetine in its hydrochloride salt form is marketed as numerous drugs, the most popular of which is Prozac. Prozac is prescribed for depression, obsessive-compulsive disorder, bulimia, agoraphobia, and premenstrual dysphoric disorder (premenstrual syndrome). Prozac and other fluoxetine medications belong to a class of drugs called selective serotonin reuptake inhibitors (SSRIs). When a nerve signal is sent, a neurotransmitter, such as serotonin, travels from a presynaptic neuron across the synaptic gap to a postsynaptic neuron. Receptors on the postsynaptic neuron capture the neurotransmitter, resulting in the transmission of the signal. After performing its function, the neurotransmitter is released back to the presynaptic cell in a process called reuptake. SSRIs slow down the return of serotonin to presynaptic neurons, allowing for a higher serotonin concentration on postsynaptic neurons. Because depression and other psychological disorders are associated with low serotonin levels, Prozac and other SSRIs help maintain serotonin levels.

Prozac was discovered by a team of chemists at the pharmaceutical company Eli Lilly. Key researchers involved in the work were Bryan B. Molloy (1939–2004), Klaus K. Schmiegel (1939–), Ray W. Fuller (1935–1996), and David T. Wong (1935–). In the middle of the 20th century, the main group of drugs for treating depression was tricyclic antidepressants (TCAs) and monoamine oxidase inhibitors (MAOIs). TCAs are named because of their three-ring chemical structure. Lilly researchers were working with TCAs in the1950s and 1960s. Prozac was developed by Eli Lilly scientists who based their work on the antihistamine diphenylhydramine; diphenylhydramine hydrochloride is marketed under the trade name Benadryl. The Lilly scientists examined diphenylhydramine because research had demonstrated that some antihistamines, including diphenylhydramine, had the ability to inhibit serotonin and serve as antidepressants. Molloy started examining diphenylhydramine-type compounds for their antidepressant properties in 1970. Molloy and his colleagues discovered fluoxetine hydrochloride had potential as an antidepressant in 1972 and it was referred to as Lilly 110140 in the first published articles on the compound, which appeared in 1974. Fluoxetine hydrochloride was no more effective than other antidepressant drugs of the time, but it produced much fewer negative side effects because it interacted specifically with the neurotransmitter serotonin but did not interfere with other neurotransmitters. TCAs inhibited the reuptake of other neurotransmitters along with serotonin. Molloy and Schmiegel applied for a patent in 1974 for the synthesis of arloxyphenylpropylamines (U.S. Patent Number 4314081). The patent named a number of compounds in this class of chemicals that could be used as antidepressants. In 1983, Eli Lilly applied to the Food and Drug Administration (FDA) for approval of fluoxetine hydrochloride as a drug used to treat depression. Prozac was first offered to the public in Belgium in 1986 and in the United States in 1988. Eli Lilly initially had a monopoly on fluoxetine hydrochloride as an antidepressant with its Prozac brand. In the mid-1990s, a lawsuit filed against Eli Lilly led to the loss of their exclusive patent rights, allowing generic fluoxetine hydrochloride antidepressants to be marketed starting in 2001.

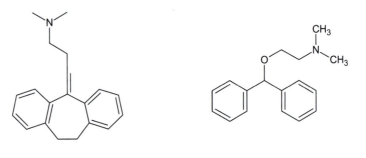

TCA stucture   diphenylhydramine

Prozac was the first SSRI antidepressant to be marketed. Because Prozac produced less severe side effects than other antidepressants, it became the drug of choice for treating depression and was made available to a wider public. Its use exploded in the 1990s, with sales peaking in 2000 when revenues from Prozac reached $2.5 billion. Eli Lilly's patent on fluoxetine hydrochloride expired in August 2001; its use continued into the 21st century but on a much smaller scale as generic fluoxetine hydrochloride products came on the market. Since its introduction in 1986, Prozac was the most prescribed drug for antidepressant until recent years when it was replaced by Zoloft, Paxil, and Lexapro as the top three antidepressants prescribed in the United States, respectively.

The health effects of Prozac and other SSRIs have been debated since these drugs were made available to the public. Foremost among health concerns is whether SSRIs increase suicidal and violent behavior. Hundreds of lawsuits have been filed against Eli Lily alleging that Prozac was responsible for all types of destructive behavior. The vast majority of these cases have been settled out of court or dismissed and only a few have gone to trial. Recent warnings by the FDA have alerted the public to potential suicidal behavior associated with antidepressants as governments, pharmaceutical companies, and researchers continue to examine the issue. There is more evidence supporting suicidal behavior in children and adolescents with psychiatric problems who use antidepressants. The FDA has approved fluoxetine in children eight years and older with major depressive disorder, but it has not approved other antidepressants for this disorder. The FDA also requires that a warning label calling attention to suicidal behavior in children be prominently published at the beginning of package inserts. The U.S. Food and Drug Administration in May 2007 proposed that makers of all antidepressant medications update warning labels to include language on increased risks of suicidal thinking and behavior in young adults ages 18 to 24 during initial treatment (generally the first one to two months).

Prozac and SSRIs have been implicated in serotonin syndrome, which occurs when two or more medications combine to elevate serotonin levels. The toxic condition results in numerous symptoms, but the most common include disorientation, irritability, neuromuscular problems, hyperthermia, and hypertension. In extreme cases serotonin syndrome results in coma or death. Prozac and SSRIs are associated with an increase in persistent pulmonary hypertension, a condition in which the fetal cardiocirculatory mechanisms remain in place after birth, causing blood to enter the heart in an oxygen poor state. Pregnant women are advised against taking SSRIs, especially during the final stages of pregnancy, because it increases the probability of persistent pulmonary hypertension.

Fluoxetine hydrochloride is most recognized as an antidepressant, but it is also used to relieve symptoms of premenstrual dysphoric disorder (PMDD) (premenstrual syndrome). These symptoms include mood swings, tension, bloating, irritability, and breast tenderness. Eli Lilly began marketing fluoxetine hydrochloride as Sarafem in 2000 for treating PMDD.

CHEMICAL NAME = methanal
CAS NUMBER = 50-00-0
MOLECULAR FORMULA = CH$_2$O
MOLAR MASS = 30.0 g/mol
COMPOSITION = C(40.0%) H(6.7%) O(53.3%)
BOILING POINT = −19.3°C
MELTING POINT = −117°C
DENSITY = 1.38 g/L (vapor density = 1.07, air =1)

Formaldehyde is a colorless, flammable gas with a distinctive pungent odor. It is the simplest aldehyde, which is a class of organic compounds with the carbonyl group bonded to at least one hydrogen atom. Formaldehyde was described by August Wilhelm von Hoffmann (1818–1892) in 1867 after the Russian Aleksandr Butlerov (1828–1886) had inadvertently synthesized it in 1857. Formaldehyde readily dissolves in water to produce a solution called formalin, which is commonly marketed as a 37% solution.

Formaldehyde is a by-product of combustion of organic compounds, metabolism, and other natural processes. Formaldehyde results from wood combustion and elevated atmospheric concentrations can result from forest fires, as well as from urban pollution sources such as transportation. Formaldehyde has been identified as a significant indoor air pollutant. Building materials such as particleboard, plywood, and paneling are major sources of formaldehyde because they incorporate formaldehyde resins as bonding adhesives. Other sources of formaldehyde in the home are carpets, upholstery, drapes, tobacco smoke, and indoor combustion products. Formaldehyde may be emitted from building materials for several years after installation. In the two decades of the 1960s and 1970s, a half million homes in the United States used urea formaldehyde foam insulation, but health complaints led to its elimination as an insulator in the early 1980s. People react differently to formaldehyde exposure, but it is estimated that between 10% and 20% of the population will experience some reaction at concentrations as low as 0.2 parts per million. Formaldehyde irritates the eyes, nose, and throats, producing coughing, sneezing, runny nose, and burning eyes. More severe reactions result in insomnia, headaches, rashes, and breathing difficulties. Some states have established indoor air quality standards ranging from 0.05 to 0.5 ppm.

The industrial preparation of formaldehyde has occurred since the late 1800s and involves the catalytic oxidation of methanol: $2CH_3OH_{(g)} + O_{2(g)} \rightarrow 2CH_2O_{(g)}$. The oxidation takes place at temperatures between 400°C and 700°C in the presence of metal catalysts. Metals include silver, copper, molybdenum, platinum, and alloys of these metals. Formaldehyde is commonly used as an aqueous solution called formalin. Commercial formalin solutions vary between 37% and 50% formaldehyde. When formalin is prepared, it must be heated and a methanol must be added to prevent polymerization; the final formalin solution contains between 5% and 15% alcohol.

Formaldehyde has hundreds of uses. Its largest use is in the production of synthetic resins. Many formaldehyde resins are hard plastics used in molding and laminates. Formaldehyde resins are used to treat textiles to make them "wrinkle-free." It is also used to produce adhesives, which are used extensively in the production of plywood and particleboard. Other common uses are as disinfectants, fungicides, and preservatives. Formalin has been the traditional embalming fluid used in the mortuary industry for the last century. Formaldehyde is also used in papermaking, textile production, and fertilizers.

More than half of the commercial formaldehyde produced is used to manufacture phenolic, urea, and melamine formaldehyde resins. Polyacetyl resins use another 5–10% of formaldehyde, and approximately 80% of formaldehyde goes into the resins and plastics industry. Phenolic-formaldehyde resins were the first synthetic plastics to be produced. The first plastic was called Bakelite. Bakelite was produced in 1906 by the Belgian-born (he immigrated to the United States in 1889) chemist Leo Hendrik Baekeland (1863–1944). Baekeland made a small fortune selling photographic paper to George Eastman (1854–1932). Using this money, Baekeland studied resins produced from formaldehyde and phenol by placing these materials in an autoclave and subjecting them to heat and pressure. Bakelite was a thermosetting phenol plastic. A thermosetting plastic hardens into its final shape upon heating. Engineers at Westinghouse used Bakelite to treat paper and canvas and compress the mixture into hard sheets to produce the first laminates. These were initially used as electrical insulators, but two of the engineers, Herbert Faber (1883–1956) and Daniel O'Conor (1882–1968), used the process to establish the Formica Corporation. The word *formica* was derived from the fact that mica was used as a common electrical insulator in the early 20th century, and the new resin could substitute "for mica."

Formaldehyde has traditionally been used as a preservative in biology and medical laboratories and in embalming fluid. Embalming fluids typically contain 5–15% formaldehyde, a significant percentage of alcohol, and other additives to perform certain functions, for example, bleaches and coloring to preserve skin color. Formaldehyde has been used to preserve dead bodies since 1900 and has several qualities that make it the preferred preservative. Foremost among these is its low cost, but it also has several biochemical advantages: it kills germs and microorganisms, destroys decomposition enzymes, retards decomposition of proteins, and hardens body tissues.

42. Formic Acid

CHEMICAL NAME = methanoic acid
CAS NUMBER = 64–18–6
MOLECULAR FORMULA = HCOOH
MOLAR MASS = 46.0 g/mol
COMPOSITION = C(26.1%) O(69.5%) H(4.4%)
MELTING POINT = 8.4°C
BOILING POINT = 100.8°C
DENSITY = 1.2 g/cm³

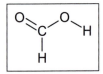

Formic acid is the simplest carboxylic acid, a group of organic compounds defined by the carboxyl group, -COOH. Formic acid is a colorless, fuming liquid with a pungent acrid odor. The common names for simple carboxylic acids come from the Latin or Greek names of their source. Formic acid is taken from the Latin word for ant, *formica*. Naturalists had observed the acrid vapor from ant hills for hundreds of years. One of the earliest descriptions of formic acid was reported in an extract of a letter written from John Wray (1627–1705) to the publisher of *Philosophical Transactions* published in 1670. Wray's letter reported on "uncommon Observations and Experiments made with an Acid Juyce to be Found in Ants" and noted the acid was previously obtained by Samuel Fisher from the dry distillation of wood ants. Formic acid is found in stinging insects, plants, unripe fruit, foods, and muscle tissue. Jöns Jacob Berzelius (1779–1848) characterized formic acid in the early 19th century, and it was first synthesized from hydrocyanic acid by Joseph Louis Gay-Lussac (1778–1850) at about the same time. A number of synthetic preparations of formic acid were found in the first half of the 19th century. Marcellin Berthelot (1827–1907) discovered a popular synthesis using oxalic acid and glycerin in 1856; he and several other chemists from his period found syntheses of formic acid by heating carbon monoxide in alkaline solutions.

The usual commercial product is marketed as a solution of 90% formic acid in water. Anhydrous formic acid must be handled carefully; it is a strong dehydrating agent and can cause severe skin burns. Formic acid is soluble in water, alcohol, ether, and glycerol. It is industrially synthesized by heating carbon monoxide and sodium hydroxide under pressure to form sodium formate (HCOONa), which is then acidified with sulfuric acid to form formic acid: $CO_{(g)} + NaOH_{(aq)}$ HCOONa$_{(aq)}$ HCOOH$_{(aq)}$. It can also be prepared by the acid hydrolysis of methyl formate (H-COOCH$_3$), which is produced from methanol and carbon monoxide or from heating oxalic acid. Formic acid can also be obtained as a by-product in the production of other chemicals such as acetic acid, acetaldehyde, and formaldehyde.

Formic acid has a number of commercial uses. It is used in the leather industry to degrease and remove hair from hides and as an ingredient in tanning formulations. It is used as a latex coagulant in natural rubber production. Formic acid and its formulations are used as preservatives of silage. It is especially valued in Europe where laws require the use of natural antibacterial agents rather than synthetic antibiotics. Silage is fermented grass and crops that are stored in silos and used for winter feed. Silage is produced during anaerobic fermentation when bacteria produce acids that lower the pH, preventing further bacterial action. Acetic acid and lactic acid are the desired acids during silage fermentation. Formic acid is used in silage processing to reduce undesirable bacteria and mold growth. Formic acid reduces *Clostridia* bacteria that would produce butyric acid causing spoilage. In addition to preventing silage spoilage, formic acid helps preserve protein content, improves compaction, and preserves sugar content. Formic acid is used as a miticide by beekeepers.

Formic acid derivatives, such as sodium formate (HCOONa) and potassium formate (HCOOK), are used as deicing compositions for airport runways and roads; they have the advantage of low environmental impact and low corrosion impact. Formates are used as brightening agents in textiles, paper processing, and laundry detergents. In the food industry, formates are used as preservatives and formate esters are used as flavoring additives. Formates are also used in pharmaceuticals, wood preservatives, perfume fragrances, resins, plasticizers, and a variety of other products.

Formic acid has several noted health effects. It is produced in the body during methanol poisoning. Methanol is oxidized in the liver to formaldehyde by the action of alcohol dehydrogenase. Formaldehyde is then converted by formaldehyde dehydrogenase to formic acid. Formic acid is responsible for the toxic effects of methanol consumption because it produces metabolic acidosis directly and indirectly by promoting lactic acidosis. Although humans continually ingest small quantities of methanol in fruit and drinks, amounting to about 10 mg per day, the kidneys eliminate the small amount of formic acid produced from this source. One controversial topic is the consumption of methanol through the artificial sweetener aspartame. Aspartame is 11% methanol. This methanol is liberated during aspartame metabolism in the body and can result in an appreciably larger methanol level (on the order of hundreds of milligrams). Therefore aspartame health effects are due to its ability to metabolize into formaldehyde and formic acid.

Exposure to formic acid is potentially dangerous. It irritates the skin, eyes, and mucous membranes and may also be toxic to the kidneys. Chronic exposure may lead to kidney damage or dermatitis. Acute exposure produces a variety of symptoms including eye irritation, blisters in the mouth and nasal membranes, headache, nausea, and breathing difficulties. Extreme exposure results in unconsciousness and death.

43. Glucose

CHEMICAL NAME = 6-(hydroxymethyl)oxane-
2,3,4,5-tetrol
CAS NUMBER = 50–99–7 (D-glucose)
MOLECULAR FORMULA = $C_6H_{12}O_6$
MOLAR MASS = 180.2 g/mol
COMPOSITION = C(40.0%) H(6.7%)
O(53.3%)
MELTING POINT = 146°C
BOILING POINT = decomposes
DENSITY = 1.54 g/cm³

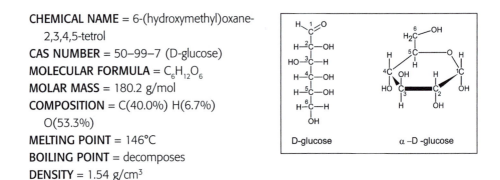

D-glucose   α–D-glucose

Glucose is one of the most important biological compounds found in nature. It is a main product in photosynthesis and is oxidized in cellular respiration. Glucose polymerizes to form several important classes of biomolecules including cellulose, starch, and glycogen. It also combines with other compounds to produce common sugars such as sucrose and lactose. The form of glucose displayed above is D-glucose. The "D" designation indicates the configuration of the molecule. The "D" configuration specifies that the hydroxyl group on the number 5 carbon is on the right side of the molecule (Figure 43.1). The mirror image of D-glucose produces another form of glucose called L-glucose.

In L-glucose, the hydroxyl group on the number 5 carbon is on the left hand side. The designations "D" and "L" are based on the configuration at the highest chiral center, which is carbon number 5. D-glucose is the common form found in nature. Glucose's biochemical abbreviation is Glc.

Glucose is the most common form of a large class of molecules called carbohydrates. Carbohydrates are the predominant type of organic compounds found in organisms and include sugar, starches, and fats. Carbohydrates, as the name implies, derive their name from glucose, $C_6H_{12}O_6$, which was considered a hydrate of carbon with the general formula of $C_n(H_2O)_n$, where n is a positive integer. Although the idea of water bonded to carbon to form a hydrate of carbon was wrong, the term *carbohydrate* persisted. Carbohydrates consist

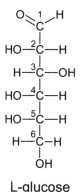

L-glucose

Figure 43.1   Numbering system for carbons in glucose.

of carbon, hydrogen, and oxygen atoms, with the carbon atoms generally forming long unbranched chains. Carbohydrates are also known as saccharides derived from the Latin word for sugar, *saccharon.*

Carbohydrates contain either an aldehyde or a ketone group and may be either simple or complex. Simple carbohydrates, known as monosaccharides, contain a single aldehyde or ketone group and cannot be broken down by hydrolysis reactions. Glucose is a simple carbohydrate containing an aldehyde group. Disaccharides consist of two monosaccharides units bonded together. Maltose and cellobiose are disaccharides formed from two glucose units. They differ in one of the glucose units having an α rather than a β configuration as discussed in the next paragraph. Sucrose and lactose are examples of disaccharides, where glucose is combined with another sugar. Sucrose is a combination of glucose and fructose; lactose is formed from glucose and galactose.

Although often displayed as an open chain structure, glucose and most common sugars exist as ring structures. The ring structure displayed is the alpha (α) form. In the α form, the hydroxyl group attached to carbon 1 and the CH_2OH attached to carbon 5 are located on opposite sides of the ring; β-glucose has these two groups on the same side of the ring. Also, glucose may exist as l and d forms. The "l" and "d" labels come from the Latin *laevus* and *dexter* meaning left and right, respectively, and refers to the ability of solutions of glucose to rotate plane-polarized light to the left or right. Rotation to the left is generally specified by using a negative sign (−) prefix and positive rotation is given by a positive (+) sign. The rotation of light in glucose is due to the existence of mirror image forms of each molecule called stereoisomers. The D and L forms are mirror images and will rotate polarized in opposite direction. The capital letters L and D that designate configuration at the highest numbered chiral center should not be confused with the small letters l and d, which refer to the direction that plane polarized light rotates.

D-Glucose is the most important and predominant monosaccharide found in nature. It was isolated from raisins by Andreas Sigismund Marggraf (1709–1782) in 1747, and in 1838, Jean-Baptiste-André Dumas (1800–1884) adopted the name glucose from the Greek word *glycos* meaning sweet. Emil Fischer (1852–1919) determined the structure of glucose in the late 19th century. Glucose also goes by the names dextrose (from its ability to rotate polarized light to the right), grape sugar, and blood sugar. The term *blood sugar* indicates that glucose is the primary sugar dissolved in blood. Glucose's abundant hydroxyl groups enable extensive hydrogen bonding, and so glucose is highly soluble in water.

Glucose is the primary fuel for biological respiration. During digestion, complex sugars and starches are broken down into glucose (as well as fructose and galactose) in the small intestine. Glucose then moves into the bloodstream and is transported to the liver where glucose is metabolized through a series of biochemical reactions, collectively referred to as glycolysis. Glycolysis, the breakdown of glucose, occurs in most organisms. In glycolysis, the final product is pyruvate. The fate of pyruvate depends on the type of organism and cellular conditions. In animals, pyruvate is oxidized under aerobic conditions producing carbon dioxide. Under anaerobic conditions in animals, lactate is produced. This occurs in the muscle of humans and other animals. During strenuous conditions the accumulation of lactate causes muscle fatigue and soreness. Certain microorganisms, such as yeast, under anaerobic conditions convert pyruvate to carbonic dioxide and ethanol. This is the basis of the production of alcohol. Glycolysis also results in the production of various intermediates used in the synthesis of other

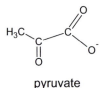

pyruvate

biomolecules. Depending on the organism, glycolysis takes various forms, with numerous products and intermediates possible.

Several common polysaccharides are derived from glucose including cellulose, starch, and glycogen. These polymeric forms of glucose differ in structure. Cellulose, the most abundant polysaccharide, forms the structural material of the cell walls of plants. Cellulose is a polymer

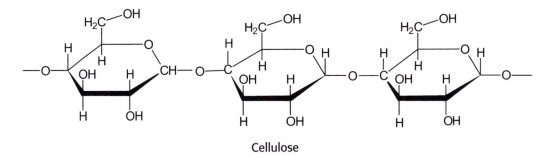

Cellulose

of glucose and consists of several thousand glucose molecules linked in an unbranched chain. Humans do not produce the enzymes, called cellulases, necessary to digest cellulose.

Bacteria possessing cellulase inhabit the digestive tracts of animals such as sheep, goats, and cows, giving these animals the ability to digest cellulose. Cellulase bacteria also exist in the digestive systems of certain insects, such as termites, allowing these insects to use wood as an energy source. Although humans cannot digest cellulose, this carbohydrate plays an important role in the human diet. Carbohydrates that humans cannot digest are called fiber or roughage. Fruit, vegetables, and nuts are primary sources of fiber. These foods have a cleansing effect on the large intestine, and such action is thought to speed the passage of cancer-causing materials through the intestine and therefore reduce the risk of colon cancer. Fiber helps retains water in the digestive system, aiding the overall digestion process. Another benefit of fiber is believed to be its ability to lower blood cholesterol, reducing heart and arterial disease.

Starch is a α-glucose polymer with the general formula $(C_6H_{12}O_5)_n$, where n can be a number from several hundred to several thousand. Plants store carbohydrate energy as starch. Grains, such as rice, corn, and wheat; potatoes; and seeds are rich in starch. Two principal forms of starch exist. One form is amylose and it accounts for approximately 20% of all starches. Amylose is a straight-chain form of starch containing several hundred glucose units. In the other form, called amylopectin, numerous glucose side branching is present. Starch is the primary food source for the human population, accounting for nearly 70% of all food consumed. Starch is broken down in the digestive system, starting in the mouth where the enzyme amylase breaks the bonds holding the glucose units together. This process ceases in the stomach because stomach acid creates a low pH environment, but it resumes in the small intestine. Glucose, maltose, and other small polysaccharides result from the digestion of starch.

In animals, most glucose is produced from starch and is not immediately needed for energy; therefore it is stored as glycogen. Because glycogen's function in animals is similar to that of starch in plants, it is sometimes referred to as animal starch. Glycogen is similar in structure to amylopectin, but it is larger and contains more glucose branching. When glucose blood levels drop, for instance during fasting or physical exertion, glycogen stored in the liver and muscles is converted into glucose and used by cells for energy. The process of conversion of glycogen to glucose is called glucogenesis. During periods of stress, the release of the hormone epinephrine (adrenaline) also causes glucose to be released from glycogen reserves (see Epinephrine). When blood glucose is high, such as immediately after a meal, glucose is converted into glycogen and stored in the liver and muscles in a process called glycogenesis. When glycogen stores are depleted and blood glucose levels are low, glucose can be synthesized from noncarbohydrate sources in a process called gluconeogenesis. Pyruvate, glycerol, lactate, and a number of amino acids are sources that can be converted to glucose.

The transformation of glucose to glycogen and glycogen back to glucose enables humans to regulate their energy demands throughout the day. This process is disrupted in individuals who have a form of diabetes known as diabetes mellitus. Diabetes mellitus occurs in individuals whose pancreas produces insufficient amounts of insulin or whose cells have the inability to use insulin (see Insulin). Insulin is a hormone responsible for signaling the liver and muscles to store glucose as glycogen. In Type 1 diabetes, called insulin-dependent diabetes mellitus, the body produces insufficient amounts of insulin. This type of diabetes occurs in youth under the age of 20 years and is the less prevalent form (about 10% of diabetics carry this form). Type 1 diabetes is controlled using insulin injections and regulating the diet. Type 2 diabetes, known as insulin-independent diabetes mellitus, is the most common form of diabetes and is associated with older individuals (generally over 50) who are overweight. In this form of diabetes, individuals produce adequate supplies of insulin, but the cells do not recognize the insulin's signal, and therefore do not capture glucose from the blood. This type of diabetes is regulated with drugs and a strictly controlled diet.

44. Glycerol (Glycerin)

CHEMICAL NAME = 1,2,3-propanetriol
CAS NUMBER = 56–81–5
MOLECULAR FORMULA = $C_3H_8O_3$
MOLECULAR MASS = 92.1g/mol
COMPOSITION = C(39.1%) H(8.8%) O(52.1%)
MELTING POINT = 17.8°C
BOILING POINT = 290°C
DENSITY = 1.26 g/cm³

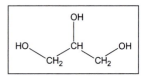

Glycerol is a colorless, viscous, hygroscopic, sweet-tasting trihydric alcohol. It is also called glycerin or glycerine, with the term *glycerol* being preferred as the pure chemical form and the term *glycerin(e)* being primarily used when the compound is used commercially in various grades. Glycerol was first isolated from olive oil and lead oxide by the Swedish chemist Carl Scheele (1742–1786) while making lead plaster soap in 1779. Scheele eventually realized that glycerol was a common ingredient in fats and oils and referred to glycerol as "the sweet principle of fats." In 1811, the French chemist Michel Eugene Chevreul (1786–1889), who was a pioneer in the study of fats and oils, proposed the name glycerine after the Greek word *glucos,* which means sweet. Chevreul decomposed soaps isolating different acids such as stearic and butyric acid and discovered that glycerol was liberated when oils and fats were boiled in a basic mixture. Théophile-Jules Pelouze (1807–1867) derived glycerol's empirical formula in 1836.

Glycerol is a by-product in the production of candles and soaps and was originally discarded in the production of these items. The process of converting a fat to soap is termed *saponification.* The traditional method of saponification involved the use of animal fats and vegetable oils. Fats and oils are esters formed when three fatty-acid molecules attach to a single glycerol molecule. When the three fatty acids attach to the three hydroxyl groups of the glycerol, a triglyceride is formed. During saponification of animal and plant products, hydrolysis of triglycerides converts triglycerides back to fatty acids and glycerol. The fatty acids then react with a base to produce a carboxylic acid salt commonly called soap.

Until 1940, the world's demand for glycerol was supplied from natural sources through the production of soaps and candles. Glycerol can also be produced through the fermentation of sugar, and this process was used to increase glycerol production during World War I. Glycerol can also be produced synthetically from propylene. The synthetic production from propylene first occurred just before World War II and commercial production started in 1943 in Germany. The synthetic process begins with the chlorine substitution of one hydrogen atom of propylene to allyl chloride: $H_2C = CH\text{-}CH_3 + Cl_2 \rightarrow H_2C = CH\text{-}CH_2Cl + HCl$. Allyl chloride is then treated with hypochlorous acid to produce 1,3-dichlorohydrin:

$$H_2C{=}CH{-}CH_2Cl \ + \ HOCl \ \longrightarrow \ \underset{\overset{|}{H_2C}{-}\overset{|}{CH}{-}CH_2Cl}{\overset{Cl \quad OH}{}}$$

Next, 1,3-dichlorohydrin is combined with a sodium or calcium hydroxide solution to produce epichlorohydrin:

$$2\ \underset{H_2C{-}CH{-}CH_2Cl}{\overset{Cl \quad OH}{}} \ + \ Ca(OH)_2 \ \longrightarrow \ 2\ \underset{H_2C{-}CH{-}CH_2{-}Cl}{\overset{O}{\triangle}} \ + \ CaCl_2 \ + \ 2\ H_2O$$

In the final reaction stage, epichlorohydrin is hydrolyzed to glycerol using a sodium hydroxide or sodium carbonate base:

$$\underset{Cl{-}CH_2{-}HC{-}CH_2}{\overset{O}{\triangle}} \ + \ NaOH \ + \ H_2O \ \longrightarrow \ \underset{H_2C{-}CH{-}CH_2}{\overset{OH \ \ OH \ \ \ OH}{}} \ + \ NaCl$$

In an alternate synthesis, the allyl chloride is hydrolyzed with NaOH to produce allyl alcohol $(H_2C = CH\text{-}CH_2\text{-}OH)$. This is then chlorinated to give monochlorohydrin and dichlorohydrin, which are hydrolyzed with sodium bicarbonate to produce glycerol.

Production of glycerol in the United States is approximately 500 million pounds annually, and it is found in thousands of products. Glycerol's properties make it useful for numerous applications. The three hydroxyl groups in glycerol allow extensive hydrogen bonding that gives glycerol its characteristic syrupy viscous texture and hygroscopic character. Approximately 40% of glycerol's use is for personal care products such as cosmetics, soaps, shampoos, lotions, mouthwash, and toothpaste. Glycerol's hygroscopic properties make it a good moisturizer in skin products. Another 25% of glycerol's annual production is used in food production. In the food industry glycerol is used as a moistening agent, as a solvent for food coloring and syrups, to prevent crystallization of sugar in candies and icings, as a preservative, and as a sweetening agent. Approximately 10% of glycerol's use goes into tobacco processing, where it is sprayed on tobacco leaves before they are shredded to serve as a moistening agent. Glycerol has the added benefit of imparting a sweet taste to chewing tobacco. The remaining 25% of glycerol's use is distributed among various industrial uses. It is used in cough syrups and elixir medicines. In industry, glycerol is found in lubricants, plasticizers, adhesives, antifreezes, resins, and insulating foams. At one time it was used almost exclusively in its nitrated form as an explosive (see Nitroglycerin), which today accounts for about 3% of its use.

45. Guanine

CHEMICAL NAME = 2-Amino-1,7-dihydro-6H-purin-6-one
CAS = 73–40–5
MOLECULAR FORMULA = $C_5H_5N_5O$
MOLAR MASS = 151.1 g/mol
COMPOSITION = C(39.7%) H(3.3%) N(46.3%)
  O(10.6%)
MELTING POINT = decomposes at 360°C
BOILING POINT = decomposes at 360°C
DENSITY = 2.2 g/cm³ (calculated)

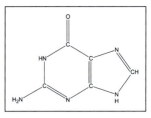

Guanine is one of the two purines comprising the five nucleic acid bases. Much of the information regarding the general role of nucleic acid bases is covered in Adenine and Cytosine. Guanine gets its name from guano, from which it was first isolated in the 1840s. Guano is nutrient-rich fecal matter deposited by sea birds, bats, and certain marine mammals. The name guano comes from the Spanish word for dung. Nutrient-rich guano is concentrated in several locales around the world. Islands off the coast of Peru were particularly rich in guano and supplied much of the world's demand for fertilizer and nitrate until the synthetic production of these were developed at the beginning of the 20th century.

Albrecht Kossel (1853–1927) determined that guanine (as well as adenine, cytosine, thymine and uracil) was a component of nucleic acid in the last two decades of the 19th century. Similar to adenine, guanine combines with ribose to form a nucleoside. The nucleoside produced is guanosine, which in turn combines with one to three phosphoryls to yield the nucleotides guanosine monophosphate (GMP), guanosine diphosphate (GDP), and guanosine triphosphate (GTP), respectively. Guanine nucleotides play an important role in metabolism including the conversion of adenosine diphosphate (ADP) to adenosine triphosphate (ATP) and carbohydrate metabolism.

Guanine is a crystalline amorphous substance that is found in guano, fish scales, and the liver of certain mammals. Guanine is responsible for the silvery iridescence of certain fish scales. Before its discovery, guanine was scraped from fish scales and used to coat beads to produce imitation pearls. Thus it was called pearl essence or pearly white in the 1700s. Guanine obtained from fish scales is used in cosmetics, especially for eye cosmetics and nail polishes.

46. Hydrochloric Acid

CHEMICAL NAME = hydrochloric acid
CAS NUMBER = 7647–01–0
MOLECULAR FORMULA = HCl$_{(aq)}$
MOLAR MASS = 36.5 g/mol
COMPOSITION = H(2.8%) Cl(97.2%)
MELTING POINT = −59°C (20%), −30°C (36%)
BOILING POINT = 108°C (20%), 61°C (36%)
DENSITY = 1.10 g/cm³ (20%), 1.18 g/cm³ (36%)

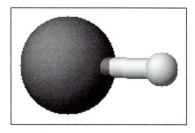

Hydrochloric acid is a strong, corrosive acid that results when the gas hydrogen chloride dissolves in water. Ancient alchemists prepared hydrochloric acid and Jabbar ibn Hayyan, known in Latin as Geber (721–815), is credited with its discovery around the year 800. The original method of preparation involved reacting salt with sulfuric acid, producing sodium hydrogen sulfate and hydrogen chloride gas. The hydrogen chloride gas is captured and dissolved in water to produce hydrochloric acid. Hydrochloric acid was formerly called muriatic acid. Terms such as muriatic and muriate were used in association with chloride substances before the discovery and nature of chlorine were fully understood. The Latin term *muriaticus* means pickled from *muri*, which is the Latin term for brine. Chlorides were naturally associated with seawater salt solutions, as chloride is the principal ion in seawater.

Hydrochloric acid is used in numerous applications, but it is generally obtained indirectly as a by-product in other chemical processes. The first large-scale production of hydrochloric acid resulted from the mass production of alkalis such as sodium carbonate (Na_2CO_3) and potassium carbonate (potash, K_2CO_3). The depletion of European forests and international disputes made the availability of alkali salts increasingly uncertain during the latter part of the 18th century. This prompted the French Academy of Science to offer a reward to anyone who could find a method to produce soda ash from common salt (NaCl). Nicholas LeBlanc (1743–1806) was credited with solving the problem. LeBlanc proposed a procedure in 1783 and a plant based on LeBlanc's method was opened in 1791. LeBlanc's method uses sulfuric acid and common salt to initially produce sodium sulfate (Na_2SO_4). Sodium sulfate is then reacted with charcoal and lime to produce sodium carbonate and calcium sulfide:

$$H_2SO_{4(l)} + 2NaCl_{(aq)} \rightarrow Na_2SO_{4(s)} + 2HCl_{(g)}$$
$$Na_2SO_{4(s)} + 2C_{(s)} + CaCO_{3(s)} \rightarrow Na_2CO_{3(s)} + CaS_{(s)}$$

Plants using the LeBlanc process were situated in areas associated with salt mines, and this naturally created locales for other industries that depended on soda ash. The alkali industry using the LeBlanc process created environmental problems near the alkali plants. The hydrogen chloride gas produced killed vegetation in the immediate vicinity of the plants. To decrease air pollution the gas was dissolved in water, creating hydrochloric acid, which was then discharged to streams, and transforming the air pollution problem into a water pollution problem. The LeBlanc process produced an oversupply of hydrochloric acid until the Solvay process supplanted it in the late 19th century. Because the Solvay process did not produce hydrochloric acid and the increase demand for hydrochloric acid, it became necessary to develop chemical plants solely for its production.

The traditional method of preparation of hydrochloric acid is the reaction of metal chlorides, especially sodium chloride with sulfuric acid (see the first reaction described). Hydrochloric acid is also produced by direct synthesis from its elements. In the chlorine-alkali industry, electrochemical reactions produce elemental chlorine and hydrogen, which can then be combined to give hydrogen chloride: $Cl_{2(g)} + H_{2(g)} \rightarrow 2HCl_{(g)}$. Hydrogen chloride is then dissolved in water to produce hydrochloric acid. By far, the most common method of producing hydrochloric acid involves its production as a by-product in chlorination reactions. The production of chlorofluorocarbons was a large source of hydrochloric acid, but the Montreal Protocol, signed in 1987 and amended in the early 1990s, restricts the production of these compounds owing to their influence on the ozone layer. This has curtailed this source of hydrochloric acid. The production of other common industrial organic chemicals such as Teflon, perchloroethylene, and polyvinyl chloride result in the production of hydrogen chloride. The production of hydrochloric acid in polyvinyl chloride production takes place when ethylene is chlorinated:

$$C_2H_{4(g)} + Cl_{2(g)} \rightarrow C_2H_4Cl_{2(g)}$$
$$C_2H_4Cl_{2(g)} \rightarrow C_2H_3Cl_{(g)} + HCl_{(g)}$$

Approximately 30% of hydrochloric acid production in the United States is used in the production of other chemicals. The second most common use is in the pickling of steel (20%), followed by oil well acidizing (19%) and food processing (17%). Pickling is a metal treatment process used to prepare metal surfaces for subsequent processing such as galvanizing or extrusion. In the iron industry, pickling involves immersing iron and steel products in vats of diluted hydrochloric acid. This removes oxides, dirt, and grease. Oil well acidizing involves injecting hydrochloric acid down well holes to dissolve limestone and carbonate formations. This expands existing fissures and creates new fissures to open channels for oil extraction.

Stomach acid is hydrochloric acid with a pH of about 1.5. The hydrochloric acid present in gastric juices helps digest food and activates specific enzymes in the digestive process. The stomach and digestive tract are protected from the hydrochloric acid secreted during digestion by a protective mucus lining and gastric buffers. Cells lining the digestive system constantly regenerate the protective mucus layer. Heartburn and stomach ulcers develop when the protective mucus lining is weakened or when excess stomach acid is generated as a result of health problems. The use of antacids are a common method to neutralize excess stomach acid. An antacid is nothing more than a base or salt that produces a basic solution when ingested.

Hydrochloric acid is also used extensively in pharmaceuticals and the food industry. When it is listed after a drug name, the drug was produced by combining a free base and hydrochloric acid to produce a hydrochloride salt. Drugs delivered as hydrochloride salts rather than free bases are more soluble in water than free forms of the drugs, tend to be more stable, are solids, and are often more compatible with the chemistry of the digestive system. In the food industry it is used in the production of gelatin and sodium glutamate, to convert cornstarch to syrup, to refine sugar, and as an acidulant.

47. Hydrogen Peroxide

CHEMICAL NAME = hydrogen peroxide
CAS NUMBER = 7722–84–1
MOLECULAR FORMULA = H_2O_2
MOLAR MASS = 34.0 g/mol
COMPOSITION = H (5.9%) O(94.1%)
MELTING POINT = −0.43°C
BOILING POINT = 150.2°C
DENSITY = 1.44 g/cm³

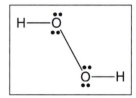

Hydrogen peroxide is a colorless liquid that is widely used as an oxidizer and bleaching agent. Hydrogen peroxide was discovered in 1818 by the French chemist Louis-Jacques Thenard (1777–1857) who called the compound eau oxygene (oxygen water). Thenard burned barium salts to form barium peroxide, BaO_2, which was dissolved in water to produce hydrogen peroxide.

Thenard's method was used for the initial commercial production of hydrogen peroxide starting in the 1870s. A popular use at this time was the bleaching of straw hats. From 1920 to 1950, the primary method of production was electrolysis. One process involved passing electric current through sulfuric acid to produce the peroxydisulfate ion ($S_2O_8^{2-}$), which was then hydrolyzed to H_2O_2: $2H_2O + S_2O_8^{2-}{}_{(aq)} \rightarrow 2H_2SO_4^-{}_{(aq)} + H_2O_{2(aq)}$. The relatively high cost of electricity of this method encouraged a search for a more economical production process. Hydrogen peroxide is currently produced on a large scale using the anthraquinone autooxidation procedure, which was developed in the 1940s. In this process, an anthraquinone, typically 2-ethyl-anthraquinone, is hydrogenated to a hydroquinone (2-ethyl-anthrahydroquinone) then reoxidized back to the anthraquinone (2-ethyl-anthraquinone) while forming hydrogen peroxide (Figure 47.1). A metal palladium or nickel catalyst is used to convert the anthraquinone to the hydroquinone, followed by autooxidation in air to generate hydrogen peroxide. The anthraquinone and hydrogen peroxide are separated; the former is recycled to repeat the process while the hydrogen peroxide is purified. There are current attempts to commercialize the direct combination of oxygen and hydrogen to H_2O_2 using catalysts, and researchers are examining new electrochemical methods.

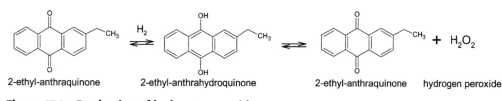

2-ethyl-anthraquinone   2-ethyl-anthrahydroquinone   2-ethyl-anthraquinone   hydrogen peroxide

Figure 47.1   Production of hydrogen peroxide.

Synthesized hydrogen peroxide is approximately 60% H_2O_2 by weight and is distilled to higher concentrations and diluted to lower concentrations for intended purposes. Different grades of hydrogen peroxide are sold that contain stabilizers and additives dependent on the end use. Food grade hydrogen peroxide comes in 35% and 50% concentrations. It is used for disinfecting purposes and also as an ingredient in cosmetics, shampoos, and medications. Reagent hydrogen peroxide for chemical and medical laboratories has a concentration of 30%. Standard grades of 35%, 50%, 60%, and 70% are used for industrial bleaching. General household hydrogen peroxide is 3% H_2O_2 and 6% is used by beauticians for hair coloring. Very high grades such as 90% are used as oxidizers in rocket propulsion.

Hydrogen peroxide decomposes over time to water and oxygen. Heat, ultraviolet light, and contaminants accelerate its decomposition, so it should be stored in cool, dark places. The foaming of H_2O_2 when a wound is cleaned is due to its catalytic decomposition from blood. Low-grade hydrogen peroxide is typically sold in brown bottles to protect the contents from ultraviolet-light exposure. Higher concentrations are stored in expandable or vented containers to accommodate the gases produced during decomposition.

World production of hydrogen peroxide is approximately 2.5 million tons, with approximately 50% used for bleaching in the pulp and paper industry and another 10% for bleaching and desizing textiles. The paper industry uses H_2O_2 in kraft bleaching. Kraft bleaching is the treatment of wood chips with sodium hydroxide and sodium sulfide to dissolve the lignin and produce a strong fibrous pulp called kraft pulp (kraft means strong in German). Kraft pulp is composed of fibers with the color of brown paper bags or cardboard. Because a number of metals catalyze the decomposition of H_2O_2, chelating agents are added to pulp mixes to deactivate metals. Bleaching with hydrogen peroxide is due to the formation of the perhydroxyl anion, HOO^- under alkaline conditions: $H_2O_2 + H_2O \rightarrow H_3O^+ + HOO^-$. The use of hydrogen peroxide to augment and replace chlorine in the paper and (textile industry) has environmental advantages and is part of the green chemistry movement.

Hydrogen peroxide has a number of environmental uses. These include water treatment, odor control, oxidation of pollutants, and corrosion control. Hydrogen peroxide is used to remove iron, manganese, and hydrogen sulfide from water supplies and wastewater. The oxidation of substances such as hydrogen sulfide reduces odors. Because H_2O_2 decomposes into oxygen and water, it has the added advantage of lowering the biological oxygen demand of wastewater.

Hydrogen peroxide is used in chemical synthesis and can function as both an oxidizing and reducing agent. Caro's acid (H_2SO_5) is made using H_2O_2. Peracetic acid ($C_2H_4O_3$) is produced by reacting acetic acid and hydrogen peroxide and is used as a disinfectant. Solid bleaching agents such as perborates and percarbonates are made using H_2O_2. It is used in epoxidation and hydroxylation reactions. Epoxidation reactions involve the breaking of double bonds in alkenes, with the carbons then bonding to the same oxygen atom to form an epoxide ring

(Figure 47.2). Hydroxylation introduces hydroxyl groups (-OH) into a compound. Hydrogen peroxide has been used since the 1930s as a propellant and oxidizer in rockets, torpedoes, submarines, and aircraft. The Germans initiated the use hydrogen peroxide for weapons and military applications. The German Messerschmitt rocket plane and V-2 rockets used H_2O_2 as a fuel. As a monopropellant (used individually), high concentration H_2O_2 (> 70%) is catalytically decomposed with a metal in a chamber at high temperature to produce gaseous products for propulsion. Rocket-grade H_2O_2 has a concentration of 90%. The X-15 rocket, which ushered in the start of the Space Age with flights starting in 1959, used hydrogen peroxide as a fuel. The Mercury, Gemini, and Apollo programs all used hydrogen peroxide for rocket propulsion of the space capsules. The lunar landing vehicle was powered and controlled by H_2O_2 rockets. As a monopropellant, H_2O_2 is much less efficient than when used as an oxidizer mixed with other fuels. The use of hydrogen peroxide as a propellant was displaced in the 1980s by other more powerful fuels, but it is still used as an oxidizer in rocket systems.

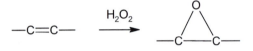

alkene   epoxide ring

Figure 47.2   Epoxidation process.

48. Hydrogen Sulfide

CHEMICAL NAME = hydrogen sulfide
CAS NUMBER = 7783–06–4
MOLECULAR FORMULA = H_2S
MOLAR MASS = 34.1 g/mol
COMPOSITION = H(5.9%) S(94.1%)
MELTING POINT = –82.3°C
BOILING POINT = –60.3°C
DENSITY = 1.55 g/L (vapor density = 1.2, air = 1.0)

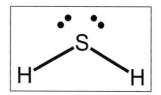

Hydrogen sulfide is a colorless, flammable, toxic gas with the characteristic odor of rotten eggs. It is produced naturally from the anaerobic bacterial decomposition of organic wastes, occurs in volcanic gases and hot springs, is a product of animal digestion, and is generated in industrial processes. Hydrogen is a natural component of natural gas and petroleum; it is only a small fraction of oil (hundreds of ppm), but may form an appreciable component of natural gas. Natural gas typically contains up to 5% hydrogen sulfide. Natural gas is considered sour if the hydrogen sulfide content exceeds 5.7 mg of H_2S per cubic meter of natural gas. The process for removing hydrogen sulfide from sour gas is referred to as sweetening the gas. Because hydrogen sulfide is associated with anaerobic respiration in sewers and swamps, it is referred to as sewer gas, swamp gas, or stink damp.

Hydrogen sulfide was known to exist in water in the 15th century and was called sulfur water or sulfur vapors. Alchemists referred to H_2S as *aer hepaticus* (hepatic air). Early chemists called it sulfuretted hydrogen, a term still used today. Carl Wilhelm Scheele (1742–1786) was the first chemist to prepare and describe hydrogen sulfide; he considered it a combination of sulfur, phlogiston, and heat. Claude Louis Berthollet (1748–1822) determined the composition of H_2S in 1789 and noted its acidic nature.

Hydrogen sulfide is produced during anaerobic respiration (fermentation). Anaerobic respiration enables organisms, primarily bacteria and other microbes, to meet their energy needs using sulfate, elemental sulfur, and sulfur compounds as electron acceptors instead of oxygen.

A simplified reaction representing anaerobic respiration is: $SO_4^{2-}{}_{(aq)} + 2(CH_2O)_{(s)} \rightarrow H_2S_{(g)} +$ $2HCO_3^{2-}{}_{(aq)}$. The hydrogen sulfide produced in this reaction can be oxidized back to sulfate or elemental sulfur if it enters the atmosphere or an oxidizing environment. Hydrogen sulfide also combines with other substances in sediments or soils to produce sulfur minerals. For example, pyrite (FeS$_2$) forms when hydrogen sulfide reacts with iron monosulfide: $H_2S_{(g)} + FeS_{(aq)} \rightarrow$ $FeS_{2(s)} + H_{2(g)}$.

The primary process for sweetening sour natural gas involves using amine solutions to remove the hydrogen sulfide. The natural gas is run through an absorption tower containing the amine solution. The two principal amines used are monoethanolamine and diethanol-amine, which absorb hydrogen sulfide. Elemental sulfur can be recovered from the solution using the Claus process. In the Claus process, the amine-extracted H$_2$S is thermally oxidized at temperatures of approximately 1,000°C to sulfur dioxide: $H_2S_{(g)} + 1.5O_{2(g)} \rightarrow SO_{2(g)} + H_2O_{(g)}$. The sulfur dioxide combines with the hydrogen sulfide to produce sulfur: $2H_2S + SO_2 \rightarrow$ $3S + 2H_2O$. The thermal conversion steps results in a 60–70% recovery of elemental sulfur. To recover most of the remaining sulfur, hydrogen sulfide and sulfur dioxide are catalyti-cally combined using activated alumina (Al$_2$O$_3$) or titanium catalysts. The catalytic process is repeated in several stages to boost sulfur recovery to greater than 95%.

Hydrogen sulfide has relatively few commercial uses. It is used to produce elemental sulfur, sulfuric acid, and heavy water for nuclear reactors. Heavy water, D$_2$O, contains the hydrogen isotope deuterium (D), which has a neutron in its nucleus, as opposed to normal hydrogen, which does not have a neutron. Heavy water is produced by the water-hydrogen sulfide exchange process (Girdler-Sulfide or G-S process). The process is based on the exchange of hydrogen and deuterium (D) in water and hydrogen sulfide at different tem-peratures. The equilibrium can be represented as $H_2O_{(l)} + HDS_{(g)} \longleftrightarrow HDO_{(l)} + H_2S_{(g)}$, where HDS is the heavy isotope variety of hydrogen sulfide and HDO is semiheavy water. The equilibrium favors formation of HDO at colder temperatures and HDS at higher temperatures. The exchange takes place in a series of towers, each with a hot lower section and a cold upper section. Ordinary water flows down the tower while hydrogen sulfide gas bubbles up through a series of perforated trays. In the cold section, the deuterium from HDS exchanges with hydrogen in cold feedwater, H$_2$O. The HDS becomes H$_2$S, and the H$_2$O is converted to HDO (D$_2$O and D$_2$S are also formed). In the hot section of the tower the process is reversed. Deuterium-enriched water is removed from the middle of the first stage tower and moves to another tower where the process is repeated. By routing water through a series of towers, the final water product is enriched to approximately 30% in deuterium. This water is sent to a distillation unit to produce reactor grade heavy water, which is 99.75% deuterium oxide, D$_2$O.

Hydrogen sulfide gas is highly toxic and caution must be taken to avoid accidental exposure, especially through inhalation. In 1713, Bernardino Ramazzini (1633–1714), an Italian physician, noted the effects of hydrogen sulfide on workers in his chapter: "Diseases of Cleaners of Privies and Cesspits" found in *De Morbis Artificum Diatribe* (A Treatise on Occupational Illness). Hydrogen sulfide is the leading cause of death from occupational inha-lation of toxic chemicals. Workers in the oil and gas industry can be exposed to dangerous concentrations, as can agricultural workers from manure sources, and wastewater treatment personnel. Concentrations as low as 0.02 ppm can be smelled, but the smell disappears at concentrations approaching 100 ppm because of disruption of the olfactory sense. Low H$_2$S concentrations ranging from several parts per million to 50 ppm irritate the eyes, nose, and throat. A 100 ppm level is classified as Immediately Dangerous to Life and Health (IDLH) by the federal government. At 250 ppm, a person may lose consciousness; experience severe stinging of the eyes, throat, and nose; and suffer pulmonary edema. An exposure to 500–700 ppm for 30 minutes or more is often fatal, and a level of 1,000 ppm causes death in a few minutes.

Hydrogen sulfide is generally not a problem in the home, but exposure may occur through faulty plumbing or enter the home through ground water sources. Intestinal bacteria in the digestive tract produce hydrogen sulfide and other gases. The odor associated with flatulence is due to H_2S. Hydrogen sulfide is also produced in the mouth and contributes to bad breath.

49. Ibuprofen

CHEMICAL NAME = 2-(4-isobutylphenyl)propanoic acid

CAS NUMBER = 15687–27–1

MOLECULAR FORMULA = $C_{13}H_{18}O_2$

MOLAR MASS = 206.3 g/mol

COMPOSITION = C(75.7%) H(8.8%) O(15.5%)

MELTING POINT = 77°C

BOILING POINT = decomposes

DENSITY = 1.0 g/cm³ (calculated)

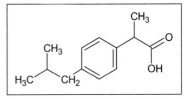

Ibuprofen is a white, crystalline anti-inflammatory drug used in numerous medications. It is the active ingredient marketed under various trade names including Advil, Motrin, and Nurofen. Ibuprofen is a nonsteroidal anti-inflammatory drug (NSAID) used as a pain reliever (analgesic), fever reducer (antipyretic), and inflammation reducer. Inflammation is a general physiological response to tissue damage characterized by swelling, pain, and heat. Ibuprofen was developed while searching for an alternative pain reliever to aspirin in the 1950s. It and related compounds were synthesized in 1961 by Stewart Adams, John Nicholson, and Colin Burrows who were working for the Boots Pure Drug Company in Great Britain. Adams and Nicholson filed for a British patent on ibuprofen in 1962 and obtained the patent in 1964; subsequent patents were obtained in the United States. The patent of Adams and Nicholson was for the invention of phenylalkane derivatives of the form shown in Figure 49.1, where R_1 could be various alkyl groups, R_2 was hydrogen or methyl, and X was COOH or COOR, with R being alkyl or aminoalkyl groups. The first clinical trials for ibuprofen were started in 1966. Ibuprofen was introduced under the trade name Brufen in 1969 in Great Britain. It was introduced in the United States in 1974. Ibuprofen was initially offered by prescription, but it became available in over-the-counter medications in the 1980s.

Ibuprofen works by inhibiting the enzyme cyclooxygenase (COX), which in turn interferes with the synthesis of prostaglandins. COX exists as several coenzyme forms that are similar in structure: COX-1, COX-2, COX-3; ibuprofen is a nonselective inhibitor of both COX-1

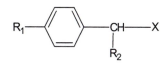

Figure 49.1 General structure of anti-inflammatory drugs.

and COX-2. COX-1 is continually produced in mammalian cells throughout the body in response to physiological stimuli. It is responsible for the production of prostaglandins, which get their name because it was originally believed they were synthesized in the prostate gland. In fact, prostaglandins are synthesized throughout the body and act like hormones by stimulating action in target cells. Prostaglandins, which are fatty acid compounds consisting of a 20-carbon chain including a 5 carbon ring, are involved in numerous physiological processes including renal function, blood clotting, and stomach mucus production. COX-2 is synthesized only in specific parts of the body (kidneys, brain, trachea) as needed and is therefore called an induced enzyme. COX-2 produces prostaglandins in response to tissue damage and inflammation. Inflammatory prostaglandins produce swelling, pain, and fever.

A common goal in the development of pain and inflammation medicines has been the creation of compounds that have the ability to treat inflammation, fever, and pain without disrupting other physiological functions. General pain relievers, such as aspirin and ibuprofen, inhibit both COX-1 and COX-2. A medication's specific action toward COX-1 versus COX-2 determines the potential for adverse side effects. Medications with greater specificity toward COX-1 will have greater potential for producing adverse side effects. By deactivating COX-1, nonselective pain relievers increase the chance of undesirable side effects, especially digestive problems such as stomach ulcers and gastrointestinal bleeding. COX-2 inhibitors, such as Vioxx and Celebrex, selectively deactivate COX-2 and do not affect COX-1 at prescribed dosages. COX-2 inhibitors are widely prescribed for arthritis and pain relief. In 2004, the Food and Drug Administration (FDA) announced that an increased risk of heart attack and stroke was associated with certain COX-2 inhibitors. This led to warning labels and voluntary removal of products from the market by drug producers; for example, Merck took Vioxx off the market in 2004. Although ibuprofen inhibits both COX-1 and COX-2, it has several times the specificity toward COX-2 compared to aspirin, producing fewer gastrointestinal side effects.

World production of ibuprofen is estimated to be 15,000 tons per year, which is approximately one-third the use of aspirin. Ibuprofen is commonly used for headaches (migraines), muscle pain, dental pain, cold, flu, fever, and menstrual cramps. Ibuprofen is most often sold as tablets, but it is also sold as liquid capsules, effervescents, chewable tablets, and liquid suspensions. Tablets commonly contain 200 mg of ibuprofen and a 1,200 mg per day limit taken over several dosages is recommended for adults. Child dosages are based on weight, but are commonly 10 mg or less per kilogram (2.2 pounds) of body weight. Ibuprofen is not recommended for pregnant women. Common side effects associated with ibuprofen are stomach pains, heartburn, constipation, headache, blurry vision, and ear-ringing. More serious side effects often associated with high prolonged doses are stomach ulcers, gastrointestinal bleeding, and renal malfunction. Increased risk of heart attack is sometimes mentioned when taking ibuprofen, but research is inconclusive as to whether ibuprofen increases the chance of a heart attack.

50. Indigo

CHEMICAL NAME = (2*E*)-2,2'-
  biindole-3,3'(1*H*,1'*H*)-dione
CAS NUMBER = 482–89–3
MOLECULAR FORMULA = $C_{16}H_{10}N_2O_2$
MOLAR MASS = 262.26 g/mol
COMPOSITION = C(73.3%) H(3.8%)
  N(10.7%) O(12.2%)
MELTING POINT = sublimes at 300°C
BOILING POINT = decomposes at 390°C
DENSITY = 1.4 g/cm³ (calculated)

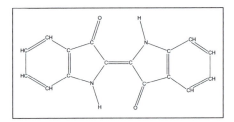

Indigo, known chemically as indigotin, is a common blue dye that has been highly valued throughout history and has played a major role in trade and commerce since ancient times. The term *indigo* is often used to describe many blue dyes produced from a number of plants. For example, woad, a blue dye obtained from the plant *Isatis tinctoria,* was used throughout the Mediterannean and Europe and is often identified as indigo. True indigo comes from the leguminous plant of the genus *Indigofera.* The *Indigofera* genus includes several hundred species, and indigo has been obtained from a number of these, but the dominant species for the dye are *Indigofera tinctoria* grown mainly in India and tropical Asia and *Indigofera suffruc-tiosa* from the tropical Americas. The name indigo comes from the Greek *indikon* and Latin *indicum* meaning "dye from India." There is evidence that indigo was used several thousand years B.C.E. Persian rugs containing indigo color exist from several thousand years B.C.E. Textile artifacts from Egyptian tombs provide evidence of indigo's use by royalty from as far back as 2500 B.C.E. The writings of Herodotus from approximately 450 B.C.E. mention indigo's use in the Mediterranean area.

Indigo was gradually introduced into Europe by traders, explorers such as Marco Polo, and Crusaders, but its widespread use did not begin until the 15th century with the opening of sea trade routes between Europe and India. As imports of indigo increased, local woad growers and the woad industry made futile attempts to reduce competition through protests that led to indigo bans in England and several other European countries. These bans temporarily delayed the use of indigo, but since they were not universally accepted throughout the continent indigo replaced woad as the dominant blue dye. Indigo became a chief trade commodity of the Dutch and East Indian Companies during the 17th century. As the demand for indigo increased, the cultivation of *Indigofera tinctoria* increased in India. Indigo plantations first developed in the western region of Gujarat in the 1600s and during the next century the main production areas shifted to the eastern regions of Bihar and Bengal. To supplement Europe's demand for indigo, plantations were established in European colonies in the Americas and Africa. South Carolina and Georgia in North America produced large quantities of indigo and it was a major crop throughout the South.

Although the processing of indigo varied somewhat between regions, the general procedure involved gathering cut *Indigofera* plants in bundles and immersing the plants in vats of clear warm water where they soaked for approximately a day. This first step resulted in the extraction of indican from the plants. Indican is the precursor to indigo and is the molecule found in all species that produce indigo. The action of bacteria on the extracted indican resulted in fermentation of indican to indoxyl, with glucose splitting off the indican (Figure 50.1). Both indican and indoxyl are colorless. After transferring the pulpy extraction to another vat, a beating process took place. This traditionally involved workers, who were often slaves, using sticks or paddles to agitate the pulpy liquid obtained from the initial soak. During beating, ashes or other basic substances were added to the mix, causing sediments and heavier particles to separate from the mixture. Lighter particles that consisted of minute flakes of indigo remained in suspension. The agitation of the mixture oxidized indoxyl, removing two hydrogen atoms while a double bond was formed between two of the oxidized indoxyls to produce indigotin,

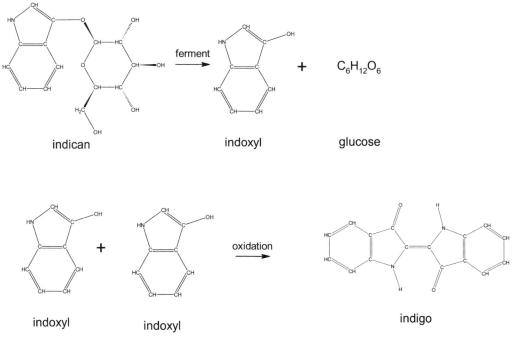

Figure 50.1 Production of indigo.

which is commonly refer to as indigo. The heavier dirt sediments were removed and the indigo-rich sediments settled to the bottom. Alternatively, the supernatant containing indigo was drawn off and further purified by processes such as filtering, boiling, or settlement in a third vat to concentrate the indigo. The final product was shaped into bricks or cakes and dried; the cakes were now suitable for transportation. The cakes could be sold directly or ground into a powder for sale.

The final indigo product was unsuitable for dyeing because it was insoluble in water. To make indigo suitable for dyeing, it must be treated with a reducing agent to produce a form of indigo called leucoindigo. Leucoindigo is soluble in a dilute alkali solution. Traditional alkali solutions consisted of lime (calcium oxide and calcium hydroxide), potash (potassium carbonate), madder, wheat bran, and aged urine were used. Today indigo is converted to leucoindigo using solutions of sodium hydrosulfite ($Na_2S_2O_4$) or thiourea dioxide (also called thiox, $CH_4N_2O_2S$) as reducing agents and sodium hydroxide to give alkali conditions. To dye a fabric it is soaked in the prepared leucoindigo solution, which has a yellow color. Leucoindigo adheres to the fiber through hydrogen bonding. On exposure to air, the leucoindigo oxidizes back to indigo imparting the blue color to the fabric (Figure 50.2). The dye adheres to the surface of the fabric, and as it wears off the original color of the fabric shows. A good example of this is seen when we observe the fading of blue jeans.

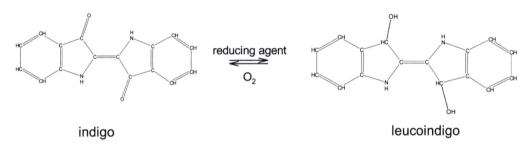

indigo   leucoindigo

Figure 50.2   Indigo reduction to leucoindigo.

At the end of the 19th century, the indigo industry was revolutionized with the synthetic production of indigo. The synthetic dye industry developed in the last half of the 19th century, with its start often attributed to William Henry Perkins's (1838–1907) accidental discovery of aniline purple in 1856. Perkins, who was only 18 at the time, was on a break from his study of chemistry at the Royal College of Chemistry in London and was attempting to synthesize quinine, $C_{20}H_{24}O_2N_2$, in his home laboratory. Quinine was of interest because its supply was insufficient to meet the demand for medicinal use as a remedy for malaria. Perkins attempted to synthesize quinine by oxidizing allyltoluidine, $C_{10}H_{13}N$: $2 C_{10}H_{13}N + 3O \rightarrow C_{20}H_{24}N_2O_2 + H_2O$.

Perkins did not obtain the white crystals that characterize quinine, but instead got a reddish-black precipitate. Perkins decided to repeat the experiment replacing allyltoluidine with aniline ($C_6H_5NH_2$). When Perkins oxidized aniline with potassium dichromate ($K_2Cr_2O_7$), he got a brown precipitate. Upon rinsing out the container containing this residue, he produced a purple substance that turned out to be the first synthetic dye. Instead of quinine Perkins had produced aniline purple. Perkins decided to leave school and commercialize his discovery. Backed by his father and a brother, Perkins established a dye plant in London and obtained a patent for aniline purple or mauve. It took several years for Perkins to perfect the manufacture of mauve, and in the process he started to convince the textile industry of the advantages of synthetic dyes.

The first synthesis of indigo is attributed to Adolf von Baeyer (1835–1917), who began his quest to synthesize indigo in 1865 but was not able to produce indigo until 1878. The synthetic production of indigo was first described by Baeyer and Viggo Drewson in 1882; Baeyer also identified the structure of indigo in 1882. The Baeyer-Drewson synthesis of indigo started with 2-nitrobenzaldehyde and acetone proceeding through a series of steps in alkali solution. Baeyer's work was not commercially viable, and it was not until 1897 that BASF (Badische Analin und Soda Fabrik) started to produce indigo commercially using a process developed by Karl von Heumann (1851–1894) that started with naphthalene. The synthetic production of indigo spelled the end of traditional methods of indigo production. By the second decade of the 20th century, nearly all indigo was produced synthetically.

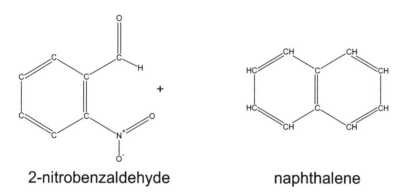

2-nitrobenzaldehyde   naphthalene

Current annual world production of indigo is approximately 18,000 tons, with BASF producing almost 40% of the world consumption. In recent years researchers have used genetic engineering using *Escherichia coli* to convert tryptophan into indigo. The desire for natural organic products has also revived traditional production methods of indigo on a small scale. Indigo's dominant use is as a textile dye, but indigo-related compounds have limited use as indicators and in food coloring. The Food and Drug Administration's FD&C Blue #2 contains indigotine (also known as indigo carmine), which is a sulfonated sodium salt of indigo.

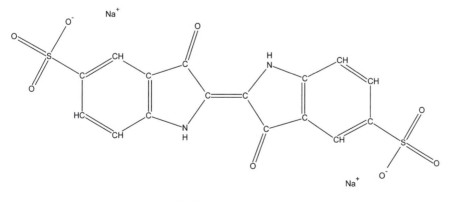

indigotine

CHEMICAL NAME = insulin

CAS NUMBER = 9004–10–8 (human)

MOLECULAR FORMULA = $C_{257}H_{383}N_{65}O_{77}S_6$

MOLAR MASS = 5,807.6 g/mol

COMPOSITION = C(53.1%) H(6.7%) N(15.7%) O(21.2%) S(3.3%)

MELTING POINT = not reported

BOILING POINT = not reported

DENSITY = not reported

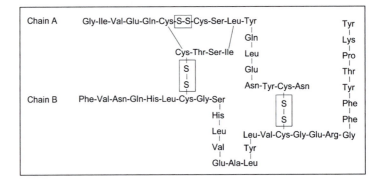

Insulin is a polypeptide hormone that consists of two peptide chains bonded by two disulfide bonds. The two chains are designated A and B. The A chain consists of 21 amino acids with a third internal disulfide bond, and the B chain contains the remaining 30 amino acids. All vertebrates produce insulin and the structure is similar in these species. For example, the insulin produced in humans and porcine species differs by only one amino acid, and humans and bovine insulin differ by three amino acids. Insulin plays a crucial role in several physiological processes. These include the regulation of sugar in the body, fatty acid synthesis, formation of triglycerides, and amino acid synthesis.

Insulin derives its name from the Latin word *insula* for island. Insulin is produced in the pancreas by β-cells in the islets of Langerhans (α-cells in the pancreas produce glucagon). Paul Langerhans (1847–1888) discovered the islets of Langerhans, which were subsequently shown to secrete insulin. Insulin binds to insulin receptors on cells and produces biochemical changes that allow the cells to take in glucose. Insulin also promotes the storage of glucose as glycogen in the liver. In this manner, insulin maintains glucose at a stable level in the bloodstream.

Insulin is most closely associated with diabetes. Diabetes results in excess blood glucose levels (hyperglycemia) and the inability of cells to absorb glucose, which in turn deprives them of energy. Instead of glucose providing energy, it is excreted excessively in the urine by diabetics. There are two main types of diabetes. Type 1 diabetes, called insulin-dependent diabetes mellitus, occurs when the body produces insufficient amount of insulin. This type of diabetes occurs in youth under age 20 and is the less prevalent form (about 10% of diabetics carry this form). Type 1 diabetes is controlled using insulin injections and regulating the diet. Type 2 diabetes, known as insulin-independent diabetes mellitus, is the most common form of diabetes and is more prevalent among older individuals (generally over 50) who are overweight. In this form of diabetes, individuals produce adequate supplies of insulin, but the cells do not recognize the insulin's signal and therefore do not capture glucose from the blood. This type of diabetes is regulated with drugs and a strictly controlled diet. It is estimated that 200 million people worldwide have diabetes, with about 16 million diabetics in the United States.

Diabetes has been prevalent in humans throughout history, but relief from the disease was not available until the 20th century. The term *diabetes* comes from the Greek words *dia bainein*, meaning to pass through or siphon. This denoted the condition of excessive urination in diabetics. Diabetes was traditionally diagnosed by a sweet taste in the patient's urine and the term *mellitus* comes from Latin meaning honey or sweet. In the late 1800s, it was known that diabetes was related to the pancreas, but it was not known exactly how. In 1901, Eugene Opie (1873–1971) determined that diabetes was related to the islets of Langerhans. During the early 20th century, knowledge was advanced on ductless glands and endocrine secretions that fed directly into the bloodstream. Diseases were treated with extracts obtained from glands. Within this framework, medical researchers attempted to treat diabetes with extracts obtained from the pancreas.

The discovery of insulin as a treatment for diabetes is primarily credited to several researchers from the University of Toronto. Frederick G. Banting (1891–1941) initiated the research in late 1920s after reading an article on diabetes and the islets of Langerhans in a medical journal. Banting had an idea on how to acquire secretions from the islets of Langerhans and wanted to attempt his method on treating diabetes. At the time, Banting had a medical practice in London, Ontario and was also assisting at the fledging medical school at Western University in London. Banting's medical practice was not very lucrative and he was searching for other areas in which to use his surgical and medical knowledge. Banting wanted to pursue research on diabetes, but the facilities at Western University did not have the necessary resources he needed. Using contacts at Western and the University of Toronto, where he received his medical degree, Banting approached John James Richard Macleod (1876–1935), who was Director of the Physiological Lab and Associate Dean of the faculty of medicine at the University of Toronto. Macleod's expertise was in glucose metabolism, but initially he was reluctant to support Banting's proposed work. Banting had little background knowledge on the research he was proposing, and Macleod was skeptical that Banting could add anything to current research in the area. Macleod eventually agreed to provide Banting with laboratory space and dogs for the research, and assigned him a student research assistant named Charles Herbert Best (1899–1978). Macleod granted Banting use of his laboratory facilities during the summer break when there was less demand for their use and Macleod himself would be in Europe for several weeks.

Banting's idea was to litigate the pancreatic ducts of dogs to cut off external secretion. This would result in the degeneration of the gland's acini leaving the islets of Langerhans from which the internal secretion could be obtained. The secretion could then be used to treat diabetic dogs. Diabetic dogs were produced by removing the pancreas of individual dogs. The research Banting proposed used standard procedures of researchers who were studying the pancreas at the time. Banting began his surgeries on dogs in May 1921 and initially was unsuccessful. Seven of the first ten dogs he operated on died during the first two weeks of the experiment. As Banting acquired more experience, better results were obtained. At the start of July, Banting and Best began using degenerate dog pancreata to prepare an extract for injection into diabetic dogs. Throughout July and into August, Banting and Best prepared extract, which they called Isletin, injected diabetic dogs, and observed the results. Cats and rabbits were also used as test subjects. Although differing results were obtained from each individual, the researchers were encouraged by the extract's ability to lower blood sugar and alleviate the diabetes in several animals. Macleod returned to the laboratory in late summer and, based on the results obtained by Banting and Best, agreed to continue to support the research.

In November and December, Banting, encouraged by data reported in the scientific literature, decided to use bovine fetuses obtained from a slaughterhouse as a source of extract. Banting also acquired the help of James Bertram Collip (1892–1965), who was a first-rate biochemist spending his sabbatical at the University of Toronto. As the work generated more excitement, Macleod took a more active role in guiding the research. Collip was able to obtain more purified extracts than Banting and Best, although as work proceeded a rift divided the researchers. Macleod and Collip collaborated building on the original work of Banting and Best. The adversarial relationship between the two pairs of scientists grew as questions of priority and ownership surfaced. In January and February 1922, doctors used extracts prepared by Collip to successfully treat human patients. An article written in March of that year proposed the word insulin for Collip's extract, and the University of Toronto applied for a patent on the drug in the spring of 1922. The Toronto group was having difficulty meeting demand for insulin in clinical trials as word spread of its efficacy in treating diabetics. To increase production and commercialize insulin, the University of Toronto group licensed insulin production to the pharmaceutical company Eli Lily in May 1922, and that summer the company invested heavily in the mass production of insulin. The challenge faced by Eli Lily and others working on insulin was to find organisms capable of producing the greatest quantity of insulin.

Insulin obtained from animal sources was the primary source of insulin until the 1980s. Porcine (pork) insulin was the principal source of insulin, as it is almost identical to human insulin; it differs by only one amino acid. Although animal insulins are effective in treating diabetes, there are several related problems with its use. These include allergenic reactions to the insulin, developing immunity to the insulin so that it is ineffective, and slower response to treatment. In the 1980s, recombinant DNA was used to manufacture human insulin by inserting the gene sequence that codes for insulin production into yeast or bacteria. Today, most insulin is made using this technique, but there is current research on using plant material to produce insulin drugs. Approximately 15 tons of insulin is produced annually worldwide.

Type 2 diabetes, also known as insulin-dependent diabetes, results from a lack of insulin. Type 2 diabetes, referred to as insulin-independent diabetes, occurs when insulin is low or cells cannot process the insulin. Diabetes impairs glucose uptake by cells, causing blood glucose levels to rise and producing a condition called hyperglycemia. This in turn results in frequent urination, dehydration, and hunger. Diabetes must be managed by monitoring blood sugar levels throughout the day, and using this information to adjust the diet and activities to keep blood glucose in an acceptable range; also, insulin must be taken for patients with Type 1 and some with Type 2. Insulin is usually administered through injections, but it can be delivered using other methods. Insulin pumps are small electronic devices that deliver insulin according to a programmed schedule throughout the day. Insulin pumps mimic the function of the pancreas to deliver a basal amount of insulin to the body. In recent years oral sprays have been developed to augment but not replace injected insulin. Another area of interest is delivery using dermal patches. Insulin cannot be delivered using conventional pills because stomach acidity denatures insulin. Although lack of insulin results in hyperglycemia, it is important when administering insulin to prevent hypoglycemia. In hypoglycemia blood sugar is too low. Hypoglycemia in diabetics can occur from too much insulin, lack of food, exercise, or lack of carbohydrates. Its symptoms include increased heart rate, nervousness, perspiration, and shakiness. When this condition is recognized, diabetics consume carbohydrates such as candy, fruit, or fruit juice to boost blood sugar levels. In extreme cases, a person can have a severe reaction and experience seizures or fall into a coma. Severe reaction to insulin is referred to as insulin shock.

52. Iron(III) Oxide

CHEMICAL NAME = iron(III) oxide
CAS NUMBER = 1317–60–8
MOLECULAR FORMULA = Fe_2O_3
MOLAR MASS = 159.7 g/mol
COMPOSITION = Fe(69.9%) O(30.1%)
MELTING POINT = 1,565°C
BOILING POINT = not reported
DENSITY = 5.24 g/cm³

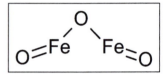

Iron(III) oxide is known in mineral form as hematite, which is the primary form of iron ore. It is also known simply as iron oxide or ferric oxide and is what is colloquially called rust when referring to the corrosion of iron objects. The use of iron marked a significant advancement in human history and was the last of the three prehistoric archaeological stages after the Stone and Bronze Ages. The period of the Iron Age is not fixed but varied as technology developed in different geographic areas. In Egypt and the Near East there is evidence that smelted iron items appeared shortly after 2000 B.C.E., although crude iron objects appeared several hundred years before, and items made from iron obtained from meteorites were present even earlier. In central Europe the early Iron Age did not occur until roughly 800 B.C.E., and in the Americas smelted iron was not used until introduced by the European explorers and colonists.

The smelting process involves reducing the iron, which is in an oxidized state in iron ore, to elemental iron. To obtain elemental iron, the iron-bearing mineral is heated with carbon so that the oxygen separates from the iron and bonds to the carbon. Smelting involves producing carbon monoxide by burning charcoal or coke to produce carbon monoxide, which is the reducing agent. The carbon monoxide then reacts with iron oxide compounds to form elemental iron. The production of iron from iron(III) oxide can be summarized as:

$$2C_{(s)} + O_{2(g)} \rightarrow 2CO_{(g)}$$

$$3CO_{(g)} + Fe_2O_{3(s)} \rightarrow 2Fe_{(l)} + 3CO_{2(g)}$$

Carbon may also act directly as a reducing agent according to the equation: $Fe_2O_{3(l)} + 3C_{(s)} \rightarrow 2Fe_{(l)} + 3CO_{(g)}$. Impurities such as silica, calcium, and aluminum are removed by adding limestone or other agents in the process. The impurities float to the top of the molten iron as slag. The crude iron produced from smelting results in pig iron. Pig iron has a carbon content of 3–5%, a high percentage of impurities, and is very brittle. It is roughly 90% iron. The name pig iron comes from the parallel sand bed molds fed by a common runner from the furnace. The parallel sand molds were thought to resemble suckling pigs. Pig iron must be further processed to reduce its carbon content and impurities to make useable iron and steel. Cast iron resembles pig iron, with a carbon content between 2% and 4%. Cast iron is hard and brittle. Steel has intermediate carbon content between 0.2% and 2%. Steel is alloyed with an array of other metals to produce many types of steels. For example, stainless steel contains chromium and nickel. High carbon steel has a carbon content of about 1.5%, is very hard, and is used for cutting tools. Wrought iron is almost pure iron, with a carbon content less than 0.2%. It is soft and malleable but has little structural strength.

Iron (III) oxide exists in mineral form as hematite. It is 70% iron and is the primary source of iron ore in the world. About 90% of the iron mined in the United States is hematite. World production of this ore is more than 1 billion tons. Magnetite and taconite are two other primary iron oxide minerals used as iron ore. The name hematite comes from the blood-red color of powdered hematite. The Greek word hematite means "blood-like." Some ancients held the belief that hematite was formed in areas where battles were fought and blood was spilled into the earth. Large deposits of hematite have been identified on Mars.

In iron production, iron ores are reduced to produce iron metal. The opposite process occurs when iron metals are oxidized to produce iron oxides or rust. Rust is primarily iron(III) oxide. Iron does not combine directly with oxygen to produce rust but involves the oxidation of iron in an electrochemical process. There are two requirements for rust: oxygen and water. The necessity of both oxygen and water is illustrated when observing automobiles operated in dry climates and ships or other iron objects recovered from anoxic water. Autos and ships subjected to these conditions show remarkably little rust, the former because of lack of water and the latter because of lack of oxygen.

Iron(III) oxide results from electrochemical reactions occurring on a piece of iron or steel. Iron metals are never uniform; there are always minor irregularities in both the composition and physical structure of the metal. These minute differences give rise to the anodic and cathodic regions associated with rust formation. During the rusting process, iron is oxidized at the anode according to the reaction: $Fe_{(s)} \rightarrow Fe^{2+} + 2e^-$. The electrons from this reaction flow through the metal to an area of the metal where oxygen is reduced according to the reaction: $O_{2(g)} + 4H^+_{(aq)} + 4e^- \rightarrow 2H_2O_{(l)}$. The hydrogen ions in this reaction are provided mainly by carbonic acid from the dissolution of atmospheric carbon dioxide in the water. The water acts as an electrolyte connecting the anode and cathode regions of the metal. Iron ions, Fe^{2+}, migrate through the electrolyte and in the process are further oxidized by dissolved oxygen in the water to Fe^{3+}. Iron (III) oxide is formed at the cathode according to the equation:

$$4Fe^{2+}_{(aq)} + O_{2(g)} + (4+2x)H_2O_{(l)} \rightarrow 4Fe_2O_3 \bullet + xH_2O_{(s)} + 8H^+_{(aq)}$$

The "x" in this equation indicates a positive whole number corresponding to the variable amount of water molecules that takes part in the reaction. The rusting process is illustrated in Figure 52.1. The area of rust formation differs from where the oxidation of iron takes place.

As noted, water serves as an electrolyte through which iron ions migrate. This explains why vehicles rust much more rapidly in regions where road salts are used to melt winter ice. The salts improve the conductivity of the electrolyte, thereby accelerating the corrosive process.

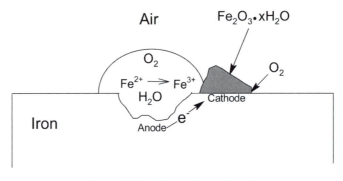

Figure 52.1   Formation of rust.

Most metals undergo corrosion of some form except for the so-called noble metals of gold, platinum, and palladium. Metals such as aluminum and zinc have an even greater tendency than iron to oxidize, but the oxide layer on these metals forms an impervious protective layer. This protects the metal below from further oxidation. Iron (III) oxide is highly porous and as a result rust does not protect the underlying metal from further corrosion.

A number of methods are used to reduce and prevent corrosion. The most common method is to paint iron materials so that the metals are protected from water and oxygen. Alloying iron with other metals is a common means to reduce corrosion; for example, stainless steel is an alloy of iron and chromium. Iron may also be protected by coating it with another

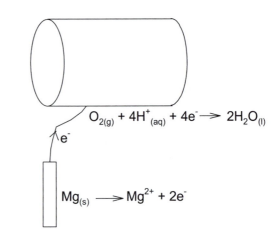

Figure 52.2   Cathodic protection, magnesium is connected to an iron object and undergoes oxidation, protecting the iron from rust.

metal. Galvanizing refers to applying a coating of zinc to protect the underlying metal. Also, because it is a more active metal, zinc oxidizes rather than iron.

Connecting iron objects to a more active metal is called cathodic protection. Cathodic protection is widely used to protect underground storage tanks, ship hulls, bridges, and buried pipes. One of the most common forms of cathodic protection is to connect the object to magnesium. When magnesium is connected to an iron object, magnesium rather than iron becomes the anode in the oxidation process. In cathodic protection bars of magnesium are connected either directly or by wire to the iron structure. Because the metal connected to the iron corrodes over time, it is called the sacrificial anode. Sacrificial anodes must eventually be replaced if they are to continue to protect the structure. The basic chemistry of cathodic protection is illustrated in Figure 52.2.

CHEMICAL NAME = 2,2,4-trimethylpentane
MOLECULAR FORMULA = C_8H_{18}
MOLAR MASS = 114.2 g/mol
COMPOSITION = C(84.1%) H(15.9%)
MELTING POINT = −107.5°C
BOILING POINT = 99.3°C
DENSITY = 0.69 g/cm³

Isooctane is a flammable liquid isomer of octane. It is best known for defining the octane number to rate the antiknock quality of gasoline, which is related to engine performance. Knock is a descriptive term used to describe the sound produced by an engine subject to inefficient combustion. Combustion in the cylinders is precisely timed to produce a smooth uniform combustion in the cylinder. The ignition spark produces a small explosion of the fuel-air mixture, producing a flame front that moves smoothly and rapidly through the cylinder. Knock occurs when a fuel-air mixture autoignites (spontaneously combusts prematurely) in a region of the cylinder before the flame front arrives. The autoignition occurs because combustion compresses the fuel mixture in localized areas of the cylinder, causing an increase in pressure and temperature in these areas. Therefore knock can be viewed as the spontaneous combustion of fuel occurring prematurely. Knocking results in inefficient engine operation and loss of power. In severe cases knocking results in overheating and engine damage. The octane rating of a gasoline is a measure of its ability to resist autoignition, which causes the knock.

The history of gasoline as a fuel parallels the development of the automobile and the internal combustion engine. At the start of the 20th century, gasoline was an undesirable by-product in kerosene production. Gasoline supply exceeded its demand, but this changed within a decade, as the automobile became a major form of transportation. Gasoline was obtained directly from distillation of crude oil (this was termed straight-run gasoline), but this process could not meet the increase demand (Figure 53.1). By 1913, cracking processes began to increase supply, and this was followed in subsequent years with advances in processes such as reforming, polymerization, and alkylation to both increase supply and improve quality. These advances continue today, with greater demand to fuel cleaner burning more efficient vehicles.

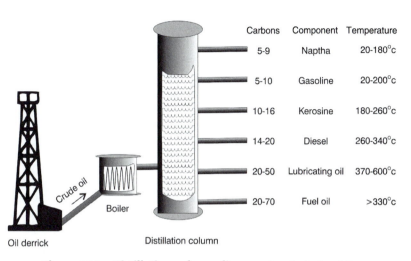

Fractionating Column for Oil

Carbons	Component	Temperature
5-9	Naptha	20-180°c
5-10	Gasoline	20-200°c
10-16	Kerosine	180-260°c
14-20	Diesel	260-340°c
20-50	Lubricating oil	370-600°c
20-70	Fuel oil	>330°c

Figure 53.1   Distillation column diagram. Drawing by Rae Déjur.

Gasoline is a mixture containing more than 500 different hydrocarbons and a variety of other compounds. The major hydrocarbon groups making up gasoline include straight-alkanes (n-paraffins), branched-alkanes (isoparaffins), alkenes (olefins), branched-alkenes (isoolefins), and aromatics (the words in parentheses indicate terms used by the petrochemical industry for classification). As early as the 1880s, it was noted that knocking in internal combustion engines was related to the type of fuel. While engineers worked on engine designs to reduce knock, chemists approached the problem by examining fuel composition and additives to reduce knock. In the early 1920s, Thomas Midgley (1889–1944) and his team of research-ers at General Motors discovered that tetraethyl lead $(Pb(C_2H_5)_4)$ added to gasoline reduced knock. This discovery enabled the car producers to increase compression ratios in engines, increasing engine performance. Tetraethyl lead was the preferred antiknock additive until the 1960s, when environmental concerns about lead pollution rose. Also, catalysts (plantinum, rhodium, palladium) used in catalytic converters to decrease air pollutants from vehicles were poisoned by lead.

To quantify the degree to which a fuel promoted or retarded knock, a committee with representatives from the gasoline and automobile industries was established in 1927. General Motors built an experimental single cylinder engine in which the compression ratio could be varied. A chemist from the Ethyl Corporation named Graham Edgar (1887–1955) experi-mented by burning different fuels in the engine and recording the knock characteristics. One of these fuels was n-heptane acquired from sap from the Jeffrey pine tree. Edgar determined that heptane performed poorly in engine tests. At this time, Edgar synthesized isooctane as a compound with good antiknock qualities. Edgar suggested that the rating of a fuel could be measured by referencing it to n-heptane, which was assigned a rating of 0, and to isooctane, which was assigned a rating of 100. By choosing these compounds, fuels of Edgar's time would have octane ratings ranging between 40 and 60.

During the next several years, a number of methods were proposed for rating knock of fuels. Different methods were proposed for different operating conditions, but two are generally used today: the Motor method and Research method. The Motor method is for an engine that operates at 900 rpm, a higher engine temperature of 149°C (300°F), and heavier load. The motor method is more reflective of high-speed freeway driving. The Research method performed at 600 rpm and variable ambient operating temperature more closely approximates conditions at low speed or starting from rest. The tests are performed with a standard one-cylinder engine made by Waukesha Engine in Wisconsin using an instrument called a knockmeter. The knock of a test fuel is compared to reference fuels that contain a blend of isooctane and n-heptane. Once the knock is matched against the reference fuel, the octane rating is given by the percentage of isooctane in the reference fuel. For example, if the fuel gives the same knock as a reference fuel with 90% isooctane and 10% n-heptane, its octane number is 90. Often the formula R+M/2 is displayed with the octane rating on gas pumps. This signifies that the octane rating is an average of ratings determined from both the Motor and Research methods. Typically, a gasoline's Research octane number (RON) is higher than its Motor octane number (MON). For fuels with octane numbers exceeding 100, reference fuels of isooctane containing tetraethyl lead are used.

The actual octane (C_8H_{18}) composition of gasoline is very low, generally comprising only a few percent by mass. The octane rating of a fuel depends on its composition. Isooctane has an octane rating of 100; normal octane has an octane rating of −18. Normal alkanes (n-paraffins) have the lowest octane ratings of the major hydrocarbons found in gasoline, and the octane number decreases as the number of carbon in the chain increases. N-butane has a RON of 113, but the RON of the next alkane, n-pentane, drops to 62, n-hexane is 19, and the reference alkane n-heptane is defined as 0. Branched compounds have higher octane ratings than straight-chain compounds. Since 1930, many chemical processes, such as alkylation and polymerization, have been developed to increase the production of branched compounds in refinery operations. High octane numbers in gasoline are those associated with the alkenes (olefins) and aromatics, especially akyl benzene compounds. For example, 2-pentene has a RON of 154. Benzene itself has a RON of 98, but 1,3,5-trimethylbenzene has a RON of 170. The highest octane numbers in gasoline are associated with cyclic alkenes, but these account for only a minute fraction of gasoline.

CHEMICAL NAME = 2-methyl-1,3
  butadiene
CAS NUMBER = 78–79–5
MOLECULAR FORMULA = C_5H_8
MOLAR MASS = 68.1
COMPOSITION = C(88.2%) H(11.8%)
MELTING POINT = −146°C
BOILING POINT = 34.1°C
DENSITY = 0.68 g/cm³

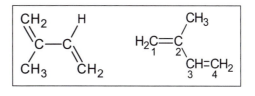

Isoprene is a volatile colorless liquid monomer that is the basic building block of rubber. Natural rubber items date back to at least 1500 B.C.E. Spanish explorers discovered native Central and South Americans using rubber in the 16th century for waterproofing, balls, and bindings. Europeans were intrigued with rubber products brought back by explorers and the substance remained a curiosity for two centuries. Joseph Priestley, one of the founders of modern chemistry, noted in 1770 that the substance could be used as an eraser if writing was rubbed with it, and so the name "rubber" came into use. In 1826, Michael Faraday (1791–1867) established the chemical formula of isoprene as C_5H_8. William Gregory (1803–1858) distilled rubber, obtaining isoprene in 1835. In 1860, Charles Hanson Greville Williams (1829–1910) isolated the monomer isoprene from rubber, showing that rubber was a polymer of isoprene.

The isoprene molecule structure shown above is just one form in which it exists. Rotation can occur around the C-C single bond, giving rise to a structure where the double bonds exist on the same side of the molecule:

Rubber results from the polymerization of isoprene to form polyisoprene. The resulting structure dictates the properties of the rubber. Natural rubber has a cis 1,4 structure (Figure 54.1). This means that the carbon atoms that form the chain attach to the same side of the chain at the 1 and 4 positions. The cis structure gives rubber its elasticity. Polyisoprene also exists in a trans 1,3 configuration. In the trans configuration, the addition takes place on opposite sides of the carbon chain. Two substances that exhibit the polyisoprene trans structure are gutta-percha and balata. Gutta-percha and balata, like rubber, are derived from the milky latex extracted from certain plants. These substances crystallize at a higher temperature than natural rubber and are partially crystallized at normal room temperatures. On cooling, gutta-percha and balata acquire a hard, leathery texture. Gutta-percha and balata have traditionally been used for sheathing of underwater cables and for the cover of golf balls.

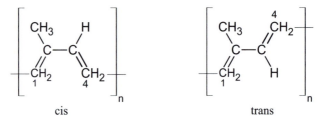

Figure 54.1 Isoprene structures.

Natural rubber is an elastomer. An elastomer is an amorphous polymer that has the ability to stretch and return to its original shape above a certain temperature. Natural rubber's properties are temperature dependent. At temperatures above 60°C, natural rubber becomes soft and sticky and at temperatures of −50°C and below, it becomes hard and brittle. To enhance natural rubber's usefulness requires that various additives be combined with the latex obtained from rubber plants. Adding ammonia to the latex prevents coagulation and allows the latex to be shipped and processed as a liquid. Charles Goodyear (1800–1860) was attempting to improve the quality of rubber in 1839 when he accidentally dropped rubber to which sulfur had been added on a hot stove. He discovered that the product had superior qualities compared to other natural rubbers of his day. The process Goodyear discovered was vulcanization. Goodyear was awarded a patent for the process in 1844 (U.S. Patent Number 3633). Vulcanized rubber is more elastic, stronger, and more resistant to light and chemical exposure. Goodyear never reaped the rewards of his discovery and died in poverty. Half a century later, the Goodyear Tire Company was named after him and is currently the largest tire producer in the United States and third in the world behind Michelin (France) and Bridgestone (Japan).

Goodyear discovered vulcanization, but he was unaware of the chemistry of the process. Rubber's elastic properties are due to the structure of isoprene. Rubber consists of coiled isoprene polymers. When rubber is stretched, the polyisoprene coils are straightened in a direction parallel to the direction of stretching. Once the stretching force is removed, the isoprene polymers return to their coiled structure. In the stretched position, the isoprene polymers can slide past each other, especially as the temperature increases. Vulcanization produces cross-links of sulfur-to-sulfur bonds between isoprene polymers. This allows elongation of rubber without sliding and gives the rubber a greater degree of elasticity over a wider temperature range. Cross-linking by adding just a few tenths of a percent sulfur greatly improves rubber's usefulness. Soft rubbers, such as that found in rubber bands, have a sulfur content of approximately 1–3%, and hard rubbers contain as much as 35% sulfur.

Natural rubber occurs in a colloidal milky suspension called latex, which is obtained from numerous plants. The most important of these is the para rubber tree, *Hevea brasiliensis*. Natural rubber is harvested by cutting a v-shape incision into a plant and allowing latex to drain into a container containing a preservative. About 50 mL of latex is obtained on a daily basis. Latex is transported to collection stations where it is processed for shipment. Processing can include preservation, coagulation, and concentrating before being sent to rubber factories. Rubber factories developed in France and England at the start of the 19th century, but it was not until the end of the century that an increased demand led to an expansion of the industry. Brazil had a monopoly on the rubber production through the later part of the 19th century. Rubber barons, who enslaved indigenous populations to gather latex from rubber trees that grew throughout the Amazon basin, ran the system (Figure 54.2). In 1876, the Englishman Henry Wickham (1846–1928) transported, some would say smuggled, 70,000 seeds from rubber trees back to Kew Gardens in England. Rubber tree seedlings from these seeds were cultivated in European herbariums and transplanted to the tropical Asian colonies of various countries. By the end of the 19th century, after numerous attempts to establish rubber in Asia, plantations developed in several regions, especially in the countries currently known as Malaysia, Indonesia, Sri Lanka, and Thailand. Brazil's monopoly of the rubber market was broken, and it was producing only about 50% of the world's natural rubber in 1910, less than 20% in 1920, and only about 1% in 1940. By 1920, more than 90% of the natural rubber production occurred on plantations. Concurrent with the rise of European plantations in Southeast Asia, the United States obtained rubber from natural sources in Central and South America.

The rise of the auto industry and an increase demand for pneumatic tires created an accelerated the demand for rubber as the 20th century unfolded. Germany was especially affected by rubber shortages in World War I as blockades prevented the importation of rubber. The increase demand motivated the search for synthetic rubber substitutes for natural rubber. A synthetic rubber termed Buna was first synthesized in Germany in 1915. Buna was a co-polymer of butadiene ($CH_2CHCHCH_2$) and acrylonitrile (CH_2CHCN), which was produced using a sodium catalyst. The name Buna came from combining "bu" for butadiene and "Na" for sodium (natrium). Historically, synthesized rubbers were more costly than natural rubber and also were of inferior quality. During the years leading up to World War II, the United States, realizing how dependent it had become on foreign sources, continued its research on synthetic rubbers. The war and Japan's control of the rubber-producing areas in Southeast Asia cut the United States' natural rubber supply by more than 90%. In response several major companies including Standard Oil, Dow, U.S. Rubber, Goodrich, Goodyear, and Firestone signed an agreement to share research to boost synthetic rubber production. The cooperative agreement resulted in improved qualities and grades of rubber and accelerated growth in the synthetic rubber industry. The United States developed a rubber called GR-S (government rubber-styrene) that was a co-polymer made from butadiene and styrene. In 1944, the United States was producing 700,000 tons of synthetic rubber. Once the war ended, the development of synthetic rubbers had displaced much of the United States' dependence on natural rubber.

Since World War II, numerous synthetic rubbers have been produced. These carry names based on the chemicals used in their production, for example styrene-butadiene rubber,

Figure 54.2 Brazilian native preparing dry rubber by smoking latex coated wooden shovel over a pot of burning palm nuts. Turtle shell bowl contains small pool of latex. *Source:* Engraving from *The Amazon and Madeira Rivers* by Franz Keller published by J. B. Lippincott and Company, Philadelphia, 1875. The Library of Congress.

polybutadiene rubber, and polychloroprene rubber. These substances are labeled as "rubber," but they differ from natural rubber. Isoprene rubber (IR) is the one synthetic rubber that is similar to natural rubber. The production of synthetic IR occurred in the mid 1950s when special organometallic catalysts for polymerizing isoprene were developed by Karl Ziegler (1898–1973) from Germany and the Italian Giulio Natta (1903–1979). Ziegler and Natta shared the 1963 Nobel Prize in chemistry for their work. Isoprene rubber has the same cis polyisoprene structure as natural rubber and has similar, but somewhat inferior, characteristics. In 2005, global consumption of isoprene rubber was approximately 165,000 tons, with about 65,000 tons of this consumption occurring in North America.

55. L-Dopa. See 33. Dopamine, L-Dopa

56. Methane

CHEMICAL NAME = methane
CAS NUMBER = 74–82–8
MOLECULAR FORMULA = CH_4
MOLAR MASS = 16.0 g/mol
COMPOSITION = C(74.87%) H(25.13%)
MELTING POINT = −182.6°C
BOILING POINT = −161.4°C
DENSITY = 0.71 g/L (vapor density = 0.55, air = 1)

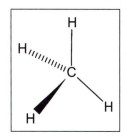

Methane is a colorless, odorless, flammable hydrocarbon gas that is the simplest alkane. The root word, *met*, in methane is derived from the Greek root word *methe* meaning wine. Methylene was used in the early 19th century as the name for methanol, which is wood alcohol, CH_3OH. Methylene comes from *methe + hydē*, the latter being the Greek word for wood, so methylene would mean wine from wood. Methanol got the names methylene and wood alcohol because it was discovered by Robert Boyle (1627–1691) in the 17th century by the destruction distillation of wood. Destructive distillation involves heating in the absence of air. Methane is the first alkane and carries the suffix "ane" denoting an alkane, thus methez + ane = methane. The structure of methane, as shown in Figure 56.1, is tetrahedral. The carbon is at the center of the tetrahedron, which can be assumed to be an equilateral pyramid, with a hydrogen atom at each of the four corners of the tetrahedron. The bond angles are all 109.5 degrees.

Methane is the principal component of natural gas, with most sources containing at least 75% methane. Methane production occurs naturally through a process called methanogenesis. Methanogenesis involves anaerobic respiration by single-cell microbes collectively called methanogens. Methanogens belong to a class of organisms called archaebacteria, which are prokaryote bacteria. Some classification systems differentiate archaeans from bacteria. The anaerobic digestion of dead plant matter under water produces methane and led to the common name of marsh gas for methane.

Methane has been used as a fossil fuel for thousands of years. The discovery of methane is attributed to the Italian physicist Alessandro Volta (1745–1827). Volta, known primarily

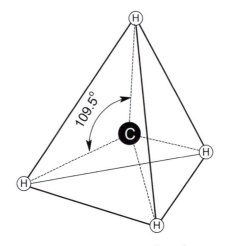

Figure 56.1   Structure of methane.

for his discoveries in electricity, investigated reports of a flammable gas found in marshes. In November 1776, Volta, while visiting the Lake Maggiore region of northern Italy, noticed that gas bubbles emanated from disturbed sediments in marshes. Volta collected the gas and began investigations on its nature. He discovered that the gas was highly flammable when mixed with air. He developed an instrument termed Volta's pistol (also called a spark eudiometer) that fired metal balls like a miniature cannon to conduct combustion experiments with methane. He also developed a lamp fueled by methane.

Methane is the principal gas found with coal and oil deposits and is a major fuel and chemical used is the petrochemical industry. Slightly less than 20% of the world's energy needs are supplied by natural gas. The United States get about 30% of its energy needs from natural gas. Methane can be synthesized industrially through several processes such as the Sabatier method, Fischer Tropsch process, and steam reforming . The Sabatier process, named for Frenchman Paul Sabatier (1854–1941), the 1912 Nobel Prize winner in chemistry from France, involves the reaction of carbon dioxide and hydrogen with a nickel or ruthenium metal catalyst: $CO_2 + 4H_2 \rightarrow CH_4 + 2H_2O$.

Methane is an important starting material for numerous other chemicals. The most important of these are ammonia, methanol, acetylene, synthesis gas, formaldehyde, chlorinated methanes, and chlorofluorocarbons. Methane is used in the petrochemical industry to produce synthesis gas or syn gas, which is then used as a feedstock in other reactions. Synthesis gas is a mixture of hydrogen and carbon monoxide. It is produced through steam-methane reforming by reacting methane with steam at approximately 900°C in the presence of a metal catalyst: $CH_4 + H_2O \rightarrow CO + 3H_2$. Alternately, methane is partially oxidized and the energy from its partial combustion is used to produce syn gas:

$$CH_4 + 2O_2 \rightarrow CO_2 + 2H_2O$$
$$CH_4 + CO_2 \rightarrow 2CO + 2H_2$$
$$CH_4 + H_2O \rightarrow CO + 3H_2$$

Hydrogen from syn gas reacts with nitrogen to produce ammonia: $N_2 + 3H_2 \rightarrow 2NH_3$. Carbon monoxide and hydrogen from syn gas can be combined to produce methanol: $CO + 2H_2 \rightarrow CH_3OH$. Methanol is primarily used for the production of formaldehyde through an oxidation process: $2CH_3OH + O_2 \rightarrow CH_2O + H_2O$ or an oxidation-dehydrogenation process: $CH_3OH \rightarrow CH_2O + H_2$.

Chlorination of methane, in which chlorine is substituted for one to all four of the hydrogens in methane, produces methyl chloride (CH_3Cl), methylene chloride (CH_2Cl_2), chloroform ($CHCl_3$), and carbon tetrachloride (CCl_4). The substitution of chlorines and fluorines in methane results in chlorofluorocarbons (see Dichlorodifluoromethane).

Methane is a fossil fuel that acts as a greenhouse gas, making it a subject of widespread interest in global warming research. Methane is much more effective at absorbing infrared radiation than is carbon dioxide. Its global warming potential is 21, which means it is 21 times more effective than carbon dioxide at trapping heat. Like carbon dioxide, concentrations of methane accelerated at the start of the Industrial Revolution and have climbed steadily throughout the 20th century, although in the last decade methane increases have plateaued for unknown reasons. Concentrations determined from ice cores place atmospheric methane concentrations at approximately 700 parts per billion (ppb) by volume in 1750, with a 150% increase to current levels of 1,750 ppb. Both natural and human sources contribute to global atmospheric methane concentrations. Estimates of the contribution from difference sources are difficult to make and vary highly among countries and global regions. Approximate contributions from major sources are summarized in Table 56.1.

Values in Table 56.1 are expressed in teragrams per year. A teragram is equivalent to 10^{12} grams. Table 56.1 indicates that approximately two-thirds of methane comes from human sources, principally related to agriculture. Animal digestion (flatulence) is chiefly from livestock such as cattle. An adult cow emits 600 liters of methane each day. Fossil fuel sources include both combustion and processing of oil, natural gas, and coal. Landfills are the primary anthropogenic source of methane in the United States.

A global system of natural gas wells, processing plants, and pipelines has been established to make natural gas available to millions of people. Exploration for conventional sources continues, but one large potential source of methane is methane hydrate. Methane hydrate is a solid crystalline structure of methane molecules surrounded by a frozen cage of water molecules. Methane hydrates form under the conditions of low temperature and high pressure. They occur in the polar regions, in cold water off the continental shelf, in permafrost, and in deep ocean sediments. The U.S. Geological Survey conservatively estimates that the amount of natural gas trapped in hydrates is twice the existing known reserves of all other fossil fuels combined.

Table 56.1   Major Sources of Atmospheric Methane in Tg/yr

Natural		Human	
Wetlands	150	Animal digestion	100
Termites	20	Rice fields	80
Oceans	15	Fossil fuels	80
Hydrates	10	Biomass burning	45
		Landfills	40
		Sewage plants	30
Total	**190**		**375**

57. Methyl Alcohol (Methanol)

CHEMICAL NAME = methanol
CAS NUMBER = 67–56–1
MOLECULAR FORMULA = CH_3OH
MOLAR MASS = 32.0 g/mol
COMPOSITION = C(37.5%) H(12.6%) O(49.9%)
MELTING POINT = –97.8°C
BOILING POINT = 64.7°C
DENSITY = 0.79 g/cm³

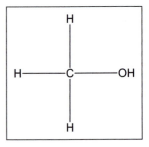

Methyl alcohol, also known as methanol or wood alcohol, is a clear, colorless, flammable liquid that is the simplest alcohol. It was first isolated in 1661 by the Irish chemist Robert Boyle (1627–1691) who prepared it by the destructive distillation of boxwood, giving it the name spirit of box, and the name wood alcohol is still used for methyl alcohol. Methyl alcohol is also called pyroxylic spirit; pyroxylic is a general term meaning distilled from wood and indicates that methyl alcohol is formed during pyrolysis of wood. The common name was derived in the mid-1800s. The name methyl denotes the single carbon alkane methane in which a hydrogen atom has been removed to give the methyl radical. The word alcohol is derived from Arabic *al kuhul* (see Ethyl Alcohol).

World production of methanol is approximately 8.5 billion gallons annually. Methanol is produced industrially, starting with the production of synthesis gas or syngas. Syngas used in the production of methyl alcohol is a mixture of carbon monoxide and hydrogen formed when natural gas reacts with steam or oxygen (see Methane). Methyl alcohol is then synthesized from carbon monoxide and hydrogen. The synthesis is conducted at high pressures, from 50 to 100 atmospheres, and in the presence of catalysts consisting of copper and oxides of zinc, manganese, and aluminum: $CO_{(g)} + 2H_{2(g)} \rightarrow CH_3OH_{(l)}$.

Methyl alcohol is poisonous and is commonly used to denature ethyl alcohol. Methanol poisoning results from ingestion, inhalation of methanol vapors, or absorption through the skin. Methanol is transformed in the body to formaldehyde (H_2CO) by the enzyme alcohol dehydrogenase. The formaldehyde is then metabolized to formic acid (HCOOH) by aldehyde dehydrogenase. Methanol poisoning is characterized by three stages. Mild poisoning may be limited to the first stage in which effects are similar to ethanol intoxication (see Ethyl Alcohol). The second stage is a latent period lasting between several hours and several days. The latent period results from the time delay from exposure until the formation of formaldehyde and formic acid in the body. During the latent period, antidotes can be administered to retard enzyme conversion to formaldehyde, allowing the kidneys to excrete methanol. Formic acid formation results in metabolic acidosis, which is the increase of acid in the body. Although numerous effects may result from methanol poisoning, it has a major effect on vision, producing conditions such as retinal edema and damage to the optic nerve. Blindness can result from ingesting as little as 3 mL of pure methanol and death can result from doses above 30 mL.

Methanol has numerous uses. Its main use is in the production of formaldehyde, which consumes approximately 40% of methanol supplies. Formaldehyde is produced from the oxidation of methanol in the presence of a copper catalyst resulting in its dehydrogenation. The main method of preparation of acetic acid is methanol carbonylation. In this process, methanol reacts with carbon monoxide to produce acetic acid (see Acetic Acid). Methanol is a common organic solvent found in many products including deicers (windshield wiper fluid), antifreezes, correction fluid, fuel additives, paints, and other coatings. A number of industrial chemicals use methanol in their production. Among these are methyl methacrylate and dimethyl terephthalate. Methanol is used to convert methylacrylamide sulfate to methyl methacrylate and ammonium hydrogen sulfate (NH_4HSO_4):

$$CH_2{=}CCONH_2 \cdot H_2SO_4 \quad \xrightarrow{\ CH_3OH\ } \quad CH_2{=}CCOOCH \quad + \quad NH_4HSO_4$$
$$\underset{CH_3}{|} \qquad\qquad\qquad\qquad \underset{CH_3}{|}$$

Methyl methacrylate is used in the manufacture of acrylic plastics and resins; the most familiar uses of acrylic plastics are for light covers, advertising signs, skylights, building panels, and Plexiglas. Methanol is used in making the ester dimethyl terephthalate from mixtures of xylene of toluene. Dimethyl terephthalate is used in the manufacture of polyesters and plastics. The polymer polyethylene terephthalate (PET) is made from dimethyl terephthalate. PET is used extensively in making plastic containers and is identified by the recycling symbol with the number 1.

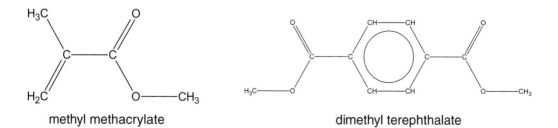

methyl methacrylate   dimethyl terephthalate

Methanol is used as a fuel additive. The common gasoline additive HEET is pure methanol and is used as a gas-line antifreeze and water remover. Methanol is used as a fuel in camp stoves and small heating devices. It is used to fuel the small engines used in models (airplanes, boats). Indianapolis Racing League and Championship Auto Racing Team require the use of methanol as a fuel and dragsters also run on methanol. In the early history of automobiles, methanol was a common fuel. The availability of cheap gasoline replaced methanol in the 1920s, but it is receiving renewed interest as an alternative fuel as the demand and cost of oil increase and oil supplies become uncertain. Methanol can be produced from coal and biomass. Methanol has a higher octane rating and generally lower pollutant emissions compared to gasoline. The relatively low flame temperature means that fewer nitrogen oxides are produced by methanol than by ethanol. One large disadvantage of methanol is that it has a lower energy density than gasoline. Using equivalent volumes of gasoline and methanol, methanol gives about half the mileage of gasoline. Another problem with methanol is its low vapor pressure, resulting in starting problems on cold days. This problem can be mitigated by using a blend of 85% methanol and 15% gasoline. This mixture is called M85 and is similar to E85 ethanol (see Ethyl Alcohol).

One energy application of methanol in its early stages of development is the direct methanol fuel cell (DMFC). A fuel cell is essentially a battery in which the chemicals are continuously supplied from an external source. A common fuel cell consists of a polymer electrolyte sandwiched between a cathode and anode. The electrodes are porous carbon rods with platinum

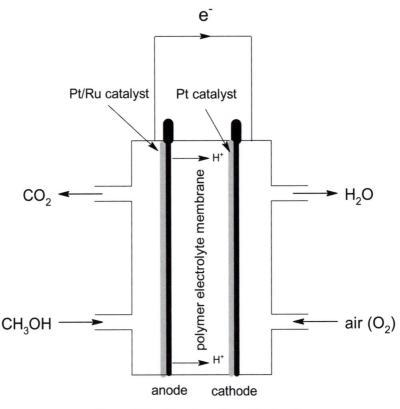

Figure 57.1 Direct methanol fuel cell.

or other catalysts. Methanol and air are fed into the cell. Oxygen from air is reduced at the cathode and methanol is oxidized to carbon dioxide and hydrogen anode:

anode reaction: $CH_3OH_{(l)} + H_2O_{(l)} \rightarrow CO_{2(g)} + 6\,H^+ + 6e^-$

cathode reaction: $1.5\,O_{2(g)} + 6\,H^+ + 6e^- \rightarrow 3\,H_2O_{(l)}$

overall cell reaction: $CH_3OH_{(l)} + 1.5\,O_{2(g)} \rightarrow CO_{2(g)} + 2\,H_2O_{(l)}$

A schematic of a DMFC is displayed in Figure 57.1. Although the first fuel cell was produced in 1839, fuel cell technology is in its early stages of development.

Methanol fuel cells are currently being produced and have limited use. Their relatively small size and use of common substances make fuel cells preferable where portability is important. Methanol fuel cells are used in military operations to provide power for communication systems, electronic devices, and general power generation. A DMFC holding 50 mL of methanol has recently been developed to power laptop computers. Another large area of research involves the use of DMFC in transportation to power electric vehicles. Governments, universities, and private companies continue to develop and increase the efficiency of fuel cells. One problem associated with DMFC is methanol crossover. This refers to the migration of methanol through the electrolyte to the cathode, reducing efficiency and inactivating the cathode catalyst. Current research focuses on improved electrolytes, catalysts, and electrodes.

58. Methylphenidate (Ritalin)

CHEMICAL NAME = α-phenyl-2-piperidineacetic
CAS NUMBER = 113–45–1
MOLECULAR FORMULA = $C_{14}H_{19}NO_2$
MOLAR MASS = 233.3 g/mol
COMPOSITION = C(72.1%) H(8.2%) N(6%)
O(13.7%)
MELTING POINT = 35°C (224°C for hydrochloride
salt)
BOILING POINT = decomposes
DENSITY = 1.1 g/cm³ (calculated)

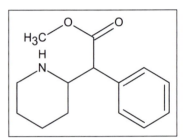

Methylphenidate is a piperidine derivative that is a central nervous system stimulant. Methylphenidate, in its hydrochloride form, is the active ingredient in the common medication Ritalin. It is prescribed for individuals, especially children, who have been diagnosed with attention-deficit hyperactivity disorder (ADHD) to help calm them and keep them focused on tasks. Methylphenidate was first synthesized from benzyl cyanide and 2-chloropyridine in Basel, Switzerland in 1944 by Ciba chemist Leandro Panizzon (1907–?). After he synthesized the compound, Panizzon's wife Marguerite, who had low blood pressure, would take methylphenidate as a stimulant before playing tennis. Panizzon named the substance Ritaline after his wife, whose nickname was Rita. Panizzon and Max Hartmann proposed an improved synthesis for methylphenidate and obtained a U.S. patent for its preparation in 1950 (U.S. Patent Number 2507631). In 1954, methylphenidate was patented for use as an agent for treating psychological disorders under the name Ritalin by the Ciba pharmaceutical company. (Ciba subsequently merged with the Geigy company to become Ciba-Geigy, and then Ciba-Geigy merged with Sandoz to form Novartis.) Methylphenidate was first used to reverse drug-induced coma. Ritalin was approved by the Food and Drug Administration (FDA) in 1955 and introduced in the United States in 1956 for several conditions including depression, senile behavior, lethargy, and narcolepsy.

Initial use of methylphenidate for treating children with ADHD was limited but increased as knowledge of the disorder and drug therapies grew. Although the condition was first recognized and described at the beginning of the 20th century, it was not until the late 1980s that consensus developed on the term attention-deficit hyperactivity disorder. ADHD is a self-descriptive condition characterized by hyperactivity, impulsivity, and inattention. There is no known cause for ADHD, and boys are diagnosed three times as frequently as girls. The use of methylphenidate for treating ADHD is based on work originating in the 1930s that demonstrated amphetamines were effective in treating hyperactive and impulsive behavior. Hyperactive children were first treated with stimulants in 1937. Since methylphenidate first appeared on the market, its use for treating ADHD has steadily increased. It is the most commonly used medication for treating children diagnosed with ADHD, with nearly 90% of all prescriptions of methylphenidate written for this condition. The use of methylphenidate has increased dramatically in the United States in recent years, with production increasing sixfold between 1990 and 2005. The United States now consumes more than 80% of the total world supply of methylphenidate.

Methylphenidate's mode of action is not completely known, but it is believed that ADHD symptoms are related to the dopaminergic areas of the brain. Animal studies indicate that methylphenidate affects several neurotransmitters to counteract ADHD behavior. Methylphenidate binds to dopamine transporters in the presynaptic neuron, blocking the reuptake of dopamine and increasing extracellular dopamine. Methylphenidate also influences norepinephrine reuptake and influences serotonin to a minor degree.

Ritalin and related generic methylphenidate drugs are available by prescription for individuals six years and older. Ritalin is distributed in 5, 10, and 20 mg tablets. In addition to ADHD, methylphenidate is used for several other medical conditions. It continues to be used for narcolepsy. It has also been used in treating depression, especially in elderly populations. Methylphenidate has been suggested for use in the treatment of brain injury from stroke or brain trauma; it has also been suggested to improve appetite and the mood of cancer and HIV patients. Another use is for pain control and/or sedation for patients using opiates.

Health concerns are associated with the use of methylphenidate. Some of the commonly reported adverse effects associated with its use are insomnia, nervousness, hypertension, headache, anorexia, and tachycardia; less common effects include weight loss, nausea, and angina. Studies have indicated that methylphenidate lead to liver tumors in mice, but limited studies on its carcinogenicity in animals have not led the FDA to change recommendations on its use. Methylphenidate is classified as a Schedule II controlled substance by the U.S. Drug Enforcement Administration. Schedule II substances have medical applications but also a high potential for abuse and addiction. Other Schedule II drugs include cocaine and morphine. Because methylphenidate is a stimulant and readily available, it has a potential for drug abuse. It is used like amphetamines as a stimulant to alter mood and help people stay awake. Although in most cases under supervised administration methylphenidate is considered safe and does not lead to dependence, there are many documented cases of abuse. Methylphenidate abuse can lead to tolerance, escalating consumption and psychological dependence. Methylphenidate medications can be diverted for illegal purposes, and there are hundreds of reports of them being stolen from pharmacies, clinics, and other licensed distributors.

Because of methylphenidate's unknown effects on long-term use and side effects, much controversy surrounds it. Many mental health workers, physicians, and education groups claim that ADHD is overdiagnosed and that methylphenidate and other stimulants are overprescribed, especially in the United States. Approximately 10 million prescriptions for methylphenidate are filled each year in the United States. The estimated number of children using methylphenidate for ADHD in the United States was about 1.5 million in 1995 and is currently estimated at about 6 million. Methylphenidate usage has plateaued in recent years. This has been attributed to increased usage of other stimulants for ADHD such as the amphetamine Adderall, which was approved by the FDA in 1996. Adderall is a longer lasting medication that has decreased the sales of Ritalin, which for many years monopolized the drug market for ADHD treatment. In the last decade, a number of other ADHD drugs have flooded the market. Concerta is a methylphenidate stimulant that is marketed as a 12-hour, once per day treatment; it was approved in 2000. Responding to its competition, Novartis introduced Ritalin LA (approved in 2002), its version of a once-per-day ADHD medication.

59. Monosodium Glutamate

CHEMICAL NAME = sodium 4-amino-5-hydroxy-5-oxo-pentanoate
CAS NUMBER = 142–47–2 (anhydrous) 6106–04–3 (monohydrate)
MOLECULAR FORMULA = CHNNaO (anhydrous) $C_5H_8NNaO_4H_2O$ (monohydrate)
MOLAR MASS = 169.1 g/mol (anhydrous) 187.1 (monohydrate)
COMPOSITION = C(35.5%) O(37.8%) H(4.8%) Na(13.6%) N(8.3%) (anhydrous)
MELTING POINT = 232°C (monohydrate)
BOILING POINT = decomposes
DENSITY = 1.6 g/cm³

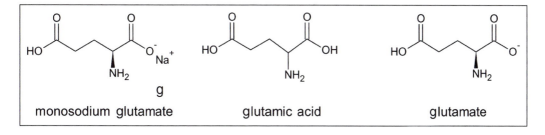

monosodium glutamate　　　　glutamic acid　　　　glutamate

Monosodium glutamate, known as MSG, is the sodium salt of the amino acid glutamic acid present in all protein. It is marketed as a white crystalline salt used to enhance flavor as the monohydrate, $C_5H_8NNaO_4•H_2O$. In this form, it is 78.2% glutamate, 12.2% sodium, and 9.6% water. When referring to MSG it is important to distinguish between several terms: MSG, glutamic acid, free glutamic acid, glutamate, and free glutamate. Monosodium glutamate, glutamic acid, and glutamate are often used interchangeably, but this presents problems when trying to distinguish different forms. Monosodium glutamate refers to the processed flavor enhancer, whereas glutamic acid and glutamate designate the amino acid produced naturally in animal and plant protein. In humans, glutamate is a nonessential amino acid. When glutamic acid (glutamate) is found in protein, it is referred to as bound glutamic acid (bound glutamate) and when not bound, as in manufactured MSG, it is referred to as free glutamic acid (free glutamate). There are two forms of glutamic acid: L-glutamic acid and D-glutamic acid. The natural glutamic acid found in humans and other higher organism is L-glutamic acid only, whereas processed monosodium glutamate contains both L-glutamic acid and D-glutamic acid along with pyroglutamic acid ($C_5H_7NO_3$) and other contaminants.

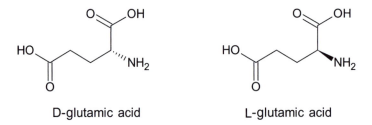

D-glutamic acid   L-glutamic acid

The use of MSG as a flavor enhancer is attributed to Kikunae Ikeda (1864–1936), a chemistry professor from Tokyo Imperial University. In 1907, Ikeda researched the unique flavor associated with the edible kelp known as konbu and the dashi (soup stock) produced from it. Japanese cooks had used konbu seaweed for centuries to enhance flavors in food, and Ikeda determined that 1-glutamic acid was responsible for the flavor. Glutamic acid was first isolated from wheat gluten by the German chemist Karl Heinrich Ritthausen (1826–1912) in 1866. Ikeda isolated the compound from the seaweed *Laminaria japonica*. He published his work in 1909 and proposed free glutamic acid as a fifth basic taste to be included with sweet, sour, bitter, and salty. He called the taste *umami,* named after the Japanese adjective *umai* meaning delicious. *Umami* can be roughly translated into the English word savory and was most apparent in meats and cheeses. MSG is odorless and has little flavor. It works synergistically with other substances found in food proteins such as the nucleotides disodium 5'-inosine monophosphate (IMP) and disodium 5'-guanosine monophosphate (GMP), to produce a pleasing savory taste. Several researchers have identified glutamic acid taste receptors since 2000, verifying *umami* as a fifth basic taste sensation.

Ikeda patented MSG as a flavor enhancer along with processes used to produce it. He started the Ajinomoto Company in 1909; the company name was taken from *Aj no moto* meaning "the essence of taste," the name Ikeda gave to his monosodium glutamate product. MSG was originally produced by hydrolysis of plant proteins using strong acids. The isolated glutamic acid was then purified and converted to MSG. In the mid-1950s, the bacterial fermentation method to produce MSG was developed in Japan. In this process bacteria strains, such as *Micrococcus glutamicus* or *Corynebacterium glutamicus*, that synthesize relatively large quantities of glutamic acid are cultivated in an aerobic broth containing a carbon source, which is typically molasses or starch. The broth also contains nutrients and a nitrogen sources such as ammonia, ammonium salt, or urea. Glutamic acid is secreted into the broth from which it can be separated. It is then concentrated, purified, and converted into MSG.

The use of MSG as a flavor enhancer began in Japan and Asia, but widespread use of the compound in the United States did not occur until after World War II. American soldiers noted MSG's ability to enhance flavors in Japanese rations and brought this knowledge back to the United States. Military personnel discovered that the Japanese used MSG in their rations. This knowledge was used in producing military rations and spread to the private food industry. Ajinomoto started manufacturing MSG in the United States in 1956 (Ajinomoto became affiliated with Kraft General Foods in 1973). Monosodium glutamate was marketed under the brand name Accent in the 1950s.

The safety of manufactured MSG has been questioned for the last several decades. The MSG controversy started in 1968 when Dr. Robert Ho Man Kwok, a Chinese-American medical researcher who immigrated to the United States, wrote a letter to the *New England Journal of Medicine* reporting adverse symptoms whenever he ate at a particular Chinese restaurant. Kwok reported numbness in the back and neck and heart palpitations. He sought answers for the condition dubbed as "Chinese restaurant syndrome." Kwok's letter elicited a response by others who experienced similar symptoms, and several ideas were advanced for the cause of the reaction. One of these implicated MSG additives, which put in motion the MSG controversy that continues today. Much of the controversy was raised by anecdotal evidence, but over the years increasing scientific research focused on the safety of MSG. A review of existing data on what was termed MSG symptom complex was performed by the Federation of American Societies for Experimental Biology (FASEB) in the mid-1990s; the review was commissioned by the Food and Drug Administration (FDA). FASEB concluded that only a small population of MSG-intolerant individuals and people with severe, poorly controlled asthma suffer MSG symptom complex when MSG is ingested in relatively high quantities. Conditions associated with MSG symptom complex included numbness in the back of the neck, radiating to the arms; warmth and fatigue in the head, upper back, neck, and arms; facial pressure or tightness; chest pain; headache; nausea; and heart palpitations.

Based on these findings, the FDA required that when MSG is added to a product, it must be identified as "monosodium glutamate" on the label. Although a label may read monosodium glutamate, the lack of this ingredient on a label does not guarantee that MSG is not being consumed. Many foods contain naturally free glutamate. For example, Parmesan cheese is among the highest free glutamate foods, with 1,200 mg of free glutamate per 100 grams. Other high free glutamate food items include peas, 200 mg/100 grams; tomatoes, 140 mg/100 grams; and corn, 130 mg/100 grams. Another source of free glutamates is hydrolyzed proteins. Hydrolyzed proteins are formed when acid or enzymes are used to digest proteins from soy meal, glutens, or other food sources. Hydrolyzed proteins are used as flavor enhancers similar to MSG. Free glutamates from hydrolyzed proteins are found in canned vegetables, canned soups, processed meats, soy sauce, seasonings, textured protein, bouillon, spices, and condiments. The FDA considers foods whose labels say "No Added MSG" or "No MSG" to be misleading if the food contains sources of free glutamates, such as hydrolyzed protein.

There does not seem to be an absolute consensus on the safety of MSG, but for the most part MSG is considered safe if consumed in normal quantities. An acceptable daily intake level is not cited by government regulators, but the average person consumes between 0.3 and 1.0 gram of MSG daily. Estimates of annual global production of MSG range between several hundred thousand tons to as high as 1 million tons. Although practically all MSG produced is used in the food industry, a small portion is used for agricultural feed supplements (for swine), pesticides, and pharmaceuticals.

CHEMICAL NAME = (5α,6α)-7,8-Didehydro-4,5-epoxy-17-
methylmorphinan-3,6-diol
CAS NUMBER = 57–27–2
MOLECULAR FORMULA = $C_{17}H_{19}NO_3$
MOLAR MASS = 285.3 g/mol
COMPOSITION = C(71.6%) H(6.7%)O N(4.9%) (16.8%)
MELTING POINT = 200°C
BOILING POINT = 254°C
DENSITY = 1.44 g/cm³

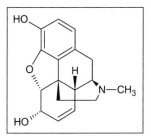

Morphine is the principal alkaloid obtained from opium. Opium is the resinous latex that exudes from the seed pod of the opium poppy, *Papver somneferum*, when it is lacerated. Alkaloids account for approximately 25% of opium, and of this 25% about 60% is morphine. Remains of poppy seeds and pods have been found in Neolithic caves, indicating that the use of opium predates written history. The opium poppy is native to the eastern Mediterranean, but today it is chiefly cultivated from the Middle East through southern Asia and into China and Southeast Asia. The first civilization known to use opium was the Sumerians, who inhabited Mesopotamia in present-day western Iraq, around 3500 B.C.E. Sumerians traded opium with other civilizations, and this led to the cultivation of opium poppies and the production of opium in many geographic areas including Egypt, India, Persia (Iran), Southeast Asia, and China.

Opium has been used medicinally throughout history. Writings of ancient physicians in many cultures espoused the virtues of opium as a remedy for all types of ailments including pain relief, cough suppression, and diarrhea. Remedies prepared by alchemists and ancient physicians often contained opium. Galen (131–200) prescribed opium for headaches, deafness, melancholy, epilepsy, asthma, and colic. The famous European physician Paracelsus (1493–1541) produced an alcoholic potion containing opium called laudanum. Varieties of laudanum were used for several hundred years as medicinal drinks and were readily available in apothecaries.

Morphine was isolated from opium at the beginning of the 19th century, and several individuals are cited for its initial isolation. Jean François Derosne, a French pharmacist, isolated an impure morphine salt in 1803. Another Frenchman, Armand Seguin (1767–1835), reported on the isolated salt of morphine in 1804, but he did not publish his findings until 1814. Most references credit the German pharmacist Friedrich Wilhelm Sertürner (1783–1841) as the discoverer of morphine. Sertürner worked on opium over a number of years and published his initial results in 1805. He continued his work on opium and administered the drug to himself and several others, producing severe narcotic effects. Sertürner published this research in 1817 and used the name morphium for his substance. The name morphium was derived from Morpheus, the Greek god of dreams. Upon reading Sertürner's work, the French chemist Joseph Louis Gay-Lussac (1778–1850) proposed that the name morphium be changed to morphine. This would be consistent with the proposal of that time that names for alkaloids carry an "ine" ending.

The process to obtain morphine from opium involves boiling a water-opium solution and adding calcium hydroxide ($Ca(OH)_2$). The calcium hydroxide combines with the morphine to form the water-soluble salt calcium morphenate. As the solution cools, other insoluble alkaloids precipitate out of solution, leaving morphine in solution. The solution is filtered and then reheated. Ammonium chloride (NH_4Cl) is added to increase the solution's pH level to 9–10, and the insoluble salt morphine hydrochloride is formed, which precipitates out upon cooling.

Once morphine had been determined to be the principal pain-killing ingredient in opium, it was substituted for opium in many treatments. Morphine's analgesic ability is due to its ability to bind and activate opioid receptors in the central nervous system (brain and spinal cord). There are different types of receptors found in the central nervous system, and the primary ones involving morphine are called μ-receptors (mu receptors). Morphine and related alkaloids have chemical structures that allow them to bind to opioid receptors. The structures follow what is known as the morphine rule. The morphine rule spells out how chemical units are bonded to form compounds so that the compounds attach to the opioid receptors. According to the morphine rule, pain-killing opioid compounds possess a phenyl ring attached to a quaternary carbon atom, which is attached to a tertiary amine group by two carbon atoms. The structure of the opioid receptor makes it possible to bind to different units of opioid analgesics. A flat part on the receptor binds to the phenyl ring, a cavity accepts the two carbon atoms, and an anionic region attracts the nitrogen in the tertiary amine group. When morphine or another opioid analgesic binds to the opioid receptors, it reduces the neurological transfer of pain signals to the brain. Morphine mimics natural pain relievers produced by the body called endorphins. When these natural pain-relieving compounds were discovered in the 1970s, the word *endorphin* was coined from the words endogenous and morphine.

Morphine was initially hailed as a miracle cure and was incorporated into many medications. It was erroneously thought that morphine use would eliminate the addictive effects of opium, and among its many uses in the 18th century was treating alcoholism. Morphine's addictive properties were quickly recognized and abuse of the drug ensued. Morphine addiction results from the drug's ability to produce an internal dreamlike euphoria, relaxation, and reduced anxiety. As addictive side effects were recognized, various governments attempted to combat opium and morphine abuse by curtailing its use. China lost two major confrontations with Britain over Britain's right to import illegal opium into China during the middle of the 19th century. The Opium Wars were fought as Britain sought to balance trade with China by using opium produced in India to supply addicted Chinese. Morphine was used extensively during wars to alleviate the pain and suffering associated with battle wounds. Thousands of Civil War survivors became addicted as a result of morphine use. Thus the use of morphine was a two-edged sword: it was an effective analgesic, but it could lead to addiction.

In a search to retain morphine's analgesic property without causing addiction, modifications of morphine were explored. In 1874, C. R. Alder Wright (1844–1894), an English pharmacist working at St. Mary's Hospital Medical School in London, boiled morphine and acetic anhydride ($C_4H_6O_3$) and produced diacetylmorphine. Wright's experiments on animal subjects produced negative effects and convinced him to abandon diacetylmorphine as a morphine substitute. Twenty years after Wright's work, the German pharmaceutical company, Bayer, examined diacetylmorphine as a remedy for tuberculosis, pneumonia, and other respiratory ailments. The work at Bayer was done simultaneously as the company developed aspirin.

Diacetylmorphine was thought to be more potent than morphine and nonaddictive. It was also believed to be an effective substitute to wean addicts off morphine. Bayer marketed diacetylmorphine under the brand name Heroine, named because subjects used in clinical trials experienced a heroic feeling when it was taken. Initially, sale and use of heroine were quite liberal. It was used as a cough suppressant and to treat morphine addiction. The American Medical Association approved heroin use in 1906. As legal heroin use increased, so did reports that heroin was addictive and produced negative side effects. It was also discovered that heroin converts into morphine in the body. Therefore heroin is a prodrug, which is a substance that does not produce an effect when taken but converts into another drug within the body. Heroin's greater fat solubility allows it to move more readily across the blood-brain barrier, resulting in a greater supply of morphine to the brain and making it more potent than morphine.

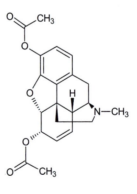

heroin

As knowledge of heroin's negative effects became known, regulations controlling the drug resulted in laws banning importation and use followed. The importation and use of heroin were banned in the United States in 1924. The regulation of heroin in most countries resulted in the illicit use and the rise of illegal syndicates to supply addicts. Today, the illicit production of opium for morphine used for heroin production is estimated to be 10 times that of licit production.

Morphine's licit use is as an analgesic and for the production of semisynthetic morphine analogs. Approximately 90% of the licit morphine produced is used to make codeine; codeine is present in very small amounts in opium. Codeine is almost identical to morphine, with an $-OCH_3$ replacing one of the $-OH$ groups in morphine. About 10% of codeine intake is metabolized to morphine in the liver. Thus, although codeine does not have analgesic properties, its ability to be converted into morphine enables its use as a pain reliever. Codeine is used extensively in over-the-counter cough suppressants. Although codeine is more toxic than morphine, it is preferred in medications because it is much less addictive than morphine.

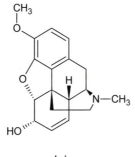

codeine

The structure of morphine was first determined in 1925 by Sir Robert Robinson (1886–1975) and John Masson Gulland (1898–1947). A total synthesis of morphine was achieved in 1952 at the University of Rochester by Marshall D. Gates (1915–2003) and his co-worker Gilg Tschudi. Since its first synthesis, a number of other processes have been used to synthesize morphine in the laboratory, but none of these is economically viable. Therefore morphine continues to be obtained through biosynthesis from poppy plants.

CHEMICAL NAME = naphthalene
CAS NUMBER = 91–20–3
MOLECULAR FORMULA = $C_{10}H_8$
MOLAR MASS = 128.2 g/mol
COMPOSITION = C(93.7%) H(6.3%)
MELTING POINT = 80.2°C
BOILING POINT = 217.9°C
DENSITY = 1.15 g/cm³

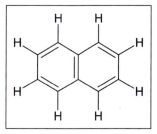

Naphthalene is a crystalline, white, flammable, polycyclic aromatic hydrocarbon consisting of two fused benzene rings. It has a pungent odor and sublimes readily above its melting point; it has been traditionally used in moth balls and is responsible for the moth balls' characteristic odor. Naphthalene is a natural component of fossil fuels and is the single most abundant component of coal tar, accounting for approximately 11% of dry coal tar. Coal tar is a thick, black, oily substance obtained by the destructive distillation of bituminous coal. In 1819, naphthalene was obtained as white crystals during the pyrolysis of coal tar by John Kidd (1775–1851), a British physician and chemist, and Alexander Garden (1757–1829), an American living in Britain. Kidd described the properties of the white crystals he obtained from coal tar and proposed the named *naphthaline* for the substance; *naphthaline* was derived from *naphtha,* a general term for a volatile, flammable, hydrocarbon liquid. Michael Faraday (1791–1867) determined the correct empirical formula for naphthalene in 1825, and Richard August Carl Emil Erlenmeyer (1825–1909) proposed the fused benzene ring structure in 1866.

Naphthalene is produced from coal tar or petroleum. It is made from petroleum by dealkylation of methylnaphthalenes in the presence of hydrogen at high temperature and pressure. Petroleum was a major source of naphthalene until the 1980s, but now most naphthalene is produced from coal tar. The pyrolysis of bituminous coal produces coke and coke oven gases. Naphthalene is condensed by cooling the coke gas and then separated from the gas. Naphthalene production in the United States is slightly greater than 100,000 tons annually.

In addition to oxidation and reduction reactions, naphthalene readily undergoes substitution reactions such as nitration, halogenation, sulfonation, and acylation to produce a variety of other substances, which are used in the manufacture of dyes, insecticides, organic solvents, and synthetic resins. The principal use of naphthalene is for the production of phthalic anhydride, $C_8H_4O_3$.

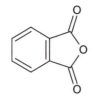

phthalic anhydride

Naphthalene is catalytically oxidized to phthalic anhydride: $2C_{10}H_8 + 9O_2 \rightarrow 2C_8H_4O_3 + 4CO_2 + 4H_2O$ using metal oxide catalysts. Phthalic anhydride is used to produce plastics, phthalate plasticizers, insecticides, pharmaceuticals, and resins. Sulfonation of naphthalene with sulfuric acid produces naphthalenesulfonic acids, which are used to produce naphthalene sulfonates. Naphthalene sulfonates are used in various formulations as concrete additives, gypsum board additives, dye intermediates, tanning agents, and polymeric dispersants. Naphthalene is used to produce carbamate insecticides such as carbaryl, which is a wide-spectrum, general-purpose insecticide. The active ingredient in the popular insecticide Sevin is carbaryl.

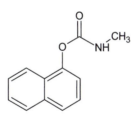

carbaryl

Naphthalene is classified by the federal government as a toxic hazardous substance. Humans are exposed to naphthalene through several sources. It is a component of gasoline, diesel fuel, tobacco smoke, vehicle exhaust, and wood smoke. Naphthalene exposure also results from its industrial use; workers involved in specific industries using naphthalene can have exposures several hundred times that of the general public. The leading cause of human exposure is residential combustion of fossil fuels and wood, followed by exposure to moth repellants. Naphthalene has been reported to disrupt red blood cells, leading to hemolytic anemia; infants are especially susceptible to this effect. The symptoms of hemolytic anemia are fatigue, loss of appetite, restlessness, and jaundice. Other effects associated with high levels of naphthalene exposure include headache, nausea, diarrhea, kidney damage, and liver damage. There is some evidence to indicate that naphthalene causes cataracts. A link between napthalene exposure and cancer has not been established. The Occupational Safety and Health Administration established the permissible exposure limit (PEL)for naphthalene at 10 ppm (parts per million) based on a time-weighted average for an 8-hour workday. The National Institute for Occupational Safety and Health has established a short-term exposure limit (not to exceed 15 minutes) of 15 ppm based on a time-weighted average.

CHEMICAL NAME = 3-[(2*S*)-1-methylpyrrolidin-2-yl]pyridine
CAS NUMBER = 54–11–5
MOLECULAR FORMULA = $C_{10}N_{14}H_2$
MOLAR MASS = 162.2 g/mol
COMPOSITION = C(74.0%) H(8.7%) N(17.3%)
MELTING POINT = −79.0°C
BOILING POINT = 247°C
DENSITY = 1.01 g/cm³

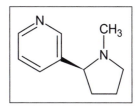

Nicotine is a colorless oily liquid alkaloid. Nicotine and its derivatives are found in all green plants, but it is most closely associated with tobacco. Tobacco leaves have particularly high concentrations of nicotine, with the leaves containing between 2% and 4% by weight; common plants such as potatoes, tomatoes, celery, and eggplant contain nicotine, but the amount is too small to have an effect as a stimulant. Tobacco is native to the subtropical and tropical Americas (there are a few species indigenous to selected areas of Africa). Columbus observed Native Americans smoking, chewing, and snorting tobacco during his voyages to the Americas. Although Columbus viewed tobacco use as a heathen practice unworthy of Europeans, subsequent explorers and their crews adopted its use, which quickly spread throughout the world. As tobacco use was adopted by different cultures, some governing bodies unsuccessfully attempted to ban its use. It was widely used to treat common ailments and diseases. Early colonists in North America recognized tobacco as a cash crop for trading with Europeans and it was widely cultivated in suitable regions.

Tobacco is a member of the nightshade (Solanaceae) family and its scientific name is *Nicotiana tabacum*. The name nicotine comes from Nicotiana after the French ambassador Jean Nicot (1530–1600). Nicot became familiar with tobacco when he was serving as ambassador to Portugal. Impressed with its use as a medicinal herb, Nicot sent seeds and cuttings back to the French Queen Catherine de Medici (1519–1589) in 1560, noting its therapeutic properties. Tobacco was called nicotiana and this was used for the scientific name.

Nicotine is the second most widely used recreational drug after caffeine. At low doses, nicotine acts as a stimulant to the central nervous system by activating acetylcholine receptors, specifically called nicotinic acetylcholine receptors, in the postsynaptic neurons during nerve transmission. Nicotine from smoking is distributed throughout the body in a few minutes and crosses the blood-brain barrier in less than 10 seconds. Physiological effects include an increase in pulse, constriction of arteries that produce elevated blood pressure, an increase in breathing rate, and loss of appetite. At higher doses nicotine acts as a depressant. Nicotine is highly addictive; it is believed that it stimulates dopamine production that activates pleasure-and-reward areas in the brain. Because nicotine produces a quick ephemeral response in the brain, the continual use of tobacco is necessary to produce the physiological reward from smoking. This makes it difficult for tobacco users to quit. Classical withdrawal symptoms are associated with people trying to "kick the habit." These include anxiety, irritability, nervousness, fatigue, and headaches.

High intake of nicotine is toxic. The lethal dose of nicotine is approximately 50 mg. The average nicotine content of a cigarette is about 8 mg, and that of a cigar is typically between 100 and 200 mg, but it may be as high as 400 mg. The toxic effect of nicotine is mitigated in smokers by its oxidation to other products when tobacco is burned. The actual delivery of nicotine to the body is only about 10% of the total nicotine content of the product. Therefore a typical cigarette will deliver about 1 mg of nicotine. The average nicotine dose from the use of a smokeless tobacco product (chewing tobacco or stuff) is about two to three times the amount from a cigarette.

The nicotine molecule consists of a pyrrolidine ring attached to a pyridine ring by a bond between carbon atoms in the two-ring systems. Nicotine was isolated in impure form from tobacco in 1809 by Louis Nicholas-Vauquelin (1763–1829). Vauquelin called the substance nicotianine. In 1826, Wilhelm Posselt (1806–1877) and Karl Ludwig Reimann (1804–1872), medical students at Heidelberg University, isolated pure nicotine and published dissertations on its pharmacology in 1828. Louis Henri Melsens (1814–1886) determined nicotine's empirical formula. Amé Pictet (1857–1937) and P. Crépieux reported the synthesis of nicotine in 1903.

Nicotine forms a number of metabolites in the body, mainly in the liver. Approximate 75% of nicotine is oxidized to cotinine, which is the primary nicotine metabolite. Cotinine can be measured in the blood, urine, and saliva and this is used as a measure of nicotine exposure in tobacco users and in those exposed to secondhand smoke. The oxidation of nicotine also produces nicotinic acid. Nicotinic acid is vitamin B_3 and has the common name niacin. Niacin deficiency results in a disease called pellagra, which is found in certain malnourished populations. Pellagra's symptoms include dermatitis, diarrhea, sensitivity to light, and dementia.

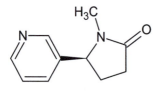

cotinine

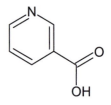

nicotinic acid or niacin

Nicotine in tobacco has always been used for medicinal purposes. Nicotine solutions made from soaking tobacco leaves in water have been used as pesticides for several hundred years. In modern times, numerous pharmaceutical companies have explored nicotine's use for treating diseases. Nicotine's most prevalent medicinal use is for smoking cessation in the form of alternate delivery systems such as gums and dermal patches. Nicotine is used medically for numerous conditions and its use is being explored in additional areas including pain relievers, attention deficit disorder medications and medications associated with Alzheimer's disease, Parkinson disease, colitis, herpes, and tuberculosis. Because of nicotine's potential therapeutic use, several large tobacco companies have developed pharmaceutical divisions.

CHEMICAL NAME = nitric acid
CAS NUMBER = 7697–37–2
MOLECULAR FORMULA = HNO₃
MOLAR MASS = 63.0 g/mol
COMPOSITION = H(1.6%) N(22.2%) O(76.2%)
MELTING POINT = –41.6°C
BOILING POINT = 83°C
DENSITY = 1.54 g/cm³

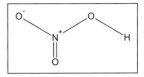

Nitric acid is a colorless, corrosive liquid that is the most common nitrogen acid. It has been used for hundreds of years. Nitric acid is a mineral acid that was called spirit of nitre and *aqua fortis,* which means strong water. Ancient alchemists first prepared nitric acid and Jabbar ibn Hayyan, known in Latin as Geber (721–815), is credited with its discovery. Geber prepared fuming nitric acid by distilling potassium nitrate (KNO_3) with sulfuric acid (H_2SO_4). The primary method of preparation of nitric acid before modern times was to react sodium nitrate ($NaNO_3$, Chilean saltpeter) with sulfuric acid and condense nitric acid from the nitric acid vapor evolved in the reaction.

Fuming nitric acid is named because of the fumes emitted by acid when it combines with moist air. Fuming nitric acid is highly concentrated and is labeled either red fuming nitric acid or white fuming nitric acid. Red fuming nitric acid, as the name implies, emits a reddish-brown fume on exposure to air. The color comes from nitrogen dioxide, which is liberated on exposure to air. The nitric acid concentration of red fuming nitric acid is approximately 85% or greater, with a substantial amount of dissolved nitrogen dioxide. White fuming nitric acid is highly concentrated anhydrous nitric acid with concentrations of 98–99%; the remaining 1–2% is water and nitrogen dioxide. Most commercial grade nitric acid has a concentration of between 50% and 70%.

Nitric acid can be prepared by several methods, but the primary method is by the oxidation of ammonia using the Ostwald method, which was named for Wilhelm Ostwald (1853–1932). The Ostwald method enabled the Germans to produce explosives in World War

I after their nitrate supplies were cut off. The first step in nitric acid production involves the oxidation of ammonia at a temperature of approximately 900°C to produce nitric oxide, NO, and water: $4NH_{3(g)} + 5O_{2(g)} \rightarrow 4NO_{(g)} + 6H_2O_{(g)}$. This process is carried out in the presence of a 90% platinum/10% rhodium catalyst in the form of wire gauze. The nitric oxide produced is further oxidized noncatalytically at a low temperature (less than 50°C) to form nitrogen dioxide and its dimer nitrogen tetroxide, N_2O_4: $2NO_{(g)} + O_2 \rightarrow 2NO_{2(g)} \rightleftharpoons N_2O_{4(g)}$. The final step involves absorbing the nitrogen dioxide-dimer in water to produce nitric acid: $3NO_{2(g)} + 2H_2O_{(l)} \rightarrow 2HNO_{3(aq)} + NO_{(g)}$. Nitric acid produced from this process has a typical concentration of between 40% and 60%. Water and nitric acid form an azeotrope so that the highest concentration obtainable from normal distillation is 68%. Concentrations higher than 68% up to almost 100% can be obtained by using extractive distillation with sulfuric acid as a dehydrating agent.

The main use of nitric acid is for the production of fertilizer, with approximately three-fourths of nitric acid production being used for this purpose. The primary fertilizer is ammonium nitrate (NH_4NO_3), produced by reacting nitric acid and ammonia: $NH_{3(g)} + HNO_{3(aq)} \rightarrow NH_4NO_{3(aq)}$. Ammonium nitrate is the preferred nitrogen fertilizer owing to its ease in production, economics, and high nitrogen content, which is 35%. Nitric acid can also be used for the acidulation of phosphate rock to produce nitrogen-phosphorus fertilizers.

Nitric acid is a strong oxidizer, which makes it useful in explosives and as a rocket propellant. Nitroglycerin and trinitrotoluene (TNT) are both produced from nitric acid. Ammonium nitrate, when mixed with small amounts of oil, can be used as an explosive that can be detonated with another explosive. Fuming nitric acid is stored as an oxidizer in liquid fuel rockets as inhibited fuming nitric. Inhibited fuming nitric acid is made by adding a small percentage (generally 0.6–4.0% depending on the acid and metal) of hydrogen fluoride. The fluoride creates a metal fluoride coating that protects metal tanks from corrosion.

Nitric acid is used for nitrating numerous other compounds to produce nitrates. Nitric acid is used to produce adipic acid ($C_6H_4O_{10}$), which is used in the production of nylon (see Nylon). In this process, cyclohexane is oxidized to a cyclohexanol-cyclohexanone mixture. Cyclohexanol and cyclohexanone are then oxidized with nitric acid to adipic acid.

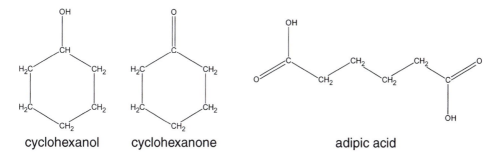

cyclohexanol   cyclohexanone   adipic acid

Additional uses of nitric acid are for oxidation, nitration, and as a catalyst in numerous reactions. Salts of nitric acid are collectively called nitrates, which are soluble in water. Nitric acid is used in the production of many items such as dyes, pharmaceuticals, and synthetic fabrics. It is also used in a variety of processes including print making.

Nitric acid is used extensively in the metal industries. Nitric acid dissolves most metals and is used to separate gold from silver in assaying and metal refining operations. Some metals such as iron and aluminum form a protective oxide coating when treated with nitric acid. *Aqua regia,* which means "royal water," is a mixture of one part nitric acid to three parts hydrochloric acid. Its name refers to its ability to dissolve precious metals such as gold and platinum. Nitric acid is used to pickle steel and brass surfaces in metal processing. Pickling is a process where metals are cleaned by immersing them in a dilute acid bath.

Nitric acid is a major component of acid precipitation (see Sulfuric Acid). It forms when nitrogen oxides, primarily NO and NO_2, dissolve in atmospheric water. The atmosphere consists almost entirely of nitrogen and oxygen. Fortunately, nitrogen and oxygen require a high temperatures such as that produced in a hot flame, to form nitrogen oxides (see Nitric Oxide). One reason it is difficult to form nitrogen oxides is that the strong triple bond present in molecular nitrogen requires a significant amount of energy to break. The bond energy of the nitrogen-nitrogen triple bond is 941 kilojoules per mol. Nitrogen oxides are a major air pollutant associated with internal combustion engines and natural gas combustion. The major source of nitrogen oxides is motor vehicles, which make up about half the source of these pollutants. Another 25% comes from power utilities and fossil fuel combustion. Nitric acid can also be produced naturally in thunderstorms, with lightning providing the energy required to form nitrogen oxides (see Nitric Oxide, Nitrogen Dioxide, Nitrous Oxide).

CHEMICAL NAME = nitric monoxide		nitrogen dioxide	dinitrogen oxide
CAS NUMBER = 10102–43–9		10102–44–0	10024–97–2
MOLECULAR FORMULA = NO		NO_2	N_2O
MOLAR MASS = 30.0 g/mol		46.0 g/mol	44.0 g/mol
COMPOSITION = N(46.7%) O(53.3%)		N(30.4%) O(69.6%)	N(63.7%)
MELTING POINT = –164°C		–11.2	–90.8°C
BOILING POINT = –152°C		21.1°C	–88.6°C
DENSITY = 1.35 g/L		2.05 g/L	.99 g/L
VAPOR DENSITY = 1.04 (air = 1)		1.58	1.53

$$N{=}O \qquad O{\nwarrow}^{N}{\searrow}O \qquad N{\equiv}N^{+}{-}O^{-} \longleftrightarrow N{=}N^{+}{=}O$$

Nitrogen oxides refer to several compounds composed of nitrogen and oxygen. This entry focuses on the three most common: nitric oxide, nitrogen dioxide, and nitrous oxide. Other nitrogen oxides not addressed include dinitrogen trioxide (N_2O_3), dinitrogen tetroxide (N_2O_4), nitrogen trioxide (NO_3), and dinitrogen pentoxide (N_2O_5). As air pollutants, nitrogen oxides generally refer to the compounds nitric oxide and nitrogen dioxide. Nitrogen oxides exist as gaseous compounds under normal conditions, although nitrogen dioxide changes phases between liquid and gas near room temperature (its liquid density at 20°C is 1.45 g/cm³). Nitrogen monoxide (NO) is commonly called nitric oxide, and dinitrogen monoxide's (N_2O) common name is nitrous oxide. Nitric oxide and nitrogen dioxide are nonflammable, toxic gases. Nitric oxide is colorless and has a sharp sweet odor; nitrogen dioxide is a reddish-brown gas (or yellow liquid) with a strong, acrid odor. Nitrogen dioxide readily dimerizes to produce N_2O_4. Dinitrogen tetroxide production is favored as temperature decreases. As N_2O_4 increases in the $NO_2 \rightleftarrows N_2O_4$ equilibrium, the color of the gas fades from brown toward colorless.

Nitric oxide is a free radical that quickly reacts in air to produce nitrogen dioxide. It is also an important biological messenger and transmitter. Nitrous oxide is a colorless, nonflammable, nontoxic gas with a slightly sweet odor and taste. Nitrous oxide is called laughing gas and has been used as a recreational inhalant, anesthetic, oxidizer, and propellant.

Nitric oxide, nitrogen dioxide, and nitrous oxide were prepared in 1772 by Joseph Priestley (1733–1804) and described in his volumes *Experiments and Observations of Different Kinds of Air* published between 1774 and 1786. Priestley called nitric oxide *nitrous air,* nitrogen dioxide *nitrous acid vapor,* and nitrous oxide *phlogisticated nitrous air,* but also referred to the latter as *diminished nitrous air.* He observed the change of clear nitric oxide to red nitrogen dioxide. Priestley prepared nitric oxide by reacting nitric acid with a metal such as copper: $3Cu_{(s)} + 8HNO_{3(aq)} \rightarrow 2NO_{(g)} + 3Cu(NO_3)_{2(aq)} + 4H_2O_{(l)}$. He prepared nitrous oxide by reducing nitric oxide using iron: $2NO_{(g)} + H_2O_{(l)} + Fe_{(s)} \rightarrow N_2O_{(g)} + Fe(OH)_{2(aq)}$. Confusion existed involving the numerous gases prepared by scientists during Priestley's era because chemical reactions produced gas mixtures. Scientists attempted to separate these mixtures into the pure components to characterize their behavior. Thus dates of discovery of gases are often ambiguous, with multiple dates being cited in the literature. For example, the year of discovery for nitrous oxide ranges between 1772 and 1793. Humphrey Davy (1778–1829) examined the physiological effects of nitrous oxide and in 1799 wrote *Researches Chemical and Philosophical, Chiefly Concerning Nitrous Oxide.* Davy noted that it had the ability to reduce pain in medical procedures and discovered that it created a temporary state of intoxicated euphoria; symptoms in people who inhaled it included disorientation, loss of motor coordination, and hysterical laughter. Thus the name laughing gas was adopted. Although Davy observed that nitrous oxide had potential as an anesthetic, its main use was as a recreational substance for several decades before it was first used as an anesthetic in 1844 (see Ether, Chloroform). Parties in which people became intoxicated on nitrous oxide, called frolics, were quite common in the early 19th century (Figure 66.1)

Nitrogen oxides are produced both naturally and by human activity. The main natural sources are volcanoes, oceans, lightning, forest fires, and biological decay. Human activities produce approximately 25 million tons of nitrogen oxides annually, which are a form of air pollution that is regulated by the federal government. Nitrogen oxides are produced when fuels are burned at high temperatures. The primary human sources of nitrogen oxides are motor vehicles, power utilities, industrial combustion, and residential combustion. Motor vehicles account for about half of the human output of nitrogen oxides, with utility plants accounting for another 30–35%. Most nitrogen oxides are emitted as NO but are quickly converted to NO_2 in the atmosphere. The federal government has established air quality standards for nitrogen dioxide at 0.053 parts per million (ppm), which equals 100 μg (micrograms) per cubic meter.

Nitrogen oxides have a number of detrimental health and environmental effects. Nitric oxide is a main ingredient in forming other pollutants such as ground-level ozone and photochemical smog. At low levels, nitrogen oxides irritate the eyes, nose, and throat, causing coughing, fatigue, and shortness of breath. Chronic exposure can lead to respiratory infections or lung damage, leading to emphysema. Breathing high concentrations of nitrogen oxides can cause spasms, burns, swelling of respiratory tissues, and fluid in the lungs. Nitrogen oxides combine with water in the atmosphere to form acids, contributing to acid precipitation. Nitrogen dioxide is highly soluble in water and forms nitric acid (HNO_3), and nitric oxide is slightly soluble and forms nitrous acid (HNO_2). Nitrogen oxides are detrimental to plants,

Laughing Gas.

"Some jumped over the tables and-chairs; some were
bent upon making speeches; some were very much inclined
to fight; and one young gentleman persisted in an attempt
to kiss the ladies."

Page 116.

Figure 66.1 Nitrous oxide frolic by George Cruikshank, 1839. From *Chemistry No Mystery,* by J. Scoffern. Caption reads: "Some jumped over the tables and chairs; some were bent upon making speeches; some were very much inclined to fight; and one young gentleman persisted in an attempt to kiss the ladies."
Source: Edgar Fahs Smith Collection, University of Pennsylvania.

causing bleaching, loss of leaves, and reduced growth rates. Nitrate particles produce haze, which impairs visibility. Nitrogen oxides react with other chemicals to form a variety of toxic products such as nitroarenes (nitro substituted benzene) and nitrosamines. Nitrogen dioxide is a strong oxidizing agent and causes corrosion. Nitrous oxide is an important greenhouse gas. Its atmospheric residence time is 120 years. A molecule of N_2O has 310 times the potential for absorbing heat compared to a molecule of CO_2. Nitrous oxide is stable and unreactive on the earth's surface, but it can be transported to the stratosphere where it absorbs energy and is converted into reactive forms of nitrogen such as nitric oxide and the nitrate radical contributing to ozone destruction. A simple model is given in the following series of reactions:

$$NO + O_3 \rightarrow NO_2 + O_2$$
$$O_3 \rightarrow O_2 + O$$
$$NO_2 + O \rightarrow NO + O_2$$
$$\text{Net: } 2O_3 \rightarrow 3O_2$$

Nitrogen oxides play a key role in ozone chemistry. Paul Crutzen (1933–) received the 1995 Nobel Prize in chemistry for research that helped define the role of nitrogen oxides in ozone depletion.

Nitric oxides have relatively few industrial uses. Nitric oxide is produced as an intermediate using the Ostwald method to make nitric acid (see Nitric Acid). Nitrogen compounds produced from nitric acid are use to manufacture fertilizers, explosives, and other chemicals. Nitrogen dioxide is used as an oxidizing agent, a catalyst in oxidation reactions, an inhibitor, as a nitrating agent for organic reactions, as a flour bleaching agent, and in increasing the wet strength of paper. Nitrous oxide is widely used as an anesthetic in dental surgery, which accounts for approximately 90% of its use. It is used by the dairy industry as a foaming agent for canned whipping creams. Pressurized nitrous oxide dissolves in the fats in the cream. When the trigger on a whipped cream can is pressed, releasing the pressure, N_2O comes out of solution producing bubbles, which instantly whips the cream into foam. Nitrous oxide is injected into fuel systems to increase the horsepower of internal combustion engines by increasing oxygen in the combustion chamber. Oxygen comes from the thermal decomposition of N_2O in the engine: $N_2O(g) \rightarrow N_2(g) + 0.5O_2(g)$. The decomposition takes place at temperatures above 297°C. Nitrous oxide used to increase engine performance, called nitrous, is stored as a liquid in a pressurized bottle at roughly 1,000 psi. An additional benefit of N_2O as a horsepower booster comes during delivery when it changes phases from a liquid to a gas. The phase change cools the intake air temperature, providing several more horsepower.

Nitric oxide is commercially produced by the catalytic oxidation of ammonia using a platinum catalyst: $4NH_{3(g)} + 5O_{2(g)} \rightarrow 4NO_{(g)} + 6H_2O_{(g)}$. Nitrous oxide is produced by the thermal decomposition of ammonium nitrate at approximately 240°C: $NH_4NO_{3(g)} \rightarrow N_2O_{(g)} + 2H_2O_{(g)}$.

During the last three decades, appreciable knowledge has been acquired on the biological importance of nitric oxide. It is synthesized in the body from the amino acid L-arginine by the action of NO synthase (NOS) enzymes. Three similar versions (isoforms) of NOS are endothelial NOS, neuronal NOS, and inducible NOS. Nitric oxide is a gaseous free radical that serves multiple functions in human physiology. Nitric oxide causes vasodilation when it is secreted from endothelial cells (the endothelium is a single layer of cells lining the interior surface of a blood vessel). As blood flow increases through an artery, nitric oxide produced by the endothelial cells relaxes the artery walls, causing it to dilate. In this role, nitric oxide is called the endothelium-derived relaxing factor. Nitric oxide also inhibits platelet aggregation. Nitric oxide plays an important role in immune systems. Macrophages produce nitric oxide using inducible NOS to directly combat bacteria and to also signal other immune responses. Nitric oxide functions as a neurotransmitter that functions by diffusing into surrounding cells rather than activating receptors like other neurotransmitters. Transmission of nitric oxide by diffusion means it can activate several surrounding neurons and does not require a presynaptic-postsynaptic pathway for nerve signal transmission. Nitric oxide also plays a role in reproduction. It functions as a vasodilator during penis erection. Nitric oxide plays a role in fertilization in some animal species, and its role in human reproduction is currently being explored.

CHEMICAL NAME = propane-1,2,3-triyl
   trinitrate
CAS NUMBER = 55–63–0
MOLECULAR FORMULA = $C_3H_5N_3O_9$
MOLAR MASS = 227.1 g/mol
COMPOSITION = C(15.9%) H(2.2%)
   N(18.5%) O(63.4%)
MELTING POINT = 13.2°C
BOILING POINT = decomposes between
   50° and 60°X
DENSITY = 1.6 g/cm³

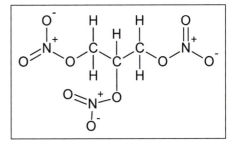

Nitroglycerin is an oily, poisonous, clear to pale yellow, explosive liquid. It was first prepared in 1846 by the Italian chemist Ascanio Sobrero (1812–1888), who nitrated glycerol using a mixture of nitric acid and sulfuric acid. Sobrero, who was injured in an explosion doing his research, realized the compound's danger and abandoned work on nitroglycerin. Twenty years after Sobrero's discovery, Alfred Nobel (1833–1896) developed its use commercially. Nobel was born in Stockholm Sweden, where his father, Immanuel Nobel (1801–1872), ran a heavy construction company. When Alfred was four, his father's company went bankrupt and Immanuel left for St. Petersburg, Russia to pursue new ventures. Immanuel rebuilt a successful business in Russia, in part because of his ability to develop and sell mines to the Russian Navy for use in the Crimean War. Alfred and the rest of his family joined his father in Russia when Alfred was nine, where he received an excellent education with private tutors. He studied in the United States and Paris, where he met Sobrero. Nobel studied chemistry, literature, and mechanical engineering as his father groomed Alfred to join him in the construction and defense industry. After another downturn in his fortune, Immanuel and his sons Alfred and Emil returned to Stockholm in 1859 to start another business. It was at this time that Alfred began to experiment with nitroglycerin, seeking a safe method for its use as an explosive. Nobel mixed nitroglycerin with other substances, searching for a safe way to transport it and make it less sensitive to heat and pressure. Several explosions at Nobel's laboratory, one of which killed Emil in 1864, prompted the city of Stockholm to ban nitroglycerin research inside the city and forced Nobel to move his studies to a barge in a lake just outside the city limits.

In 1864, Nobel had developed an explosive that was almost eight times as powerful as gunpowder on a weight basis and began mass producing nitroglycerin. Early shipments and use of nitroglycerin were made with nitroglycerin in its unstable liquid state. It was shipped in zinc cans, which were often packed in crates using sawdust as packing material. Nitroglycerin could detonate when disturbed, and the probability of an explosion increased with temperature and the presence of air bubbles in the liquid nitroglycerin. Furthermore, impurities in the form of residual acids used in nitroglycerin's production could corrode the zinc shipping containers and also produce gases that could trigger an explosion. Numerous deaths were attributed to nitroglycerin when it was first marketed, and Nobel continued to experiment with methods to make nitroglycerin safer. One of these was mixing nitroglycerin with materials to make a solid form of nitroglycerin. Nobel discovered that when nitroglycerin was mixed with a silica-based diatomaceous earth material called kieselguhr, a relatively stable product resulted. The mixture produced a paste that Nobel could pack into cardboard tubes; these could then be inserted into holes drilled into rock structures and detonated. In 1867, Nobel patented his mixture and called it dynamite, a name derived from the Greek word *dunamis*, meaning power. Nobel also perfected a blasting cap made from mercury fulminate ($Hg(ONC)_2$) and potassium chlorate ($KClO_3$) to detonate the nitroglycerin. Nobel's business expanded rapidly, as dynamite was increasingly used for construction and defense purposes. Although Nobel's fame and fortune were based on his invention of dynamite, he was an able inventor and chemist. Over the years he received 355 patents, including ones for synthetic rubber and artificial silk. His will requested that the bulk of his fortune, which approached $10 million, be used to fund annual prizes in the areas of chemistry, physiology or medicine, physics, literature, and peace.

Nitroglycerin is made by nitrating glycerol. Early industrial processes used a batch process in which glycerol was added to a mixture with approximately equal volumes of nitric acid and sulfuric acid. The sulfuric acid serves to ionize the nitric acid and removes water formed in the nitration process. Removing the water formed in nitration increases the yield of nitroglycerin. Acids and water must be removed from the desired nitroglycerin through a washing process. The production of nitroglycerin is highly exothermic, and it is important to keep the temperature below room temperature to prevent an explosion. Early production methods used cooling coils in the nitration vessels to regulate the temperature. During the latter half of the 20th century, safer continuous production methods replaced batch processes. In these methods much smaller reactors are required, as glycerol is reacted with the acids.

The explosive power of nitroglycerin is related to several factors. Each molecule of nitroglycerin contains three nitrate groups, which serve as oxidizing agents for the hydrocarbon groups to which they are bonded. The hydrocarbon groups provide the fuel and the nitrate groups provide the oxidizer necessary for combustion. In typical combustion reactions involving hydrocarbons, oxygen does not come from the molecule but from air or another source external to the molecule. Glycerin contains both the fuel and oxidizer bonded together. The detonation of nitroglycerin can be represented by the reaction: $4C_3H_5(ONO_2)_{3(l)} \rightarrow 12CO_{2(g)} + 10H_2O_{(g)} + 6N_{2(g)} + O_{2(g)}$. Four moles of liquid nitroglycerin produces 29 moles of gaseous products in this reaction. Also, the production of stable products results in a highly exothermic reaction. The hot gaseous products produce a rapid expansion in volume, producing a shock wave that propagates throughout the material, causing an almost instantaneous reaction. This is contrasted with combustion where the flame moves through the material as the fuel and oxygen react.

Nitroglycerin has medicinal use as a vasodilator. Workers in the original nitroglycerin plants developed headaches, which led to the discovery that nitroglycerin is a vasodilator. The main medical use of nitroglycerin is to treat angina pectoris. Angina pectoris is a condition in which the heart does not receive sufficient blood (oxygen) supply, producing a tight sensation in the chest. This lack of oxygen supply may be due to arthrosclerosis, thickening of the arteries. Nitroglycerin was first used to treat this condition in the late 19th century. It is prescribed today in various forms (tablet, ointment, patches, and injection) for patients who suffer from angina pectoris. Nitroglycerin is marketed under various trade names: Nitro-Dur, Nitrostat, Nitrospan, Nitro-Bid, and Tridil. When used in medications, the name glyceryl trinitrate is often used instead of nitroglycerin.

CHEMICAL NAME = [17α]-17-Hydroxy-
19-norpregn-4-en-20-yn-3-one
CAS NUMBER = 68–22–4
MOLECULAR FORMULA = $C_{20}H_{26}O_2$
MOLAR MASS = 298.4 g/mol
COMPOSITION = C(80.5%) H(8.8%)
O(10.7%)
MELTING POINT = 204°C
BOILING POINT = decomposes
DENSITY = 1.15 g/cm³

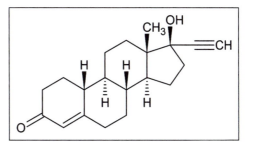

Norethindrone is a progestin (a synthetic substance with properties similar to progesterone) that is best known as the first female oral contraceptive, or the "pill." Norethindrone's global impact on society and culture has made it one of the most important inventions in history. The development of norethindrone as a female oral contraceptive took place indirectly over 30 years as a result of steroid research. This research accelerated in the 1930s when structures and medical applications of steroidal compounds were determined. Steroids are lipids, which include cholesterol, bile salts, and sex hormones, that are characterized by a structure of three fused six-carbon rings and a five-carbon ring (Figure 68.1).

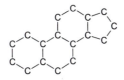

Figure 68.1 Steroid structure.

Contraceptives and other birth control methods have been used throughout human history. Ancient methods included a plethora of potions, condoms made from animal skins, withdrawal, abstinence, cervical caps, and spermicidal concoctions. During the 20th century women groups and national organizations such as Planned Parenthood developed that specifically sought a safe contraceptive for females. The use of norethindrone to fulfill this role was due to the efforts of several key individuals; foremost among these was Russell Earl Marker (1902–1995). Marker received his undergraduate and master's degrees from the University of Maryland, but he left before completing his doctorate because he refused to take additional coursework in physical chemistry. After leaving the University of Maryland, Marker worked in hydrocarbon chemistry for several years, leading to a position at the Rockefeller Institute. There he developed an outstanding research record and left for Penn State University to pursue steroid research. Marker's position at Penn State was supported by the pharmaceutical company Parke Davis. By the end of the 1930s, Marker was successful in extracting progesterone from animal urine. At that time progesterone was being examined for treating menstrual cramps and miscarriages. Although Marker was able to produce progesterone from animal sources, he was aware that numerous steroid compounds existed in plants and decided to focus his work on obtaining progesterone from plant sources. Marker's work focused on the hypothesis that sarsasapogenin, a sapogenin plant steroid from the sarsaparilla plant, could be degraded by splitting off the end ring structure as shown in Figure 68.2, leaving a steroidal structure from which progesterone could be synthesized. Marker was successful in finding a process to remove the end ring group, and this procedure came be known as the Marker degradation. The process involved acetylation with acetic anhydride followed by oxidation with chromic acid (CrO_3) and hydrogenation.

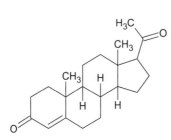

progesterone

sarsasapogenin

Figure 68.2　Progesterone and sarsasapogenin.

The sarsasapogenin derived from the sarsapilla plant was not an economically viable source for synthesizing progesterone, so Marker began searching for other suitable plant steroids. In 1941, he discovered that using wild yams from the species *Dioscorea* found in Mexico supplied the compound diosgenin, which was almost identical to sarsasapogenin. Using diosgenin, Marker was able to produce progesterone economically. On returning to the United States from Mexico, Marker was unsuccessful in convincing Parke Davis and Penn State to pursue his discovery. He resigned his position at Penn State in 1943 and returned to Mexico in 1944, where he partnered with Mexican hormone producers to form the company Syntex. Marker's relationship with Syntex and his Mexican partners lasted only one year because of disagreements over patents and profits. Marker left Syntex and took his progesterone production knowledge with him. Marker formed a new company, and Syntex hired a chemist named George Rosenkranz (1916–) as its new research director. Part of Rosenkranz's job was to continue the work Marker had started.

Rosenkranz's arrival at Syntex started the next chapter in the development of "the pill." Marker left chemical research several years after leaving Syntex and became involved in

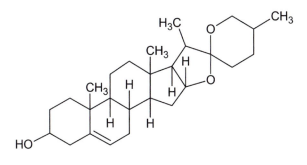

diosgenin

other endeavors. Meanwhile, Rosenkranz recruited chemists to Syntex to develop hormonal research. One of these was Carl Djerassi (1923–). Djerassi's worked on synthesizing cortisone and other hormones and then turned his attention to progestin. Progesterone was administered by painful injection directly into the cervix and Djerassi sought a suitable oral progesterone analog. Djerassi and Syntex were not interested in making an oral contraceptive initially, but they were interested in making a progesterone drug that could be used to treat menstrual problems such as cramps, irregular periods, and endometriosis, as well as cervical cancer and infertility. Luis Ernesto Miramontes (1925–2004), a researcher working under Djerassi, first synthesized norethindrone in 1951, but Djerassi and Rosenkranz shared the credit as inventors of "the pill." A group at Searle had synthesized a similar compound called norethynodrel, which was converted to norethindrone by stomach acid.

The next chapter in the use of norethindrone as an oral conception involved Luis Pincus (1903–1967). Pincus was a Harvard researcher whose controversial work on fertility and reproduction forced him to leave Harvard. He moved to Clark University in Worcester, Massachusetts, and while there he helped establish the Worcester Foundation for Experimental Biology. Early in 1950, interest in the use of progesterone as a contraceptive developed as a result of concerns about the burgeoning world population and calls from birth control

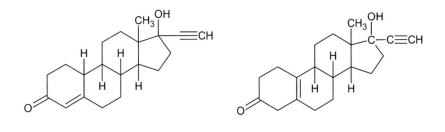

norerthindrone   norethynodrel

advocates. Pincus was approached by Margaret Sanger (1879–1966) in 1951 to find a suitable oral contraceptive for women. Sanger founded the American Birth Control League, which developed into Planned Parenthood. Sanger's request was supported by funds from Katherine McCormick (1875–1967). McCormick was one of the first female graduates of MIT and held a degree in biology. She married the son of Cyrus McCormick (1809–1884) who had patented the Cyrus reaper, which eventually led to the development of the International Harvester Corporation. Katherine McCormick was a friend of Sanger and a philanthropist who had the finances to back Sanger's request to Pincus. Pincus examined numerous compounds as possible progestins. Norethindrone and norethynodrel were two of the many compounds Pincus examined. After trials with animals indicated positive results, Pincus teamed with a Harvard gynecologist named John Rock (1890–1984). Rock had worked with progestins therapeutically in treating menstrual problems and infertility. Because controversy surrounded the use of progestins as contraceptives, norethindrone and norethynodrel were clinically tested for their therapeutic effects. Much of the testing took place outside the United States. Pincus, backed by Searle, conducted trials on norethynodrel in Haiti, Puerto Rico, and Brookline Massachusetts. Concurrently, Syntex tested norethindrone with subjects in Mexico City, Los Angeles, and San Antonio. In 1957, both norethindrone and norethynodrel were approved by the Food and Drug Administration (FDA) for treating menstrual problems and infertility. Under much controversy, the use of norethindrone and norethynodrel as oral contraceptives took three more years for approval. In 1960, the FDA approved Searle's norethynodrel under the trade name Enovid. Norethindrone was approved as an oral contraceptive in 1962 under the trade name Ortho-Novum. By 1965, 5 million American women were using the pill, and by 1973 the number had doubled. Today it is estimated that 100 million women use the pill worldwide, with about 10% of these in the United States.

Progestin contraceptives work by producing pregnant-like conditions in a female to prevent ovulation. During pregnancy, progesterone is released by the placenta during development of the fetus. This in turn suppresses development of egg follicles and ovulation. Progestins mimic this condition and thus prevent or delay ovulation. Oral contraceptives currently use progestin and estrogen in combination to prevent ovulation and thicken cervical mucus. The latter make it harder for sperm to enter the uterus and for an egg to implant on the uterine wall.

CHEMICAL NAME = 2S,5R,6R)-3,3-dimeth-
yl-7-oxo-6-[(phenylacetyl)amino]-4-thia-1-
azabicyclo[3.2.0]heptane-2-carboxylic acid
CAS NUMBER = 61–33–6
MOLECULAR FORMULA = $C_6H_{18}N_2O_4S$
MOLAR MASS = 334.4 g/mol
COMPOSITION = C(57.5%) H(5.4%)
N(8.4%) O(19.1%) S(9.6%)
MELTING POINT = 209–212°C (for Penicillin
G sodium)
BOILING POINT = decomposes
DENSITY = 1.4 g/cm³

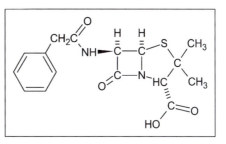

Penicillin was the first natural antibiotic used to treat bacterial infections and continues to be one of the most important antibiotics. The name comes from the fungus genus *Penicillium* from which it was isolated. *Penicillus* is Latin for "brush" and refers to the brushlike appearance of filamentous *Penicillium* species. Species of this genus are quite common and appear as the bluish-green mold that appears on aged bread, fruit, and cheese. The term *penicillin* is a generic term that refers to a number of antibiotic compounds with the same basic structure. Therefore it is more appropriate to speak of penicillins than of penicillin. The general penicillin structure consists of a β-lactam ring and thiazolidine ring fused together with a peptide bonded to a variable R group (Figure 69.1). Penicillin belongs to a group of compounds called β-lactam antibiotics. Different forms of penicillin depend on what R group is bonded to this basic structure. Penicillin affects the cell walls of bacteria. The β-lactam rings in penicillins open in the presence of bacteria enzymes that are essential for cell wall formation. Penicillin reacts with the enzymes and in the process deactivates them. This in turn inhibits the formation of peptidoglycan cross-links in bacteria cell walls. Peptidoglycans consist of a network of protein-carbohydrate chains that form a strong outer skeleton of the cell. Thus penicillin weakens the cell wall and causes it to collapse. Humans and other animal cells have membranes rather than walls. Because animal cells do not have corresponding cell wall enzymes that are present in bacteria, penicillins do not harm human cells.

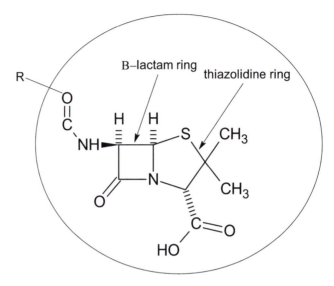

Figure 69.1   General structure of penicillin.

The discovery of penicillin is generally credited to Alexander Fleming (1881–1955) in 1928, but the development of penicillin as an antibiotic took place sporadically over the last decades of the 19th century and first half of the 20th century. Joseph Lister (1827–1912) noted in 1871 that bacteria growth was inhibited by the presence of *Penicillium* molds. Lister used *Penicillium* to treat patients but never published his results. Other researchers published articles that implied that molds could be used to kill bacteria. In 1896, Ernest Duchesne (1874–1912) discovered the antibiotic properties of penicillin while working on his dissertation. He noted that *Penicillium glaucum* prevented the growth of *Escherichia coli* when incubated in the same culture dishes. Other researchers reported before Fleming that extracts from *Penicillium* species inhibited the growth of bacteria, but the difficulty in working with *Penicillium* and mixed results led early scientists to abandon the research.

Alexander Fleming was a medical doctor working in the Inoculation Department at St. Mary's Hospital in London. Fleming studied agents to combat bacteria. In 1922, he discovered the enzyme lysozyme using mucus from his own nose. Lysozyme destroys bacterial cell walls. Fleming's work involved culturing numerous plates, which consisted of Petri dishes, with bacteria and isolating subcultures for study. His discovery of penicillin occurred when Fleming was working on staphylococci for a reference manual on bacteria. During his work at the end of the summer in 1928, Fleming took a month-long vacation. Rather than discard his Petri dishes, he piled them in the dark corner of his small laboratory. When Fleming returned in September to continue his staphylococci work, he retrieved his stored Petri dishes to make observations and to create subcultures of interesting variants. Soon after his return, he was visited by a former co-worker. Fleming had a pile of Petri dishes near a small vessel of Lysol. Lysol was used for rinsing and disinfecting the dishes. Fleming and his former co-worker observed that one Petri dish was contaminated with mold and that staphylococci growth was absent in the vicinity of the mold. Fleming cultured the mold in a tube of broth and in succeeding weeks showed his plate to scientists who visited his laboratory. Fleming's plate did not raise much interest. He worked on penicillin for several months after his discovery in 1928 and published an article in 1929 reporting his 1928 observation. He then worked sporadically on penicillin over the next several years and published a second article in 1932, but the difficulties that plagued earlier researchers on *Penicillium* also prevented Fleming from making significant progress with the mold.

The commercial development of penicillin can be traced to 1938, when Ernest Boris Chain (1906–1979), a German biochemist who fled Germany for England in 1933, and Howard Walter Florey (1898–1968) expanded their work on lysozyme to search for other antibacterial agents. Chain was part of Florey's research team at the Dunn School of Pathology at Oxford University. He reviewed the literature and Fleming's 1929 article stimulated his interest on penicillin because of its similarity to lysozyme. Chain convinced Florey that penicillin had potential as an antibacterial agent, and laboratory work on *Penicillium notatum* (currently called *Penicillium chrysogenum*) began in 1938. Florey assembled a research team to work on penicillin, but lack of research funds from England and Oxford University initially hampered his efforts. England's entrance into World War II also made difficult the task of acquiring funds for a project with unclear potential.

The first challenge of the research teams was purifying penicillin by extracting the compound from the mold. Norman Heatley (1911–2004) devised extraction and purification methods to obtain quantities of penicillin sufficient for study. It was determined that penicillin was nontoxic when injected in animal subjects. In May 1940, a significant experiment took place in which eight mice injected with streptococci bacteria and half of these were treated with penicillin. The mice injected with penicillin survived and the untreated mice died after a day. Similar results were obtained with subsequent experiments as the researchers refined the amount and frequency of penicillin injections needed to combat infections in different animal subjects.

The next step in penicillin's use for humans involved clinical trials on humans. The large problem that Florey's team faced was producing enough penicillin to treat human subjects rather than small animal subjects such as mice. The penicillin initially used in Florey's Oxford laboratory was produced by cultivating *Penicillium* in hundreds of clay containers modeled after bedpans. The labor-intensive process produced limited quantities of penicillin. Heatley devised a continuous process for extracting penicillin from mold filtrate, and this increased the quantity and purity of penicillin so that limited trials could be performed on several human subjects. The amount of penicillin for human trials was so limited that researchers recycled penicillin by extracting it from subjects' urine. The initial human trials were performed on five subjects and were successful to the extent that penicillin cured infections without harmful side effects, but one subject died when the supply of penicillin ran out and another died of complications unrelated to the treatment. Florey's team published and presented their results in 1941, and Florey was convinced that the evidence for penicillin's efficacy would enable him to obtain financial backing to conduct the human trials necessary to commercialize penicillin. Florey was unsuccessful in convincing government health officials or pharmaceutical companies in Britain to provide financial backing to scale up production, so he arranged a trip to the United States to seek backing. Florey used several close professional colleagues and friends in the United States to arrange a tour of pharmaceutical companies, foundations, and agencies; Florey and Heatley arrived in the United States in July 1941, carrying with them a small quantity of *Penicillium* cultures. During their tour, they were able to arrange a pilot study to increase mold production with brewing techniques used in fermentation at the Bureau of Agricultural Chemistry Lab in Peoria, Illinois. At the Peoria laboratory Heatley worked with Andrew J. Moyer (1899–1959) to develop methods to increase production. These methods led to using a broth of corn steep liquor and lactose rather than glucose for culturing *Penicillium* and deep-culture fermentation, in which the mold was grown throughout the volume of a vat rather than on the surface. While Heatley collaborated with Moyer on production techniques, Florey continued to promote penicillin to pharmaceutical companies. Several of these companies were already working on penicillin and Florey was able to increase their efforts.

Florey returned to Oxford in late September, and Heatley remained in the United States to continue work on increasing production, first in Peoria and then at Merck's factory in New Jersey. Florey succeeded in increasing interest on penicillin research in the United States, but he was unsuccessful in obtaining more penicillin for his own clinical trials. Florey's team continued to produce penicillin using laboratory techniques and to perform limited human trials. Meanwhile, U.S. pharmaceutical firms, stimulated by the country's sudden entry into World War II on December 7, 1941, made commercial penicillin production a priority. Mass production of penicillin by U.S. firms started in 1943, and it was used immediately to treat wounded soldiers. Penicillin reduced suffering, prevented amputations, cured pneumonia, and saved thousands of lives during the war and was hailed as a "miracle drug." The United States increased production throughout the war years and after the war widespread civilian use commenced. U.S. firms took out patents on the production methods. Chain had urged Florey to patent production processes, but Florey and his supporters thought that this was unethical. Ironically, British firms had to pay American firms royalties for methods that originated in their country. There was also disagreement over credit for the work on penicillin between Florey and Fleming. Although Fleming had written the article that motivated Florey to examine penicillin as an antibiotic, Florey and his team's work led to its commercialization. Fleming, Florey, and Chain shared the Nobel Prize in physiology or medicine in 1945 for their work on penicillin.

Three years after mass production of penicillin began, the first evidence of resistance appeared. Because bacteria exist as large populations that rapidly reproduce, the widespread use of an antibiotic inevitably leads to biological resistance. Bacteria have evolved to produce a number of resistance mechanisms to antibiotics. One is the production of β-lactamase enzymes, which breaks down the β-lactam ring, rendering it ineffective. Another mechanism is modification of the penicillin-binding proteins. These proteins are essential in building the peptidoglycan cross-links of the cell wall.

Because of the ability of bacteria to develop resistance to penicillin, pharmaceutical companies must continually develop different penicillin compounds for continued use as an antibiotic. Different forms are also used depending on the type of infection, delivery method, and individual. The form discovered by Fleming and used by Florey was benzylpenicillin or Penicillin G. Today there are numerous compounds that are classified as penicillins that are marketed under various trade names. Early penicillins were biosynthetic compounds obtained from molds. Modern penicillins are semisynthetic in which penicillin obtained from natural sources is further synthesized to impart specific properties to the compound.

From 1940 to 1945, researchers sought the structure of penicillin. Several forms of penicillin were discovered. In 1945, Dorothy Mary Crowfoot Hodgkin (1910–1994) used x-ray crystallography to determine the structure of penicillin. John C. Sheehan (1915–1992) synthesized a penicillin called penicillin V in 1957 at Massachusetts Institute of Technology. Although Sheehan's work showed that penicillin could be synthesized, the method was not practical on a commercial level. Currently, about 45,000 tons of penicillins are produced annually worldwide.

CHEMICAL NAME = phenol
CAS NUMBER = 108–95–2
MOLECULAR FORMULA = C$_6$H$_5$OH
MOLAR MASS = 94.1 g/mol
COMPOSITION = C(76.6%) H(6.4%) O(17.0%)
MELTING POINT = 43.0°C
BOILING POINT = 181.8°C
DENSITY = 1.07 g/cm³

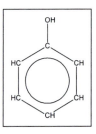

Phenol is a colorless or white crystalline solid that is slightly soluble in water. Phenol is the simplest of the large group of organic chemicals known as phenols, which consist of compounds where a carbon in the phenyl aromatic group (C$_6$H$_5$) is directly bonded to hydroxyl, OH. Phenol was probably first prepared by Johann Rudolph Glauber (1604–1668) in the middle of the 17th century. Its original name was carbolic acid, a name given by Friedlieb Ferdinand Runge (1795–1867), who isolated phenol from the destructive distillation of coal tar in 1834. The name phenol was first used in 1841 by Charles Frédéric Gerhardt (1816–1856).

Phenol's first prominent use was by Joseph Lister (1827–1912) as an antiseptic. Throughout human history, infection often resulted in death, even when the wound could be surgically treated. A broken bone piercing the skin, which today is a painful but not life-threatening injury, historically resulted in infection and possible amputation or death. Lister was inspired by Louis Pasteur's (1822–1895) germ theory of disease, and he began to use antiseptic methods during routine surgery during the 1860s. Lister treated wounds directly with phenol and cleaned surgical instruments in phenol solutions. He also devised a carbolic sprayer that could be used to mist phenol into operating and recovery rooms. Lister was able to reduce surgical infectious mortality rates from 50% to approximately 15% using his antiseptic techniques, and this practice spread as Lister educated others on the use of antiseptic practices. Lister used phenol diluted in water and oils in his surgical practice. Phenol is toxic and causes white blisters when applied directly to the skin. Because it is a powerful bactericide, phenol can be found in numerous consumer products including mouthwashes, antiseptic ointments, throat lozenges, air fresheners, eardrops, and lip balms.

Phenol was prepared before World War I through the distillation of coal tar. The first synthetic process involved the sulfonation of benzene followed by desulfonation with a base. In this process, benzene sulfonic acid is prepared from the reaction of benzene and sulfuric acid:

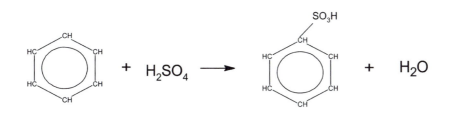

Benzene sulfonic acid is then reacted with base to produce phenol:

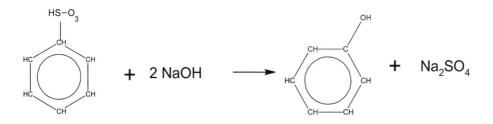

Another method developed by Dow chemical used the hydrolysis of chlorobenzene using NaOH:

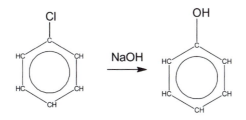

The most common current method of phenol production is from the cumene hydroperoxide rearrangement process. In this process, benzene reacts with propylene to produce cumene. Cumene is oxidized to cumene hydroperoxide. When cumene hydroperoxide is treated with dilute sulfuric acid, it rearranges and splits into phenol and acetone. Because the reactants are inexpensive and the process is simple, the acidic oxidation of cumene is used to produce more than 95% of the world's supply of phenol.

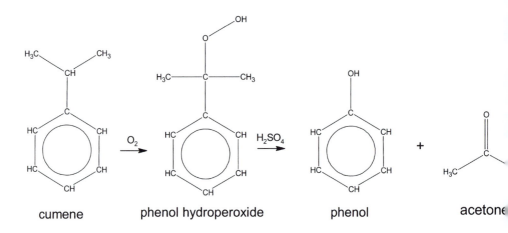

cumene   phenol hydroperoxide   phenol   acetone

The predominant use of phenol today is for phenolic resins (see Formaldehyde). Phenol was a chief component of the first synthetic plastic produced at the beginning of the 20th century. The material was produced in 1906 by a Belgian immigrant to the United States named Leo Hendrik Baekeland (1863–1944). Baekeland made a small fortune selling photographic paper to George Eastman (1854–1932), the founder of Eastman Kodak. With this money, Baekeland studied resins produced by combining phenol and formaldehyde in an autoclave and subjecting them to heat and pressure. Baekeland called the resin he produced Bakelite. Bakelite was a thermosetting phenol plastic. A thermosetting plastic hardens into its final shape on heating.

Phenol continues to be a primary chemical used to make thermoset resins. These resins are made by combining phenol with aldehydes such as formaldehyde. More than 4 billion pounds of phenolic resins are used annually in the United States. Phenolic resins find their widest use in the construction industry. They are used as binding agents and fillers in wood products such as plywood, particleboard, furniture, and paneling. Phenolic resins are impregnated into paper, which, after hardening, produces sheets that can be glued together to form laminates for use in wall paneling and countertops. Decking in boats and docks are made from phenolic resin composites. Phenolic resins are used as sealing agents and for insulation. Because phenolic resins have high heat resistance and are good insulators, they are used in cookware handles. Because they are also good electrical insulators, they are used in electrical switches, wall plates, and for various other electrical applications. In the automotive industry, phenolic resins are used for parts such as drive pulleys, water pump housings, brakes, and body parts.

In addition to the construction industry, phenol has many other applications. It is used in pharmaceuticals, in herbicides and pesticides, and as a germicide in paints. It can be used to produce caprolactam, which is the monomer used in the production of nylon 6. Another important industrial compound produced from phenol is bisphenol A, which is made from phenol and acetone. Bisphenol A is used in the manufacture of polycarbonate resins. Polycarbonate resins are manufactured into structural parts used in the manufacture of various products such as automobile parts, electrical products, and consumer appliances. Items such as compact discs, reading glasses, sunglasses, and water bottles are made from polycarbonates.

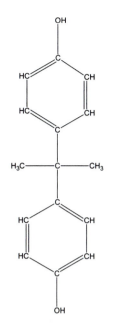

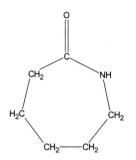

caprolactam

bisphenol A

71. Phosphoric Acid

CHEMICAL NAME = phosphoric acid
CAS NUMBER = 7664–38–2
MOLECULAR FORMULA = H_3PO_4
MOLAR MASS = 98.0 g/mol
COMPOSITION = H(3.1%) P(31.6%) O(65.3%)
MELTING POINT = 42.4°C
BOILING POINT = decomposes at 260°C
DENSITY = 1.88 g/cm³

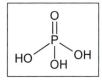

Pure phosphoric acid, also called orthophosphoric acid, is a clear, colorless, mineral acid with moderate strength. It is normally marketed as an aqueous solution of 75–85% in which it exists as a clear, viscous liquid. Although phosphoric acid is considered the compound H_3PO_4, phosphoric acids consist of a number of acids that may be formed from phosphoric acid. Heating phosphoric acid results in the combination of two or more H_3PO_4 molecules to form polyphosphoric acids by the elimination of water through condensation reactions. For example, diphosphoric ($H_4P_2O_7$), also called pyrophosphoric acid, is produced by heating phosphoric acid to about 225°C as shown in Figure 71.1. Tripolyphosphoric acid ($H_5P_3O_{10}$) and tetrapolyphosphoric acid ($H_6P_4O_{13}$) are made in a similar manner. Further heating and dehydrating can also create metaphosphoric, HPO_3, and phosphoric anhydride, P_4O_{10}. Metaphosphoric acid can be considered as a phosphoric acid molecule minus a water molecule. Metaphosphoric acid units combine to produce cyclic metaphosphoric acids. Phosphorus anhydride is expressed using the molecular formula P_2O_5 (diphosphorus pentoxide) and is also called superphosphoric acid. The compound P_2O_5 is used as the basis of rating the phosphorus content of fertilizers. Superphosphoric acid used by industry consists of a mixture of superphosphoric acid and polyphosphoric acids.

Phosphoric acid (orthophosphoric acid), as well as the different phosphoric acids, can ionize to form corresponding phosphates. When phosphoric acid loses its three hydrogen ions, it becomes the phosphate or orthophosphate ion, PO_4^{3-}. Likewise, when the polyphosphoric acids lose their hydrogen ions, the corresponding polyphosphate ion is produced. For example,

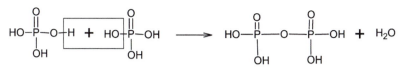

diphosphoric acid

Figure 71.1   Formation of diphosphoric acid.

pyrophosphoric acid loses four hydrogen ions to produce the pyrophosphate ion $P_2O_7^{4-}$ and PO^{3-} is the metaphosphate ion. The phosphoric acids can ionize partially to form intermediate ions.

Phosphoric acid was produced but not identified by alchemists in ancient times. It derives its name from the element phosphorus, which was discovered in 1669 by Henning Brand (1630–1710). Brand vaporized concentrated urine in the absence of oxygen and condensed the vapors obtaining a white waxy powder, which glowed in the dark. Phosphorus got its name from Latinized Greek words *fos* for light and *phéro* to carry, so phosphorus meant "to carry light." Calcium phosphate, $Ca_3(PO_4)_2$ was discovered in bone ash in 1769 by Karl Wilhelm Scheele (1742–1786) and Johann Gahn (1745–1718). Scheele subsequently isolated phosphorus from bone ash and produced phosphoric acid by reacting phosphorus and nitric acid. Scheele's method replaced bone as the main source of phosphorus rather than urine.

In the early 19th century the importance of fertilizers in agriculture was scientifically demonstrated, and bones became the main source of phosphorus for soil nutrients. Bone meal applied to soils to enrich their phosphorus content had limited efficacy because bone meal, which was largely $Ca_3(PO_4)_2$, had low solubility in water. Phosphates tend to precipitate out with calcium and other divalent and trivalent ions. A number of individuals were looking for methods to convert less soluble phosphate found in bones to more soluble forms. John Bennett Lawes (1814–1900) patented a process in 1841 of making superphosphate from bones and later extended his process to phosphates obtained from rock. Superphosphates are made by treating $Ca_3(PO_4)_2$ with sulfuric acid to make more soluble calcium hydrogen phosphates: $Ca_3(PO_4)_2 + 2H_2SO_4 \rightarrow Ca(H_2PO_4)_2 + 2CaSO_4$. In this reaction $Ca(H_2PO_4)_2$ is monobasic calcium phosphate, which is also called superphosphate. Calcium hydrogen phosphates (superphosphates) are more water soluble and therefore more readily available to plants.

Based on the work of Lawes and others, the commercial fertilizer industry was launched in the mid-19th century. The production of phosphate fertilizer starts with the making of phosphoric acid, which is then converted into phosphate salts. The first phosphoric acid plant was built in Germany around 1870 and the first plant in the United States in 1890. Phosphate rock is converted into phosphoric acid by either the wet process or thermal (dry) process, with the former accounting for more than 90% of world production. The wet process involves reacting dried crushed phosphate rock with concentrated (93%) sulfuric acid in a reactor to form phosphoric acid and dihydrate calcium sulfate (gypsum) or hemihydrate calcium sulfate (see Calcium Sulfate): $Ca_3(PO_4)_{2(s)} + 3H_2SO_{4(aq)} + 6H_2O_{(l)} \rightarrow 2H_3PO_{4(aq)} +$ $3[CaSO_4 \bullet 2H_2O]_{(aq)}$. The phosphoric acid and gypsum are separated by filtration, and a portion of the acid is recycled back to the reactor to increase the P_2O_5 concentration of the acid. Purification of the phosphoric acid is performed using organics, which is then separated from the organics in extraction columns. The phosphorus content of the final product is generally between 30% and 40%.

The thermal process yields much higher purity H_3PO_4 needed for pharmaceuticals, high-grade chemicals, and the food and beverage industry. The thermal process starts by smelting phosphate rock with silica (SiO_2) and coke in an electric furnace. Elemental phosphorus is then combusted with air at temperatures between 1,700°C and 2,700°C to produce P_2O_5, which is then hydrated with a dilute phosphoric acid spray in a chamber to produce a H_3PO_4 mist Concentrated phosphoric acid is recovered from the mist in the chamber. The thermal process produces a concentrated phosphorus solution with a P_2O_5 content of approximately 85%. The chemical reactions representing the thermal process are: $P_{4(l)} + 5O_{2(g)} \rightarrow P_2O_{5(g)}$ and $2P_2O_{5(g)} + 6H_2O_{(l)} \rightarrow 4H_3PO_{4(aq)}$.

Phosphoric acid is second only to sulfuric acid as an industrial acid and consistently ranks in the top 10 chemicals used globally. The amount of phosphate rock processed globally in 2005 was 160 million tons, with about 40 million tons processed in the United States. Florida is the leading producer of phosphate rock in the United States. The production of fertilizers dominates the use of phosphoric acid, accounting for more than 95% of its use in the United States, but it is used in a number of other applications. The use of phosphate and polyphosphates increased steadily after World War II as detergents replaced laundry soaps. Phosphates were used as builders and water softeners. A builder is a substance added to soaps or detergents to increase their cleansing power. The increased use of detergents with phosphate resulted in increasing loads of phosphorus to water bodies. Because phosphorus was a limiting factor in many aquatic ecosystems, the increased loads often resulted in algal blooms, loss of oxygen, fish fills, and cultural eutrophication. To mitigate the detrimental effects of phosphate on water bodies, many states and local governments took action to eliminate or reduce phosphate in detergents starting in the 1970s. After many states had established phosphate detergent bans, the manufacture of phosphate detergent for household laundry was voluntarily stopped in the mid-1990s.

Phosphoric acid is used as an intermediate in the production of animal feed supplements, water treatment chemicals, metal surface treatments, etching agent, and personal care products such as toothpaste. It is used as a catalyst in the petroleum and polymer industry. Phosphoric acid is used in food as a preservative, an acidulant, and flavor enhancer; it acidifies carbonated drinks such as Coca Cola and Pepsi, giving them a tangy flavor. Phosphoric acid is used as a rust remover and metal cleaner. Naval Jelly is approximately 25% phosphoric acid. Other uses for phosphoric acid include opacity control in glass production, textile dyeing, rubber latex coagulation, and dental cements.

72. Piperine

CHEMICAL NAME = 1-[5-(1, 3-Benzodiox-
ol-5-y1)-1-oxo-2, 4-Pentdienyl
CAS NUMBER = 94–62–2
MOLECULAR FORMULA = $C_{17}H_{19}NO_3$
MOLAR MASS = 285.3 g/mol
COMPOSITION = C(71.6%) H(6.7%)
N(4.9%) O(16.8%)
MELTING POINT = 130°C
BOILING POINT = decomposes
DENSITY = 1.19 g/cm³

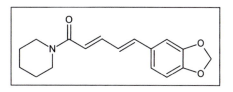

Piperine is an alkaloid responsible for the taste and flavor of pepper. In pure form it exists as a yellow to pale-yellow crystal, with a burning taste and pungent order. It is found naturally in plants in the Piperaceae family, particularly *Piper nigrum* (black pepper) and *Piper longum* (Indian long pepper). *Piper nigrum* is a perennial woody vine that grows to approximately 30 feet. It produces spikelike flowers that hold berries called peppercorns. Black pepper is obtained form the dried, immature berries of *Piper nigrum*; white pepper comes from the mature red berries of *Piper nigrum* after removing the outer skin and inner pulp to obtain the seed. Pepper is processed by harvesting the berries (peppercorns), using a threshing process to remove the berries from the spikes. Berries for black pepper are picked before they are ripe while still green. Berries are then cleaned and dried. During the drying process the berries ferment, turning to dark-brown-black. In small-scale operations, the berries are sun dried; in more elaborate operations ovens are used. Peppercorns are often sorted and graded before being ground manually or mechanically to produce the final pepper product.

The word *pepper* comes from the Indian Sanskrit word *pippali,* which became the Greek *péperi* and Latin *piper. Piper nigrum* is native to southwest India and Sri Lanka, where it has a long history of traditional use. Pepper was traded as a spice and used as a medicinal substance for at least 4,000 years. It was first cultivated in the present Indian state of Kerala, which is located at the southern tip of India along its southwest coast. Pepper was freely traded between

India, Southeast Asia, the Middle East, and northern Africa centuries before it was used by ancient Greeks and Romans. In ancient times, it was used as a spice, preservative, and medicine. Pepper was such a valuable commodity that it was often used as a medium of exchange and as a form of money. Centers of commerce such as Alexandria, Constantinople (Istanbul), Calicut, and Venice arose from the trade of pepper. Arab monopolies for pepper (and other spices) kept prices high, making pepper a spice for the wealthy. To break the Arab monopolies, European powers explored for direct trade routes to India. Portugal built on Vasco da Gama's (1469–1524) trade route completed in 1488 to India to establish its own pepper monopoly in the 16th century. This in turn prompted Spain to search for western routes to India.

Hans Christian Oersted (1777–1851), the Danish scientist best associated with his work on electromagnetism but trained as a pharmacist, studied plant alkalis in the first two decades of the 19th century. He reported work on pepper as early as 1809, isolating piperine from black pepper in 1819 and publishing his findings in 1820. Oersted extracted a resin from pepper plants with alcohol, formed a soluble salt by adding hydrochloric acid with alcohol, and then separated piperine from solution by precipitation and distillation. In 1882, Leopold Rügheimer (1850–1917) synthesized piperine from piperinic acid chloride and piperidine. The complete synthesis of piperine was reported in 1894 by Albert Ladenburg (1842–1911) and M. Scholtz. Ladenburg and Scholtz used piperonal ($C_8H_6O_3$) and acetaldehyde (CH_3CHO) to produce piperinic acid ($C_{12}H_{10}O_4$), which was then reacted with thionyl chloride ($COCl_2$) and piperidine ($C_5H_{11}N$) to produce piperine.

Pepper is the most widely used spice in the world. Its use as a seasoning was especially valued to mask offensive odors and tastes before modern preservation techniques that reduced spoilage were available. Pepper use is not limited to food; it is also added to alcoholic beverages. Pepper flavorings are used in some soft drinks. The spicy pepper taste is also obtained from other compounds; the most familiar of these is capsaicin. Capsaicin is responsible for the hot taste associated with various peppers: jalapeño, bell, chili, etc. The ability of piperine, capsalcin, and other compounds to produce a hot taste has to do with their structures. Piperine and capsaicin are classified as vanilloids because they contain a vanillyl or vanillyl-like group along with an amide group and a hydrocarbon tail. The vanilloid structure allows them to bind and activate vanilloid receptors in sensory neuron membranes. The receptors function as ion channels that selectively pass ions when vanilloids bind to them. The passage of ions produces a change in ion concentration inside and outside the neuron that produces a signal to the brain interpreted as a hot, spicy sensation. The vanilloid receptor responsible for this sensation is called TRPV1, which stands for transient receptor potential vanilloid (type) 1; TRPV1 regulates the passage of Ca^{2+} ions into and out of the nerve cell.

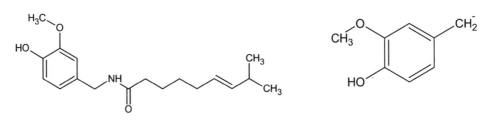

capsaicin   vanillyl

Capsaicin is 70 times hotter than piperine. The hotness of food has traditionally been quantified by the use of a scale called the Scoville Organoleptic Scale, named after Wilbur L. Scoville (1865–1942), a pharmacist who derived the scale in 1912. Scoville's original method was based on progressively diluting a solution of a pepper or another food extract with water sweetened with sugar and then having a panel of tasters rate when the hot taste could no longer be sensed. The amount of dilution needed to eliminate the hot taste gives the Scoville rating. Therefore Tabasco sauce with a rating of 5,000 requires a 5,000:1 dilution to eliminate the hot taste. The original Scoville scale was highly subjective. Modern methods use high-performance liquid chromatography (HPLC) to determine the heat rating of a substance. Specific concentrations of various compounds are determined and equated to capsaicin standard. The HPLC method calculates the degree of heat in ASTA (American Spice Trade Association) pungency units, which can be converted to Scoville Units (a Scoville unit equal to 15 times the ASTA value). Scoville ratings range from 0 to 16,000,000 for pure capsaicin. Scoville ratings are highly variable, with food ingredients being rated over a range; for example, jalapeños are rated between 3,000 and 8,000 Scoville Heat Units (SHU); cayenne pepper rates between 30,000 and 80,000 SHU. The hottest chilies have SHU ratings between 500,000 and 900,000.

The antimicrobial activity of black pepper is generally weak and limited to specific organisms. Because of this, it has limited use as a preservative, especially when compared to other substances such as salt, mustard, and cinnamon. It is used in combination with other spices as a preservative and is effective against specific bacteria; for example, several species in the genus *Lactobacillus* (lactic acid bacteria found in dairy products). Piperine is used in granular formulations as a pesticide against small animals such as dogs, cats, skunks, raccoons, and squirrels. Piperine pesticides are nontoxic and operate as a repellant when animals smell or test them. Piperine is also incorporated with other compounds to make insecticides; it is effective against houseflies, lice, and various other pests.

Piperine has been used medicinally for thousands of years and this continues today. It is used to treat asthma and chronic bronchitis. It also has putative analgesic and anti-inflammatory properties that stem from its antioxidant properties. Research is examining the use of piperine in treating malaria. Antiepilepsirine is a derivative of piperine that is used to treat different types of epilepsy, especially in China. A current area of interest is the efficacy of piperine in increasing the bioavailability of certain nutrients and drugs. It is thought to possibly aid digestion and increase the absorption in the digestive tract of numerous drugs used as cancer treatments, antihistamines, steroidal inflammation reducers, and antibiotics. Studies have shown that the administration of piperine leads to higher blood serum levels of many drugs and nutrients, but the exact mechanism is not understood and under investigation.

73. Potassium Carbonate

CHEMICAL NAME = potassium carbonate
CAS NUMBER = 584–08–7
MOLECULAR FORMULA = K_2CO_3
MOLAR MASS = 138.2 g/mol
COMPOSITION = K(56.6%) C(8.7%) O(34.7%)
MELTING POINT = 891°C
BOILING POINT = decomposes
DENSITY = 2.3 g/cm³

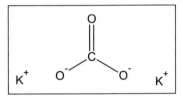

Potassium carbonate is a white, crystalline, salt that forms basic aqueous solutions used in the production of fertilizer, glass, ceramics, explosives, soaps, chemicals, and wool treatments. It was the main compound once referred to as potash, although the term today is not reserved exclusively for potassium carbonate, but for several potassium salts. In the fertilizer industry potash refers to potassium oxide, K_2O, rather than potassium carbonate. Pearlash is a purer form of potash made by heating potash to remove impurities. The name potash comes from the traditional method of making potassium carbonate, which has been performed since ancient times. Preparing potassium carbonate involved burning wood or other plant material, leaching the ashes in a wooden barrel covered on its bottom with straw, and then evaporating or boiling away the water in the leachate in clay or iron pots to recover potassium and sodium alkalis.

Potash was a valuable commodity used for items such as glass, soap, dyes, and gunpowder. Pearlash was used as a leavening agent before sodium bicarbonate was used. Russia and Sweden forests were the traditional sources of European potash before the middle of the 1700s. As the forests of Europe became depleted and colonists acquired knowledge on potash production, Europeans turned to North American colonists to supply potash. The vast forests of North America supplied the wood needed to support potash production. England used large quantities of potash in colonial times for glass making, soap, dyeing, and saltpeter; but the main need was to scour wool, which involved cleaning the wool to remove dirt, grease, and contaminants. By the mid-18th century, England used several thousand tons of potash each year. As American colonists spread west and cleared land for farming, burning the wood and stumble to produce potash became a valuable source of income. Up to 100 bushes of ashes worth roughly $5 could be obtained from an acre of land; potash distributors could make $75 from a 500-pound barrel of potash. Hardwoods such as elm, ash, sugar maple, hickory, beech, and basswood were particularly good sources of high-grade potash. By the end of the 18th century, North America's coastal region was largely settled and its forests were already being depleted. Efforts were made to increase potash production as the preferred woods for its production became scarcer. The first patent ever issued in the United States was granted in 1790 to Samuel Hopkins (1743–1818) of Philadelphia (many references mistakenly attribute the patent to Samuel Hopkins of Pittsford, Massachusetts) for "the making of Pot ash and Pearl ash by a new Apparatus and Process." Hopkins's method burned the ashes a second time in a redesigned furnace to more completely oxidize the carbon and increase the carbonate yield. Hopkins also recycled the residue left after leaching the raw ashes to increase the potash yield. In exchange for using his process, Hopkins required $50 down and another $150 over a five-year period; alternatively, the user of his patent could put 50 pounds of potash down with 150 pounds due over the next 5 years.

The United States was a primary supplier of the world's potash for the first half of the 19th century. In 1851, deposits of soluble potassium salts were discovered in northern Germany near Stassburg and in 1861 production began. The Stassburg deposits became the primary source of world potash, but this consisted of a mix of potassium salts, rather than potassium carbonate, with high concentrations of potassium chloride (KCl) and potassium sulfate (K_2SO_4). Also, the introduction of industrial processes to produce less expensive soda ash (Na_2CO_3), which could be used in place of potash for many applications, ended the traditional method of potash production using wood ashes (see Sodium Carbonate). The development of electrolytic chemical production methods at the end of the 19th century established the modern method for potassium carbonate. This involves the electrolytic conversion of a potassium chloride (KCl) solution into potassium hydroxide (KOH), chlorine gas (Cl_2), and hydrogen gas (H_2). Potassium hydoxide is then reacted with carbon dioxide to produce sodium carbonate: $2KOH_{(aq)} + CO_{2(g)} \rightarrow K_2CO_{3(aq)} + H_2O_{(l)}$.

Today potassium carbonate is still used for numerous applications. Its primary use is in the production of specialty glasses and ceramics. It is used to make optical glass, glass used for video screens in televisions and computers, and laboratory glassware. Its is used in certain glasses rather then cheaper sodium carbonate owing to its better compatibility with lead, barium, and strontium oxides incorporated in these glasses. These oxides lower the melting point of glass and produce a softer glass. Potassium carbonate has a higher refractive index than sodium carbonate producing a more brilliant glass. Also, K_2CO_3 glass has greater electrical resistivity and the ability to withstand greater temperature changes. Potassium carbonate is a common flux combined with titanium dioxide to produce frits used in ceramics. A frit is a calcined mixture of fine silica, a pigment, and a flux that is ground a specific particle size and used to produce glazes, enamels, and additives in glass making.

Potassium carbonate is used in agriculture and food production. Agriculturally, K_2CO_3 is used as a drying agent in agriculture for alfalfa hay and other legumes. The potassium carbonate is sprayed on the crop stems at cutting to reduce field curing time. It functions by modifying the waxy cuticle layer of the plants so that it is more permeable to water. It is also applied to grapes to decrease drying time in the production of raisins. Potassium carbonate is used as a spray or drip fertilizer and also as a constituent of compound fertilizers. Its high water solubility and alkaline property make it useful for supplying potassium to acidic soils, especially in vineyards and orchards. Dutch-processed cocoa uses potassium carbonate as an alkalizing agent to neutralize the natural acidity of cocoa. It is used to produce food additives like potassium sorbate and monopotassium phosphate.

Potassium carbonate is used in the chemical industry as a source of inorganic potassium salts (potassium silicates, potassium bicarbonate), which are used in fertilizers, soaps, adhesives, dehydrating agents, dyes, and pharmaceuticals. Potassium carbonate used to make potassium lye produces soft soaps, which are liquids or semisolids rather than solids. Other uses of potassium carbonate includes use as a fire suppressant in extinguishers, as a CO_2 absorbent for chemical processes and pollution control, an antioxidant in rubber additives, and in pharmaceutical formulations.

74. Potassium Nitrate

CHEMICAL NAME = potassium nitrate
CAS NUMBER = 7757–79–1
MOLECULAR FORMULA = KNO_3
MOLAR MASS = 101.1 g/mol
COMPOSITION = K(38.7%) N(13.8%) O(47.5%)
MELTING POINT = 334°C
BOILING POINT = decomposes at 400°C
DENSITY = 2.1 g/cm³

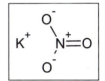

Potassium nitrate is a solid, colorless, crystalline ionic compound that exists as the mineral niter. Potassium nitrate is also known as saltpeter. The name saltpeter comes from the Latin *sal petrae*, meaning salt of stone or salt of Petra. Petra is an area in Jordan where large deposits of camel dung heaps provided a source for saltpeter production. The term *saltpeter* or Chilean saltpeter is also used for sodium nitrate, $NaNO_3$; a major source of this compound was Chile. In this discussion, saltpeter is considered to be potassium nitrate and not sodium nitrate. Other names associated with potassium nitrate included nitre and nitre of potash. Potassium nitrate is used as a fertilizer, to preserve meat, for preparation of nitrogen compounds, and as an oxidizer in propellants and explosives. Saltpeter's most prominent use in human history is as the principal ingredient in gunpowder.

The exact date of the invention of gunpowder is unknown, but it is generally thought to have occurred around 800 c.e. in China or possibly in the Arab world. Ancient alchemists' search for elixirs resulted in the discovery of the first gunpowder known as black powder. The use of black powder in weapons first appeared in Chinese writings dating from the 11th century. Roger Bacon did extensive studies of black powder and is often credited with introducing Europe to its uses in the 13th century. Black powder is a mixture of saltpeter, charcoal, and sulfur. Although the exact proportions of ingredients in black powder vary, a mixture of approximately 75% saltpeter, 15% charcoal, and 10% sulfur produces a typical powder.

The potassium nitrate used in gunpowder was originally obtained from natural mineral deposits of niter. Small quantities formed as efflorescence deposits on damp stone walls were identified as early as 2000 B.C.E. in Sumerian writings. As the use of black powder expanded with the development of weapons, the demand for saltpeter exceeded supply. This was exacerbated during times of war. To meet the demand for saltpeter to produce black powder, a saltpeter industry developed that followed prescribed methods to produce large quantities of saltpeter. The method depended on processing dirt obtained from areas where nitrates would naturally form (Figure 74.1). These were areas in which animal waste had accumulated such as the dirt floors of barns, stables, herding pens, caves, or cellars. The ammonia compounds in the urine and fecal wastes in these areas underwent nitrification to produce nitrates, which combined with potassium in the soil to form saltpeter. Nitrification involves the conversion of ammonia (NH_3) and ammonium (NH_4^+) compounds into nitrite (NO_2^-) and nitrate (NO_3^-):

$$NH_3 + O_2 \rightarrow NO_2^- + 3H^+$$
$$NO_2^- + H_2O \rightarrow NO_3^- + 2H^+$$

Figure 74.1   Saltpeter production in the Middle Ages. Soil saltpeter is leached with water in a tub (B) that contains a layer of sand in the bottom and a plug (C). Filtrate is collected in another tub (D) and is reduced in a copper vat (A), which is heated over a hearth. Reduced solution s placed in crystallization containers (E) where saltpeter crystallizes on copper sticks. From *De Re Metallica* (On the Nature of Metals) by Georgius Agricolla (1494–1555), published in 1556.

The process of saltpeter production involved mixing the dirt with ashes to keep the pH neutral and then refining and separating the saltpeter.

As the demand for saltpeter increased, large saltpeter processing areas called saltpeter plantations developed. The large-scale production of saltpeter involved creating nitre beds, which were boxes in which dung and plant organic matter were piled. The nitre beds required addition of lime or other basic materials to neutralize acids that formed in the process. Beds were treated with water to keep them moist, but they had to be protected from the elements. The latter was important to keep nitrates from washing away. The watering process often used urine or dung water from the cleaning of pens and stables to promote production. The beds were frequently stirred and the incorporation of plant material increased the organic matter and the porosity of the heap. The latter was important to make sure the waste had adequate exposure to oxygen. The nitre beds were similar to composting. After a period lasting from several months to several years, the treated wastes were leached, and the leachate was further processed to increase the production of potassium nitrate as opposed to other nitrates such as magnesium or sodium nitrate. This step involved treatment with ashes, which were a source of potash (potassium carbonate). The final step involved separating the saltpeter out of solution from other salts. Because saltpeter is much more soluble than common salt (NaCl) in hot water, the solution was boiled and saltpeter was concentrated in the liquid portion while NaCl precipitates were removed. Saltpeter's solubility decreases quickly as temperature decreases, but common salt's solubility is fairly constant with temperature. Therefore once the initial separation process was over, the liquid was cooled and saltpeter precipitated. Even after the separation process, the saltpeter contained impurities and a typical product might consist of only 75% saltpeter. This amount was generally sufficient for general purposes, but this product could be further refined using various techniques to obtain high-grade saltpeter. Such techniques were used when high-quality black powder was desired such as that produced by the French under Antoinne Lavoisier (1743–1794) in the late 18th century and his apprentice, Eleuthè Irénée du Pont (1771–1834), who fled to America. The du Pont gunpowder plant developed into the largest chemical company in the United States.

One of most common locations for producing saltpeter in the United States was in limestone caves. The soils in these caves contained bat guano that had accumulated over thousands of years, and the caves themselves contained natural alkaline soils and were protected from the elements. Both small and massive saltpeter production centers were scattered throughout areas that contained these caves, such as Kentucky, Tennessee, and Indiana. Saltpeter production in these areas played a significant role during times of war such as the War of 1812 and the Civil War. Countries that lacked sufficient saltpeter production to support their needs had to rely on importation of saltpeter from an ally or colony. Chilean saltpeter from the Atacama Desert supplied much of the world's saltpeter starting in the 1830s. Chilean saltpeter was sodium nitrate and not potassium nitrate. Sodium nitrate could be converted to potassium nitrate by reacting it with potassium chloride, $KCl_{(aq)}$: $NaNO_{3(aq)} + KCl_{(aq)} \rightarrow KNO_{3(aq)} + NaCl_{(aq)}$. The saltpeter was obtained as it precipitated as the solution cooled.

Although the most prominent use of saltpeter is for the production of black powder, potassium nitrate is also used as fertilizer. In the first half of the 17th century, Johann Rudolf Glauber (1604–1668) obtained saltpeter from animal pens and discovered its use to promote plant growth. Glauber included saltpeter with other nutrients in fertilizer mixtures. Glauber's work was one of the first to indicate the importance of nutrient cycling in plant nutrition.

The Germans' need to supplant Chilean saltpeter supply, which could be cut off by enemy blockades, led to the search for methods to synthesize nitrates. The reaction required a supply of ammonia, which was economically synthesized by Fritz Haber (1868–1934) before World War I (see Ammonia). Ammonia could then be converted to nitric acid through the Ostwald process and then nitric acid can be reacted with bases to produce nitrates (see Nitric Acid):

$$KOH_{(aq)} + HNO_{3(aq)} \rightarrow KNO_{3(aq)} + H_2O$$

75. Propane

CHEMICAL NAME = propane
CAS NUMBER = 74–98–6
MOLECULAR FORMULA = C_3H_8
MOLAR MASS = 44.1 g/mol
COMPOSITION = C(81.7%) H(18.3%)
MELTING POINT = –187.6°C
BOILING POINT = –42.1°C
DENSITY = 1.30 g/L (vapor density = 1.52, air = 1)

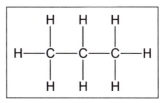

Propane is a colorless, odorless, flammable gas that follows methane and ethane in the alkane series. The root word *prop* comes from the three-carbon acid propionic acid, CH_3CH_2COOH. Propionic acid comes from the Greek words *protos* meaning first and *pion* meaning fat. It was the smallest acid with fatty acid properties. Propane is the gas used to fuel barbecues and camp stoves giving it the common name bottled gas. It is marketed as liquefied petroleum gas (LPG) or liquefied petroleum; it should be noted that LPG is often a mixture that may contains butane, butylene, and propylene in addition to propane. In addition to cooking, propane can be used as an energy source for space heating, refrigeration, transportation, and heating appliances (clothes dryer). Propane is a $10 billion industry in the United States, and the United States consumes approximately 15 billion gallons of propane annually.

Propane can be stored as liquid in pressurized (approximately 15 atmospheres) storage tanks and/or at cold temperatures and vaporizes to a gas at atmospheric pressure and normal temperatures. This makes it possible to store a large volume of propane as a liquid in a relatively small volume; propane as a vapor occupies 270 times the volume of propane in liquid form. This makes liquid propane an ideal fuel for transport and storage until needed.

Bottled gas was sold as early as 1810 in England, but propane was discovered at the start of the 20th century with the development of the natural gas industry. Problems in natural gas distribution were attributed to condensation of particular fractions of natural gas in pipes. This led to the removal of these higher boiling point gases and a search for applications for these gases. Concurrent with the development of the natural gas industry, society's increasing use of automobiles began. Volatile gasoline used in early cars vaporized appreciable amounts of hydrocarbons leading to a continual expense to early car owners. Walter Snelling (1880–1965), a research chemist working for the U.S. Bureau of Mines in 1910, investigated emissions from Ford Model-T gasoline tanks. Snelling developed a still for separating liquid and gaseous fractions of gasoline as he investigated methods to control emissions. Thus Snelling was able to separate propane (and butane) from natural gas. Snelling discovered that propane would condense at moderate pressures and developed a method for bottling liquid propane. By 1912, Snelling and others found that propane could be used as a fuel for cooking and lighting and as a source for cutting torches. Snelling joined with associates to found the American Gasol Company, which was the first commercial provider of propane. A year later, Snelling sold his patent for propane for $50,000 to Frank Phillips (1873–1950), founder of Phillips Petroleum.

Propane has been used as a transportation fuel since its discovery. It was first used as an automobile fuel in 1913. It follows gasoline and diesel as the third most popular vehicle fuel and today powers more than half a million vehicles in the United States and 6 million worldwide. The widespread use of propane is hampered by the lack of a distribution system, but it has been used to fuel fleets of buses, taxis, and government vehicles. Also, it is heavily used to power equipment such as forklifts. Propane is cleaner burning than gasoline or diesel and has been used to reduce urban air pollution. Compared to gasoline it emits 10–40% of the carbon monoxide, 30–60% of the hydrocarbons, and 60–90% of the carbon dioxide. An advantage of cleaner burning propane is that engine maintenance is improved because of lower engine deposits and fouling. Propane's octane ratings range between 104 and 110. The lower emissions are somewhat compromised by propane's lower energy value; propane has about 75% of the energy content of gasoline when compared by volume. Propane is separated from natural gas and is also produced during petroleum processing. Approximately 53% of the propane produced in the United States comes from the small fraction (less than 5%) found in natural gas and the remainder comes petroleum refining.

Propane's greatest use is not as a fuel but in the petrochemical industry as a feedstock. As an alkane, it undergoes typical alkane reactions of combustion, halogenation, pyrolysis, and oxidation. Pyrolysis or cracking of propane at several hundred degrees Celsius and elevated pressure in combination with metal catalysts result in dehydrogenation. Dehydrogenation is a primary source of ethylene and propylene:

$$CH_3CH_2CH_3 \rightarrow C_2H_4 + H_2$$
$$CH_3CH_2CH_3 \rightarrow C_3H_6 + H_2.$$

Because propane contains three carbon atoms, halogenation produces two isomers when propane is reacted with a halogen. For example, when propane is chlorinated the chlorine can bond to one of the terminal carbons to produce 1-chloropropane: $CH_3CH_2CH_3$ $\xrightarrow{Cl_2}$ $CH_3CH_2CH\text{-}Cl$ or the chlorine can bond to the central carbon to give 2-chloropropane:

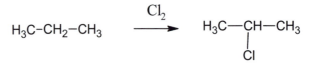

Propane demonstrates that the carbon atoms have different characteristics in alkanes with more than two carbon atoms. The terminal carbon atoms in propane are bonded to three hydrogen atoms and one carbon atom. A carbon atom bonded to only one other carbon atom is referred to as a primary or 1° carbon. The central carbon atom in propane is bonded to two other carbon atoms and is called a secondary or 2° carbon. A hydrogen atom has the same classification as the carbon atom to which it is attached. Thus the hydrogen atoms attached to the terminal carbon atoms in propane are called primary (1°) hydrogens, whereas the central atom has secondary (2°) hydrogen. The difference in bonds leads to differences in reactions and properties of different isomers. For example, breaking a primary bond requires more energy than breaking a secondary bond in propane. This makes formation of the isopropyl radical $CH_3CHCH_3\bullet$ easier than the n-propyl radical, $CH_3CH_2CH_2\bullet$. Even though the formation of the isopropyl is more favorable energetically, the greater number of primary hydrogen atoms leads to approximately equal amounts of n-propyl and isopropyl radicals formed under similar reaction conditions.

Oxidation of propane can produce various oxygenated compounds under appropriate conditions, but generally alkanes are relatively unreactive compared to other organic groups. Some of the more common oxidation products include methanol (CH_3OH), formaldehyde (CH_2O), and acetaldehyde (C_2H_4O). Propane can be converted to cyclopropane by conversion to 1,3 dichloro-propane using zinc dust and sodium iodine $ClCH_2CH_2CH_2Cl \xrightarrow{\text{Zn, NaI}}$ cyclopropane.

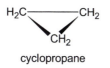

cyclopropane

76. Propylene

CHEMICAL NAME = propene
CAS NUMBER = 115-07-1
MOLECULAR FORMULA = C_3H_6
MOLAR MASS = 42.1 g/mol
COMPOSITION = C(85.6%) H(14.4%)
MELTING POINT = −185.2°C
BOILING POINT = −47.6°C
DENSITY = 1.95 g/L (vapor density = 1.49, air = 1)

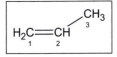

Propylene is a colorless, flammable gas that follows ethylene as the second simplest alkene hydrocarbon. It has an odor similar to garlic and has wide use in the chemical industry as an intermediate in the synthesis of other derivatives such as polypropylene, propylene oxide, isopropyl alcohol, acetone, and acrylonitrile. The production of propylene is similar to ethylene and is obtained through steam cracking of hydrocarbon feedstocks. Steam cracking is a process used to break molecules into smaller molecules by injecting the catalysts with steam.

Accelerated production and use of polypropylene began in the late 1950s when the discovery of Ziegler-Natta catalysts made large-scale polymerization of propylene economically feasible. The polymerization of propylene leads to several different structures that vary in their properties based on their tacticity. Tacticity, derived from the Greek word *tactos* meaning ordered, refers to how groups are arranged in a polymer. The general structure of the polypropylene molecule can be pictured as polyethylene in which a methyl (CH_3) group has replaced a hydrogen atom in each monomer. Three general structures for polypropylene are termed isotactic, syndiotactic, and atactic. These are depicted in Figure 76.1. The structures are shown using conventional symbols to depict the three-dimensional character of the molecules. Solid wedges represent bonds that lie above or out from the plane of the page, and dashed lines are used to represent bonds that lie behind the page. The isotactic structure shows a regular pattern, with the methyl groups on every other carbon in the chain coming out from the page. This means that all methyl groups lie on the same side of the molecule. In the syndiotactic structure, the methyl groups alternate regularly above and below the plane of the page, which means they are on opposite sides of the chain. The atactic structure displays a random pattern of methyl groups above and below the plane of the page.

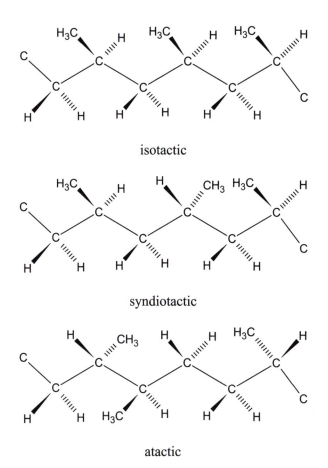

isotactic

syndiotactic

atactic

Figure 76.1   Tacticity forms of polypropylene.

The different forms of polyethylene, such as low-density polyethylene (LDPE) and high-density polyethylene (HDPE), dictate its physical properties. In the polymerization process, the isostatic structure forms helical coils that allow tight packing, resulting in a highly ordered crystalline structure and producing a hard, strong, stiff plastic with a high-melting-point. Conversely, the random atactic configuration prevents a tight structure resulting in an amorphous, soft substance. Polypropylene production accelerated around 1960 with the advent of Ziegler-Natta catalyst to control the polymerization process. In the last 20 years, a new group of catalysts called metallocene catalysts have resulted in significant advances in the propylene industry. Metallocene catalysts consist of a transition metal, such as titanium or zirconium, sandwiched between carbon rings.

Metallocene catalysts have allowed greater control and advances in polymerization. The polymerization of isostatic propylene up to about 1995 resulted in a structure with approximately 5% atactic polypropylene. Metallocene catalysts made it possible to produce 100%

isostatic or syndiotactic polypropylenes. The catalysts have also allowed chemists to control the chain length of polypropylene tacticities in polymers to produce various polypropylenes with a range of physical and chemical characteristics. For example, rubbery elastomer polypropylene results by producing atactic polyethylene chains with regions of isostatic polypropylene interspersed along the chain. The isotactic regions are areas along the chain where greater attraction and packing between molecules takes place, resulting in cross-linking of the chains. This is similar to the vulcanization process in rubber. Thus a soft flexible polypropylene is produced. Polypropylene is also co-polymerized with polyethylene to expand its applications.

Polypropylene products can be identified with the recycling symbol containing the number 5 with the letters PP underneath the symbol. Several important characteristics of polypropylene are compared to polyethylene in Table 76.1.

Table 76.1   Comparison of Polypropylene and Polyethylene

	Polypropylene	**LDPE**	**HDPE**
Density (g/cm³)	0.905	0.920	0.950
Melting point (°F/°C)	327/164	230/110	260/125
Hardness (Shore D)	D75	D45	D70
Tensile strength (psi)	4,800	2,000	4,500

Low-density polyethylene, LDPE; high high-density polyethylene, HDPE.

Polypropylene is closer to HDPE in its properties. Polypropylene is more heat resistant than polyethylene, and its higher melting point makes it preferable for items subjected to heat such as dishwashers. It is also used extensively for containers of dairy projects. Familiar plastic containers holding yogurt, butter, margarine, and spreads are generally made of polypropylene. Another advantageous property of polypropylene is that it is resistant to many solvents, acids, and bases. This makes it an ideal for several common applications: the casing of car batteries, truck bed liners, outdoor carpet and welcome mats, tops for plastic bottles, storage tanks, car trim and paneling, and toys.

Polypropylene is also used extensively in fiber form in textile applications. One third of polypropylene's production in the United States is used as fiber and the worldwide use of fiber polypropylene was approximately 2.5 million tons in 2005. Major uses of fiber polyethylene are carpeting, upholstery, paper and packaging, construction fabric liners, diapers, and rope.

Propene is used as a starting material for numerous other compounds. Chief among these are isopropyl alcohol, acrylonitrile, and propylene oxide. Isopropyl alcohol results from the hydration of propylene during cracking and is the primary chemical derived from propylene. Isopropyl alcohol is used as a solvent, antifreeze, and as rubbing alcohol, but its major use is for the production of acetone. Acrylonitrile is used primarily as a monomer in the production of acrylic fibers. Polymerized acrylonitrile fibers are produced under the trade names such as Orlon (DuPont) and Acrilan (Monsanto). Acrylonitrile is also a reactant in the synthesis of dyes, pharmaceuticals, synthetic rubber, and resins. Acrylonitrile production occurs primarily through ammoxidation of propylene: $CH_3 - CH = CH_2 + NH_3 + 1.5 \, O_2 \rightarrow CH_2 = CH - C \equiv N + 3 \, H_2O$.

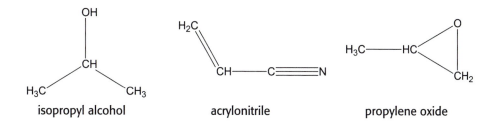

isopropyl alcohol

acrylonitrile

propylene oxide

Propylene oxide is produced from the chlorohydrination of propene similar to the process used to make ethylene oxide (see Ethene). A major use of propylene oxide involves hydrating propylene oxide to produces propylene glycol, propylene polyglycols, and other polyether polyols. These products are used to produce both rigid and flexible polyurethane foams, but they are also used to produce polyurethane elastomers, sealants, and adhesives.

77. Quinine

CHEMICAL NAME = (2-ethenyl-4-azabicyclo[2.2.2]
    oct-5-yl)-(6-methoxyquinolin-4-yl)-methanol
CAS NUMBER = 130–95–0
MOLECULAR FORMULA = $C_{20}H_{24}N_2O_2$
MOLAR MASS = 324.4 g/mol
COMPOSITION = C(74.0%) H(7.5%) N(8.6%) O(9.9%)
MELTING POINT = 177°C
BOILING POINT = decomposes
DENSITY = 1.21 g/cm³

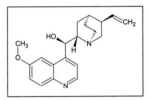

Quinine is a white crystalline alkaloid best known for treating malaria. Quinine is derived from the bark of several species of trees in the genus *Cinchona* in the Rubiaceae family. Cinchona trees grow on the eastern slopes of the Andes Mountains at elevations of several thousand feet. Early explorers and missionaries noted that South American natives used the ground bark of Cinchonas to treat fever, cramps, and the chills. Because these symptoms were associated with malaria, Cinchona bark powder was recognized as a possible treatment in the 1600s by Jesuit missionaries. Jesuits used the powder to prevent and treat malaria. Missionaries returning to Europe brought back the powder from the "fever tree," which was colloquially called "Jesuit powder," and its use as a medicine was endorsed by Vatican officials. Jesuit powder mixed with water or alcohol became a prevalent remedy for fever and malaria in Catholic countries, but in many Protestant areas its use was shunned.

Malaria is a parasitic disease that has been prevalent throughout human history. The word *malaria* comes from the Italian words *mala aria* meaning "bad air." It was originally associated with air emanating from swamps and wet places. Malaria results in high fever, chills, headache, and weakness in its victims. The disease is caused by four species in the protozoa genus *Plasmodium.* The exact cause of malaria was not determined until the 1880s when Charles Louis Alphonse Laveran (1845–1922) discovered the protozoa responsible for the disease. Subsequently, Ronald Ross (1857–1932) determined the entire life cycle of the parasite that produced malaria. Ross was awarded the Nobel Prize in physiology or medicine in 1902, and Laveran received the same recognition in 1907.

People are exposed to malaria through contact with the female *Anopheles* mosquito. When a person is bitten by an infected mosquito, it injects asexual *Plasmodium* sporozoite cells (along with anticoagulant and saliva) into the blood of the victim. The sporozoites travel to the liver where they divide asexually into merozoites. The merozoites multiply in the liver and enter the bloodstream, invading the red blood cells where they enlarge and reproduce, rupturing the cell. This process releases toxins into the body, producing the symptoms associated with malaria. Today malaria is estimated to infect between 300 and 500 million people, with more than 1 million deaths occurring annually. Malaria is most prevalent in tropical and subtropical Africa and Asia where the Anopheles mosquito thrives. Malaria has largely been eradicated in temperate and developed countries.

Although malaria continues to be a major disease affecting millions, it was even more widespread in recent history. As the demand for Cinchona bark powder used for treating malaria in Europe and European colonies increased, Cinchona trees disappeared in many of its growth areas. Spain cornered the market for Cinchona bark and its European rivals attempted to establish the trees in their own countries or colonies. Smuggled Cinchona seedlings resulted in many unsuccessful attempts by the English, Italian, Dutch, and French to establish Cinchona plantations. The Dutch finally established successful plantations in Java in the 1860s.

As merchants tried to increase the supply of Cinchona, scientists sought to isolate the active ingredient in Cinchona responsible for its antimalarial properties. After decades of work by numerous investigators, quinine was finally isolated in 1820 by Pierre-Joseph Pelletier (1788–1842) and Joseph-Bienaimé Caventou (1795–1877). The name quinine originates from the native word for the Cinchona tree *quina quina,* which became the Spanish word quino for cinchona. The development of organic synthesis in the middle of the 19th century and the limited supply of quinine stimulated attempts to synthesize it. William Henry Perkins's (1838–1907) attempt to synthesize quinine in 1856 led to his discovery of mauve, which was a significant discovery in the dye industry (see Indigo). Quinine's synthesis was still incomplete as World War II commenced. The war provided further motivation to synthesize quinine owing to the large number of Allied troops fighting in the tropical Pacific and Japanese control of Cinchona plantations in the Indo-Pacific. The first complete synthesis is often credited to Robert Burns Woodward (1917–1979) and William von E. Doering (1917–) working at Harvard in 1944. Woodward and Doering synthesized the compound D-quinotoxine from which quinine could supposedly be synthesized from an already documented process, but there was no evidence that the correct stereoisomer of quinine had been isolated. The total synthesis of quinine with stereochemical control is credited to Gilbert Stork (1921–) in 2001.

Quinine's use as an antimalarial agent spans several hundred years, but it has been replaced in recent years by other substances such as chloroquine. Because some *Plasmodium* strains have developed resistance to several malaria medications, quinine use is being revived. About 60% of quinine production is used for medicinal purposes, and the drug is available by prescription. In addition to its use as an antimalarial agent, quinine medications are used to treat leg cramps, muscle cramps associated with kidney failure, hemorrhoids, heart palpitations, and as an analgesic. At higher concentrations it is toxic and causes a condition known as cinchonism. Conditions associated with cinchonism include dizziness, hearing loss, visual impairment, nausea, and vomiting.

Nonmedicinal use of quinine, accounting for about 40% of its use, is primarily as a flavoring agent in condiments and liqueurs. The most common food use of quinine is tonic water. Tonic water originated in India where English colonists drank carbonated water mixed with quinine to prevent malaria. The bitter taste of quinine was often masked by mixing it with alcoholic beverages; one result of this practice was the drink gin and tonic. Current Food and Drug Administration regulations in the United States limit the amount of quinine in tonic water to 83 parts per million (83 mg per liter). This level is significantly less than that required for therapeutic purposes, so the use of commercial tonic waters to combat malaria is not practical.

78. Saccharin

CHEMICAL NAME = 1,2-benzisothiazol-3(2*H*)-one 1,1-dioxide
CAS NUMBER = 81–07–2
MOLECULAR FORMULA = $C_7H_5NO_3S$
MOLAR MASS = 183.2 g/mol
COMPOSITION = C(45.9%) H(2.7%) N(7.7%) O(26.2%) S(17.5%)
MELTING POINT = 229°C
BOILING POINT = decomposes
DENSITY = 0.83 g/cm³

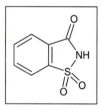

Saccharin is the oldest and one of the best-known artificial sweeteners. It was accidentally discovered in 1878 by Ira Remsen (1846–1927) and his postdoctoral research fellow Constantin Fahlberg (1850–1910) at Johns Hopkins University. Constantin, who was working on toluene derivatives from coal tar, noticed a sweet taste on his hand one evening at dinner. He traced the taste back to the oxidized sulfonated chemicals he was working with and determined it was a sulfonated amide benzoic acid compound. Remsen and Fahlberg jointly published their findings on the compound in 1879 and 1880 in American and German journals. During the next several years, Remsen continued his academic work as one of the world's leading chemists, and Fahlberg perfected methods for commercialization of saccharin. Fours year after they published their work, Fahlberg and his uncle, Adolf List, applied for a United States patent for the compound, which was granted in 1885 (U.S. patent number 319082). Fahlberg named the compound saccharine from the Latin word *saccharum* for sugar. Fahlberg did not inform Remsen of his plan to commercialize saccharine and never mentioned Remsen in the patent. Remsen was furious with Fahlberg and referred to him as a scoundrel and considered his actions unethical. Remsen never benefited financially from the discovery of saccharin, but Fahlberg prospered. Fahlberg's defense was that he had done all the work on saccharine and Remsen showed no interest in its commercial applications.

Saccharin was first introduced to the public in 1885. It was initially promoted as an antiseptic and food preservative. Fahlberg and his uncle started Fahlberg, List, & Company near Magdeburg Germany to produce saccharin in 1886. The use of saccharin as a sweetener started around 1900 when it was marketed for use by people with diabetes. Because saccharin was a cheap sugar substitute, it was viewed as a threat to the sugar industry. Sugar manufacturers in Europe, Canada, and the United States lobbied for laws restricting saccharin's use; for example, laws enacted in some countries limited its use to pharmaceuticals. At the start of the 20th century, saccharin was not produced in the United States but was imported from Germany. This fact led to the founding of Monsanto Chemical Company in 1901. John Francis Queeny (1859–1933), a pharmaceutical distributor, could not persuade his employer to manufacture saccharin. He decided to start his own production plant. He chose his wife's maiden name to conceal his intentions from his current employer. Queeny hired a Swiss chemist named Louis Veillon (1875–1958) who knew how to produce saccharin. The company established itself by selling saccharin exclusively to Coca-Cola. It then expanded into supplying vanilla and caffeine to Coca-Cola.

Calls to regulate saccharin in foods have been present throughout its history. Early in the 20th century, the political climate promoted legislation and government oversight to ensure that food was safe. In 1906, the passage of the Federal Food and Drug Act gave government regulatory authority concerning the safety of food. The Department of Agriculture's Bureau of Chemistry, the predecessor of the Food and Drug Administration (FDA), performed research and made recommendations with respect to food additives. In 1907, a study by the newly created Board of Food and Drug Inspection made claims (latter refuted) that saccharin damaged the kidneys and other organs. The leader of the Bureau of Chemistry, Harvey W. Wiley (1844–1930), was a member of the Board and held the view that saccharin (and other chemicals such as benzoates) was dangerous. This was seen as a threat to food producers who were increasingly using additives as preservatives and sweeteners. These producers appealed directly to President Theodore Roosevelt, who himself was using saccharin to guard against diabetes. In a meeting between Roosevelt, representatives from several food producers, and Board members, Wiley claimed that consumers were being deceived and in danger from eating canned corn with saccharin. Roosevelt questioned Wiley as to whether he believed that saccharin was dangerous to his health and Wiley affirmed it was. Roosevelt responded to this with: "Anybody who says saccharin is injurious to health is an idiot." Thus saccharin was both a scientific and political issue.

In 1911, a board of federal scientists issued a report claiming saccharin caused digestive problems when more than 0.3 gram per day were consumed. Food containing more than 0.3 gram was considered adulterated. The scientists recommended that using more than 0.3 gram of this amount should be reserved for drugs for people needing to limit their sugar intake. Saccharin used as a sweetener was limited until World War I created sugar shortages and rationing led to loosening saccharin restrictions. Saccharin was in even greater demand during World War II.

In the war's aftermath, the use of saccharin as a dietary supplement gained popularity. It was often used as a sodium or calcium salt, which was highly soluble in water. Saccharin was introduced in the sugar formulation of Sweet'N Low in 1957. Different Sweet'N Low formulations exist, depending on where it is sold and in what form it is sold; some Sweet'N Low products do not use saccharin, but rather other artificial sweeteners. Saccharin is approximately 300 times as sweet as sucrose based on weight. Thus very little saccharin is required for sweetening. Sweet'N Low individual packets produced for the United States contain 3.6% calcium saccharin, with most of the product being dextrose; cream of tartar and calcium silicate for anticaking are also used in Sweet'N Low. In addition to using saccharin and saccharin salts as a sugar substitute, it is used in personal care products such as tooth paste, lip balms, and mouthwash. It is also used in pharmaceuticals. Approximately 30,000 tons per year of saccharin and saccharin salts are used globally each year, with about 5,000 tons of this used in the United States.

Questions on saccharin's safety has followed its usage to the present day. Saccharin is banned in Canada (except in special cases), several European countries, and many other countries. Countries where it is legal place restrictions on its use. Saccharin has been regulated in the United States since the beginning of the century. A Canadian study in 1977 that reported saccharin caused bladder cancer in rats led to a highly publicized debate concerning its safety. Based on this study, the FDA proposed to ban saccharin, but public response against an outright ban prompted Congress to pass the Saccharin Study and Labeling Act on November 23, 1977. This act placed an 18-month moratorium on FDA's ban and required foods that contained saccharin to have the following label warning: "Use of this product may be hazardous to your health. This product contains saccharin which has been determined to cause cancer in laboratory animals." The moratorium was extended on several occasions through the 1990s, and during this period studies focused on the potential toxicity of saccharin. After a review in 1997, the National Institute of Health recommended the continued listing (as it had been since 1981) of saccharin as an anticipated human carcinogen in its official report of cancer-causing substances. Additional saccharin studies and reevaluation of the original Canadian rat study led the federal government to delist saccharin in 2000. The Canadian study was thought to have limited relevance to humans because of the excessively high doses of saccharin given to rats, the rat species used in the study, and sodium's role in producing tumors. Legislation, signed into law on December 21, 2000, repealed the warning label requirement for products containing saccharin. The National Cancer Institute's position is that there is no clear evidence linking saccharin to cancer in humans.

Saccharin is synthesized using two methods: the Remsen-Fahlberg process and the Maumee or Sherwin-Williams method. The Remsen-Fahlberg synthesis of saccharin starts by reacting toluene with chlorosulfonic acid to give ortho and para forms of toluene-sulfonic acid (Figure 78.1). The acid can be converted to sulfonyl chlorides by treating with phosphorus pentachloride. The ortho form, o-toluene-sulfonyl chloride, is treated with ammonia to give o-toluene-sulfonamide, which is then oxidized with potassium permanganate to produce o-sulfamido-benzoic acid. On heating, the latter yields saccharin. Another synthesis was developed at Maumee Chemical Company in Toledo, Ohio, and it came to be known as the Maumee process. This process starts with phthalic anhydride, which is converted into anthranilic acid. Anthranilic acid is then reacted with nitrous acid, sulfur dioxide, chlorine, and ammonia to give saccharin. The Maumee process was further refined by the Sherwin-Williams Company and is therefore now referred to as the Sherwin-Williams process.

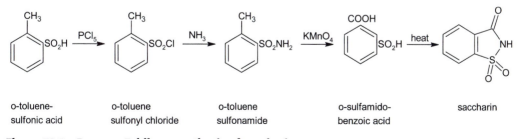

Figure 78.1 Remsen-Fahlberg synthesis of saccharin.

CHEMICAL NAME = silicon dioxide
CAS NUMBER = 14808–60–7
MOLECULAR FORMULA = SiO_2
MOLAR MASS = 60.1 g/mol
COMPOSITION = Si(46.7%) O(53.3%)
MELTING POINT = 1,713°C
BOILING POINT = 2,230°C
DENSITY = 2.6 g/cm³

$$\ddot{O}\!=\!Si\!=\!\ddot{O}$$

Silicon dioxide, generally known as silica, is a colorless solid that exists in numerous crystalline forms, the most common of which is quartz. Oxygen and silicon are the two most common elements in the earth's crust, and silicon dioxide is the principal component of sand. Silica is used biologically, most notably by phytoplanktonic diatoms and the zooplankton radiolarians in their shells. The words silica and silicon come from the Latin *silex* meaning hard stone or flint. Silica, SiO_2, should not be confused with silicates or silicones. Silicates contain the basic tetrahedral unit SiO_4^{4-} bonded to metal ions such as aluminum, iron, sodium, magnesium, calcium, and potassium to form numerous silicate minerals. Silicones are synthetic polymers made of monomers with at least two silicon atoms combined with an organic group and generally containing oxygen.

Silica can exist in either a crystalline or noncrystalline form. In quartz, SiO_2 exists in the natural crystalline state and possesses long-range order, with the silicon atom covalently bonded to oxygen atoms in a tetrahedral arrangement in a regular repeating pattern. Glass is an example of noncrystalline silica. Although natural glasses exist, silica glasses are produced when silica is heated to an elevated temperature and then rapidly cooled. The rapid cooling does not allow the SiO_2 to form a regular crystalline structure with long-range order. The result is a solid that behaves like a viscous liquid when heated. Glass is sometimes called a solid solution and flows at a very slow rate. This can sometimes be seen in old window glass where the bottom is slightly thicker than the top. The difference in crystalline and noncrystalline forms is shown as a two-dimensional representation in Figure 79.1. The actual structures form a three-dimensional tetrahedral pattern. Silica is sold as sand and its main uses are for glass; ceramics; foundry sand, a source of silicon in the chemical industry; as a filtration media; a filler/extender; an abrasive; and as an adsorbent.

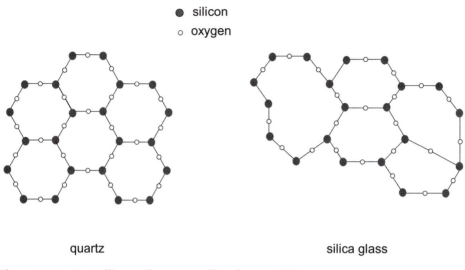

quartz   silica glass

Figure 79.1   Crystalline and noncrystalline forms of SiO$_2$.

The use of silica sand in glass is thousands of years old. The first known silica glass was produced around 3500 B.C.E. in Mesopotamia (present day Syria and Iraq), although there is some evidence of earlier production in Egypt and Phoenicia (Lebanon). The melting point of pure silica is 1,713°C, but the mixing of other substances with the SiO$_2$ lowers its melting point. Egyptians added natron (sodium carbonate) to SiO$_2$. Some glasses are produced at temperatures as low as 600°C. As the art of glass making developed, individuals discovered how to produce different glasses by adding various substances to the silica melt. The addition of calcium strengthened the glass, and other substances imparted color to the glass. Iron and sulfur give brown glass, copper produces a light blue color, and cobalt a dark blue color. Manganese was added to produce a transparent glass, and antimony to clear the glass of bubbles. Most modern glass produced is soda-lime glass and consists of approximately 70% SiO$_2$, 15% Na$_2$O (soda), and 5% CaO (lime). Borosilicate glass is produced by adding about 13% B$_2$O$_3$. Borosilicate glass has a low coefficient of thermal expansion and is therefore very heat resistant. It is used extensively in laboratory glassware and in cooking where it is sold under the brand name Pyrex.

Because of silica's high melting point, it is ideal for making molds for metal casting. It is regularly used to form iron, aluminum, and copper items. Silica is the primary filter medium used in wastewater treatment. Filtration systems often modify silica physically and chemically to produce activated silica formulations. Besides water treatment, activated silica gels used for chromatography in chemistry laboratories. In the construction industry, silica glass is used as fiber glass insulation, silica sand is a basic ingredient in cement and concrete, and is used indirectly in building products. Silica is used as filler in paints, adhesives, rubber, and coatings. It is added to personal care products such as tooth polishes.

Exposure to silica can result in the disease called silicosis. Silicosis is a disabling, nonreversible, and sometimes fatal lung disease caused by overexposure to respirable crystalline silica. In silicosis, silica particles enter the lung where they become trapped, producing areas of swelling. The swelling results in nodules that become progressively larger as the condition worsens. Silicosis is defined at several levels of severity: chronic silicosis, accelerated silicosis, and acute silicosis. Chronic silicosis results from long-term (20 years) exposure to low concentrations of silica, whereas acute silicosis is the result of a short-term exposure (a year or less) to high concentrations. Symptoms may not be obvious in cases of chronic silicosis and x-ray screening is recommended for at-risk groups. These include sand-blasters, miners, laborers who regularly saw, drill, and jack-hammer concrete, and general construction such as tunnel drilling. In advanced stages of silicosis, individuals have difficulty breathing, especially when active.

80. Sodium Bicarbonate

CHEMICAL NAME = sodium hydrogen carbonate
CAS NUMBER = 144–55–8
MOLECULAR FORMULA = NaHCO₃
MOLAR MASS = 84.0 g/mol
COMPOSITION = Na(27.4%) H(1.2%) C(14.3%) O(57.1%)
MELTING POINT = 50°C
BOILING POINT = converts to Na_2CO_3 at 100°C, decomposes at 149°C
DENSITY = 2.2 g/cm³

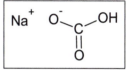

Sodium bicarbonate, which is the compound commonly called baking soda, exists as a white, odorless, crystalline solid. It occurs naturally as the mineral nahcolite, which derives its name from its chemical formula by replacing the "3" in NaHCO₃ with the ending "lite." The world's main source of nahcolite is the Piceance Creek Basin in western Colorado, which is part of the larger Green River formation. Sodium bicarbonate is extracted using solution mining by pumping hot water through injection wells to dissolve the nahcolite from the Eocene beds where it occurs 1,500 to 2,000 feet below the surface. The dissolved sodium bicarbonate is pumped to the surface where it is treated to recover NaHCO₃ from solution. Sodium bicarbonate can also be produced from the trona deposits, which is a source of sodium carbonates (see Sodium Carbonate).

Most sodium bicarbonate in the United States is made synthetically by the reaction of sodium carbonate solution (Na₂CO₃) with carbon dioxide: $Na_2CO_{3(aq)} + H_2O_{(l)} + CO_{2(g)} \rightarrow$ 2NaHCO₃₍ₐ_q₎. It can also be produced using the Solvay process, which uses ammonia, carbon dioxide, and salt to produce sodium bicarbonate according to the following series of reactions:

$$2NH_{3(g)} + CO_{2(g)} + H_2O_{(l)} \rightarrow (NH_4)_2CO_{3(aq)}$$

$$(NH_4)_2CO_{3(aq)} + CO_{2(g)} + H_2O_{(l)} \rightarrow 2NH_3HCO_{3(aq)}$$

$$NH_4HCO_{3(aq)} + NaCl_{(aq)} \rightarrow NaHCO_{3(s)} + NH_4Cl_{(aq)}$$

Sodium carbonate was imported into the United States until the 1840s when Austin Church (1799–1879), a physician, and his brother-in-law, John Dwight (1819–1903), opened a baking soda factory in New York. The company started as John Dwight & Co. in 1846. The Arm & Hammer brand name was devised by Church's son, James A. Church, who owned a mill called Vulcan Spice Mills. Vulcan was the Roman god of fire and, according to mythology, fashioned tools and weapons for gods and heroes in his workshop on Mt. Olympus. Church had devised the arm and hammer logo to symbolize products from his spice mill. When he joined his father's baking soda business in 1867, he brought the arm and hammer as a brand symbol for their baking soda. The descendants of Church and Dwight marketed baking soda under the brand name Arm & Hammer, which is currently used by the Church and Dwight Company for baking soda, as well as other products they produce.

Total production of sodium bicarbonate in 2004 was about 1.6 million tons, with the United States, Western Europe, and Japan accounting for more than 80% of its production. Sodium bicarbonate's greatest use is for the food industry; about one-third of its production is consumed in the United States for food applications. Sodium bicarbonate, used in the form of baking soda and baking powder, is the most common leavening agent. When baking soda, which is an alkaline substance, is added to a mix, it reacts with an acid ingredient to produce carbon dioxide. The reaction can be represented as: $NaHCO_{3(s)} + H^+ \rightarrow Na^+_{(aq)} + H_2O_{(l)} + CO_{2(g)}$, where H^+ is supplied by the acid. Baking powders contain baking soda as a primary ingredient along with acid and other ingredients. Depending on the formulation, baking powders can produce carbon dioxide quickly as a single action powder or in stages, as with a double-action powder. Baking soda is also used as a source of carbon dioxide for carbonated beverages and as a buffer.

In addition to baking, baking soda has numerous household uses. It is used as a general cleanser, a deodorizer, an antacid, a fire suppressant, and in personal products such as toothpaste. Sodium bicarbonate is a weak base in aqueous solution, with a pH of about 8. The bicarbonate ion (HCO_3^-) has amphoteric properties, which means it can act as either an acid or a base. This gives baking soda a buffering capacity and the ability to neutralize both acids and bases. Food odors resulting from acidic or basic compounds can be neutralized with baking soda into odor-free salts. Because sodium bicarbonate is a weak base, it has a greater ability to neutralize acid odors.

The second largest use of sodium bicarbonate, accounting for approximately 25% of total production, is as an agricultural feed supplement. In cattle it helps maintain rumen pH and aids fiber digestibility; for poultry it helps maintain electrolyte balance by providing sodium in the diet, helps fowl tolerate heat, and improves eggshell quality.

Sodium bicarbonate is used in the chemical industry as a buffering agent, a blowing agent, a catalyst, and a chemical feedstock. Sodium bicarbonate is used in the leather tanning industry for pretreating and cleaning hides and to control pH during the tanning process. Heating sodium bicarbonate produces sodium carbonate, which is used for soap and glassmaking. Sodium bicarbonate is incorporated into pharmaceuticals to serve as an antacid, a buffering agent, and in formulations as a source of carbon dioxide in effervescent tablets. Dry chemical type BC fire extinguishers contain sodium bicarbonate (or potassium bicarbonate). Other uses of bicarbonate include pulp and paper processing, water treatment, and oil well drilling.

81. Sodium Carbonate

CHEMICAL NAME = sodium carbonate
CAS NUMBER = 497–19–8
MOLECULAR FORMULA = Na_2CO_3
MOLAR MASS = 106.0 g/mol
COMPOSITION = Na(43.4%) C(11.3%) O(45.3)
MELTING POINT = 851°C
BOILING POINT = decomposes
DENSITY = 2.5 g/cm³

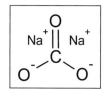

Sodium carbonate is known as soda ash or washing soda and is a heavily used inorganic compound. Approximately 45 million tons of soda ash are produced globally both naturally and synthetically. Soda ash is obtained naturally primarily from the mineral trona, but it can also be obtained from nahcolite ($NaHCO_3$) and salt brine deposits. Trona is a freshwater sodium carbonate-bicarbonate evaporite, with the formula $Na_3CO_3HCO_3 \cdot 2H_2O$. The largest known deposit of trona is located in the Green River area of Wyoming, and other large deposits are found in Egypt's Nile Valley and California's Searles basin around the city of Trona. Soda ash is produced from mined trona by crushing and screening the ore and then heating it. This produces a soda ash mixed with impurities. Pure soda ash is obtained by dissolving the product and precipitating impurities combined with filtering processes.

Sodium carbonate, Na_2CO_3, has been used historically for making glass, soap, and gunpowder. Along with potassium carbonate, known as potash, sodium carbonate was the basis of the alkali industry, which was one of the first major chemical industries. Throughout history, alkalis were obtained from natural sources. There is evidence of Egyptians using soda ash to make glass as early as 3500 B.C.E. Natron ($Na_2CO_3 \cdot 10H_2O$) imported from Egyptian lakes was a major source of soda ash for Europe. Soda ash was also produced by burning wood and leaching the ashes with water to obtain a solution that yielded soda ash when the water was boiled off. The name soda ash originates from the barilla plant, which was used to produce soda ash. The scientific name of this plant is *Salsola soda,* but it goes by the common names of sodawort or glasswort because the soda produced from it was used in making glass. Barilla is a common plant found in saline waters along the Mediterranean Sea in Spain and Italy. Barilla was dried and burned to produce soda ash. The depletion of European forests and international disputes made the availability of alkali salts increasingly uncertain during the latter part of the 18th century. This prompted the French Academy of Science in 1783 to offer a reward to anyone who could find a method to produce soda ash from common sea salt, NaCl. Some of the leading chemists of the day sought a solution to the soda ash problem, but it was Nicolas LeBlanc (1743–1806) who was credited with solving the problem. LeBlanc proposed a procedure in 1783, and a plant based on LeBlanc's method was opened in 1791. Unfortunately, LeBlanc's association with French Royalty led to the confiscation of the plant at the time of the French Revolution. Furthermore, conflicting claims for LeBlanc's method were made by several other chemists and he never received the reward. LeBlanc became disheartened and destitute and committed suicide in 1806.

LeBlanc's method uses sulfuric acid and common salt to initially produce sodium sulfate, Na_2SO_4. Sodium sulfate is then reacted with charcoal and limestone to produce sodium carbonate and calcium sulfide:

$$H_2SO_{4(l)} + 2NaCl_{(aq)} \rightarrow Na_2SO_{4(s)} + 2HCl_{(g)}$$
$$Na_2SO_{4(s)} + 2C_{(s)} + CaCO_{3(s)} \rightarrow Na_2CO_{3(s)} + CaS_{(s)}$$

The sodium carbonate and calcium sulfide were separated by mixing with water. Because sodium carbonate was soluble in the water and calcium sulfide was insoluble, the former would be suspended in solution.

LeBlanc's process increased the demand for sulfuric acid, and the alkali and acid industries were the first modern large-scale chemical industries. Plants using the LeBlanc process were situated in areas associated with salt mines, and this naturally created locales for industries that depended on soda ash. The alkali industry using the LeBlanc process created environmental problems near the alkali plants. The hydrogen chloride gas killed vegetation in the immediate vicinity of the plants. To decrease air pollution, the gas was dissolved in water, creating hydrochloric acid that was then discharged to streams, but this just turned the air pollution problem into a water pollution problem. Another problem was created by the solid calcium sulfide product. Calcium sulfide tailings stored around alkali plants reacted with air and water, creating noxious substances such as sulfur dioxide and hydrogen sulfide. Landowners adjacent to alkali plants sought relief from the environmental damage resulting from soda production. The situation got so bad in England that the Parliament's House of Lords enacted the first of the "Alkali Acts" in 1863. These laws regulated the production of soda ash and required producers to reduce their environmental impacts on the surrounding countryside.

The LeBlanc process was the principal method of producing soda ash until 1860 when the Belgian Ernest Solvay (1838–1922) developed the process that bears his name. The Solvay process, sometimes called the ammonia method of soda production, uses ammonia, NH_3, carbon dioxide, and salt to produce sodium bicarbonate (baking soda), $NaHCO_3$. Sodium bicarbonate is then heated to give soda ash. The series of reactions representing the Solvay process are:

$$2NH_{3(g)} + CO_{2(g)} + H_2O_{(l)} \rightarrow (NH_4)_2CO_{3(aq)}$$
$$(NH_4)_2CO_{3(aq)} + CO_{2(g)} + H_2O_{(l)} \rightarrow 2NH_4HCO_{3(aq)}$$

$$NH_4HCO_{3(aq)} + NaCl_{(aq)} \rightarrow NaHCO_{3(s)} + NH_4Cl_{(aq)}$$

$$2NaHCO_{3(g)} \xrightarrow{heat} Na_2CO_{3(s)} + H_2O_{(l)} + CO_{2(g)}$$

Natural sources supplies the need for soda ash use in the United States, whereas countries without a natural source depend more heavily on synthetic soda ash and imports. One disadvantage of synthetic soda ash production is the production of environmental pollutants. Industrial soda ash comes in two main grades: light and dense. Light soda ash has a larger grain size. The primary use of soda ash is in the production of glass. Approximately 30% of the soda ash production in the United States is used for this purpose. Decahydrate soda ash, $Na_2CO_3 \cdot 10H_2O$, is referred to as washing soda because it was traditionally used with laundry soap as a softening agent. Today it is used with laundry soaps and detergents to maintain pH and as a builder (an additive that enhances cleaning). Soda ash is used in the paper-making industry to soften wood chips in pulp production. It is used as a base in the chemical industry to increase pH, and it is also a source of sodium ions in chemical processes. Sodium bicarbonate or baking soda ($NaHCO_3$) is produced by reacting carbon dioxide with a solution of sodium carbonate.

82. Sodium Chloride

CHEMICAL NAME = sodium chloride
CAS NUMBER = 7647–14–5
MOLECULAR FORMULA = NaCl
MOLAR MASS = 58.4 g/mol
COMPOSITION = Na(39.3%) Cl(69.7%)
MELTING POINT = 801°C
BOILING POINT = 1,465°C
DENSITY = 2.16 g/cm³

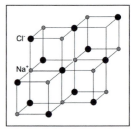

Sodium chloride is the familiar compound commonly referred to as salt or table salt. The mineral form of sodium chloride is halite and is found in natural deposits throughout the world. It accounts for approximately 2.7% by weight of the dissolved minerals in seawater. Sodium chloride is an ionic compound existing as a white crystalline cubic structure of alternating sodium and chloride ions. Sodium chloride is essential for life, with the average adult requiring about 1 to 2 grams per day. Salt supplies sodium and provides numerous essential functions such as maintaining water balance in cells, taking part in nerve signal transmission and muscle contraction.

The use of sodium chloride as salt dates back to prehistoric times. Throughout much of human history salt consumption was low and has only increased relatively recently. Prehistoric people obtained salt from eating meat and fish, which supplied the diet with ample salt. Herbivores sought out natural salt sources to supply their needed nutrients and prehistoric humans were attracted to salt licks for game and the salt itself. As prehistoric people moved from a hunter-gathering form of existence to agricultural societies, changes in diet reflected a less nomadic existence as plant matter formed a larger part of the diet. Consuming more plant material meant less salt was being consumed. Domesticated animals required salt and early farmers discovered that vegetables could be seasoned with salt. The use of salt for preserving meat was one of the major developments that supported permanent settlements. Salt preservation works by drawing water out of food, creating an environment that does not support the growth of microbes responsible for spoilage. In addition to preservation, humans discovered that salt could be used for other applications such as curing hides, making pottery, and preparing medications.

The earliest written records contain references to salt. A Chinese treatise on pharmacology called the Peng-Tzao-Kan-Mu dates from 4,700 years ago cites 40 different types of salt along with methods for extracting and processing salt. Egyptian art depicts salt making and it is being used for various procedures such as curing and mummification. Numerous references in the Bible's Old Testament refer to salt.

As settlements grew and uses of salt increased, various methods were used by cultures to obtain salt. These consisted of mining from shallow underground deposits, evaporating seawater and saline lake water, mining evaporites from dried lake deposits, and trade. Salt has been traded throughout history and was responsible for establishing numerous cities, both as producing and/or trading centers. The Greek historian Herodutus (484–425 B.C.E.) wrote extensively about Sahara salt and its importance in African commerce. Salt was extracted from pans near Ostia, the ancient port city of Rome located at the mouth of the Tiber, since 1400 B.C.E. Ostia was the main source of salt for Rome as it grew into a major city of the ancient world. The *Via Salaria* was a salt supply route that ran between the Adriatic Sea and Rome. Venice established itself as a leading center of salt production and trade as early as the 6th century. Timbuktu in the western Africa country of Mali was founded as a major trade center for Saharan salt and gold in the Middle Ages. Salt was transported along trade routes to the area by camel. Salt played an important role in the economics and politics of many other cities, demonstrated by the word salt in their names, for example, Salzburg, Hallein, and Hallstatt (salz is the German word for salt and Hall comes from the old Celtic word for salt).

The universal need for salt and its use throughout the world made salt a common exchange currency. The word salary comes from the Latin word *salarium* derived from *sal* for salt. A *salarium* was the compensation provided in provisions and money to Roman officials and soldiers. Because one of the most essential provisions was salt, it was supplied directly or an allowance for salt was provided. Adam Smith (1723–1790) makes a number of references to salt in *The Wealth of Nations*. Throughout most of history, salt was a sign of wealth and provided a basis for taxes, monopolies, and conflicts. As cities and towns developed, salt commerce was increasingly regulated and based on a multitude of commercial contracts, taxes, and licenses controlled by royalty and merchants. France's salt tax and requirement to purchase a set amount of salt, which was known as the gabelle, was applied nationally and in various forms throughout its provinces symbolized state oppression. Gabelles were inconsistently applied throughout regions by collectors called farmers in an intricate set of regulations as salt moved across regional boundaries. France's salt taxes were not unique; rather they were common sources of revenue for local governments throughout the world. Local populations despised oppressive salt taxes, which in extreme cases led to revolts; France's gabelle contributed to the French Revolution and was abolished in 1790 only to be reinstated by Napoleon (1769–1821) to fund his military campaigns. Many long-standing salt taxes existed into the 20th century. For example, Mahatma Gandhi's (1869–1948) famous salt march to Dandi in 1930 to obtain salt illegally defied British rule and was a major event in India's move toward independence. France did not abolish its gabelle until after World War II.

The three main industrial methods used to produce salt are the solar evaporation method, mining of rock salt, and solution mining. The solar evaporation method is the oldest process used to obtain salt. This method is applied in geographic areas with high solar input and low rainfall, which are typically arid or desert regions such as the Middle East, Baja Mexico, and the Mediterranean. Solar evaporation consists of capturing seawater or saline lake water in shallow ponds or natural pans where the sun and wind evaporate water to concentrate the brine. The brine is concentrated in a series of ponds until it reaches a high enough concentration to crystallize. As evaporation continues, the brine becomes supersaturated, resulting in crystallization of sodium chloride and other salts. The brine can be moved between ponds and processed to increase sodium chloride yield, but impurities, which mainly consist of calcium, magnesium, and potassium sulfates, will still be present. A salt mixture builds up on the bottom of crystallization ponds over months until it is thick enough to be harvested. Harvesting of a pond typically occurs after a year or several years of crystallization. The harvested salt is drained, dried, and washed. This removes more of the impurities, producing a salt with about 90–95% sodium chloride. Sea salt is often marketed as a more natural product because it still contains traces of other sea salts. The final processing stages involves refining the salt to produce almost pure (more than 99%) sodium chloride. Refining involves various separation processes. A common procedure is to dissolve the salt in water to produce a brine and use chemicals to precipitate impurities out of the brine. The brine is then heated in special crystallizers using steam and water evaporates under a vacuum to recrystallize sodium chloride. This may then be further concentrated using centrifuges.

Sodium chloride is produced using hard rock mining located from salt deposits located in many places on the planet. All salt deposits were once covered by the sea or salt lakes. They consist of highly concentrated sodium chloride formations laid down when ocean or saline lakes dried, leaving the salt behind. North American salt deposits were formed several hundred million years ago when major seawater intrusions occurred over the continent. Over time, salt deposits were covered by marine deposits and other rock formations. Halite has a relatively low density and can intrude upward through overlying deposits. Salt deposits may be just a few feet to thousands of feet thick and can consist of almost pure halite or be interspersed with other rock formations. Major North American salt deposits occur along the Gulf of Mexico coast, around the Great Lakes, and a region across the central United States in parts of Kansas, Oklahoma, and Texas (Figure 82.1).

Underground salt mining consists of sinking shafts and using a "room and pillar" system where large volumes called rooms of salt are excavated, leaving pillars to support the roof of the mine. Rooms are typically 10 to 40 feet high, with horizontal dimensions of tens to hundreds of feet. As the mine is developed, more rooms are added to produce a checkerboard pattern. Salt is extracted using explosives, modern machinery, and conveyor belts and subjected to washing and refining methods similar to sea salt.

A third major process for extracting salt is solution mining; this is especially useful for concentrated deep deposits. In solution mining injection wells are drilled into a salt deposit and water is pumped into it. The salt dissolves in the water to produce brine that is pumped back to the surface where it is processed to obtain salt.

Sodium chloride has numerous uses; one major producer lists more than 1,400 uses for its salt. Global production of salt is about 230 million tons annually; about 50,000 tons are produced in the United States. The largest consumer of salt is the chemical industry, which uses approximately 60% of total production. The major chemical industry that uses salt is the chlor-alkali industry to produce soda ash (in countries that do not obtain it from natural deposits), caustic soda ($NaOH$), and chlorine (see Sodium Carbonate and Sodium

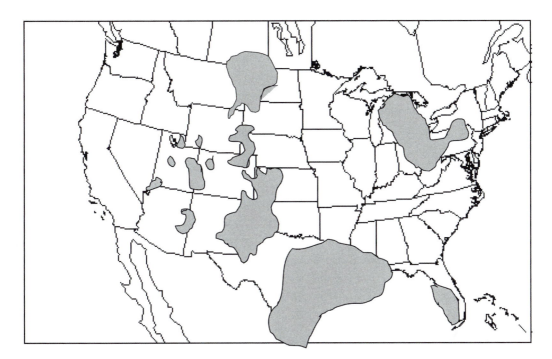

Figure 82.1 Salt deposits in the continental United States.

Hydroxide). Approximately 40% of the sodium chloride used in the United States is used in the chlor-alkali industry. Nearly all (98%) of the chlorine and sodium hydroxide produced in the United States comes from the electrolysis of sodium chloride solution. Salt is also used as a feedstock to obtain sodium, sodium chlorate ($NaClO_3$), and calcium chloride ($CaCl_2$). Other industries that use sodium chloride are oil and gas, textiles and dyeing, metal processing, paper and pulp, leather treatment, and rubber; the largest of these is the oil and gas industry where salt is used as a constituent of drilling fluids and as an additive to saturate salt formations. A total of 30% of global sodium chloride production is used in the food industry. This is the second largest global use of salt, but in the United States, road deicing is the second largest use. About as much salt is used for road deicing as in the chlor-alkali industry, which means that almost 80% of salt usage is for chlor-alkali production or deicing in the United States. The remaining 10% of global production is used for numerous applications such as water softening, agriculture feed supplements, and deicing. Sodium chloride is used in water softening where sodium is exchange for magnesium and calcium ions responsible for hard water.

The most common use of salt is as part of daily diets. Although it is added directly to food, 75% of the salt consumed in the United States is a result of eating processed foods The National Academy of Sciences has determined that a minimum daily requirement of 500 mg of sodium is safe, which equates to 1,300 mg of salt . The Academy and the federal government recommend that sodium consumption be no more than 2,400 mg per day, which equals 6,100 mg of salt. Most Americans consume levels higher than this, and many health organizations recommend decreasing salt intake. Excess salt can lead to health problems such as elevated blood pressure, although recent research seems to indicate that normal or moderately high salt consumption does not result in hypertension in most people. Some individuals are salt sensitive and have trouble excreting excess salt. Excessive salt intake and retention by the kidneys raise blood pressure because increased water retention raises blood volume. Excessive salt also has been associated with other health problems. The loss of calcium in high-salt diets increases the risk of osteoporosis. Salt has also been linked to asthma, stomach cancer, and kidney stones.

83. Sodium Hydroxide

CHEMICAL NAME = sodium hydroxide
CAS NUMBER = 1310–73–2
MOLECULAR FORMULA = NaOH
MOLAR MASS = 40.0 g/mol
COMPOSITION = Na(57.5%) O(40%) H(2.5%)
MELTING POINT = 318°C
BOILING POINT = 1,390°C
DENSITY = 2.1 g/cm³

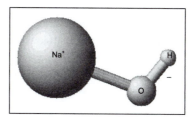

Sodium hydroxide is the most widely used strong base in the chemical industry. It exists as a white, odorless, crystalline solid at room temperature. Sodium hydroxide is a highly corrosive and toxic substance and is called caustic soda or lye. Sodium hydroxide is used as an alkali in the production of numerous products including detergents, paper, synthetic fabrics, cosmetics, and pharmaceuticals. It is the ingredient in many common cleaners, degreasers, drain cleaners, and personal care products.

Sodium hydroxide began to be produced on a large scale in the mid-19th century using soda ash and lime. The process is known as the lime causticization method and is no longer used. It involved the reaction of slaked lime ($Ca(OH)_2$) and soda ash (Na_2CO_3): $Ca(OH)_{2(aq)} + Na_2CO_{3(aq)} \rightarrow 2NaOH_{(aq)} + CaCO_{3(s)}$. During the same period, Charles Watt demonstrated that sodium hydroxide could be produced through electrolysis of brine. Watt passed a current through a brine solution in an electrolytic cell and obtained sodium hydroxide and hydrogen at the anode of the cell and chlorine at the cathode: $2NaCl_{(aq)} + 2H_2O_{(l)} \rightarrow Cl_{2(g)} + 2NaOH_{(aq)} + H_{2(g)}$.

Watt's work laid the foundation for the current electrolysis method used to produce sodium hydroxide. This method was developed independently during the 1880s by the American chemist Hamilton Young Castner (1858- 1899) and the Austrian chemist Karl Kellner (1850–1905). Castner left Columbia University before graduating and developed a method to produce sodium more economically for use in the aluminum industry. Until the late 1800s, aluminum was considered a precious metal because it was difficult to obtain it in pure form. The process involved the reduction of aluminum chloride using sodium. Castner moved to England where his method for producing sodium was used for a brief period. In 1866, Charles Martin Hall (1863–1914) in the United States developed a method to produce aluminum electrochemically (U.S. Patent Number 400766), and Paul L. T. Héroult (1863–1914) developed a similar process in France. The Hall-Héroult method for producing aluminum replaced the old sodium method (see Aluminum Oxide).

Castner turned his interest to gold extraction, which required high-quality sodium hydroxide. Castner developed a three-chambered electrolytic cell. The two end chambers contained brine and graphite electrodes. The middle chamber held water. The cells were separated excepted for a small opening on the bottom, which contained a pool of mercury that served as the cell's cathode. When current flowed through the cell and the cell was rocked, sodium reduced from the brine came into contact with water in the middle cell to produce a sodium hydroxide solution. As Castner built his mercury cell, Kellner was working on a similar design. Rather than compete with each other, Castner and Kellner joined forces to establish the Castner-Kellner Alkali Company to produce sodium hydroxide, which competed with soda ash and potash as an industrial base, and chlorine, which was used primarily to make bleach.

As the demand for chlorine increased during the 20th century, the electrolysis method for making sodium hydroxide slowly replaced the lime causticization method. By 1970, sodium hydroxide was produced exclusively by electrolysis. Numerous types of electrolytic cells are used to make sodium hydroxide, but most can be classified into two general categories: diaphragm and mercury. Most sodium hydroxide is made using diaphragm cells. In this type of cell, a diaphragm separates chambers with each chamber containing multiple electrodes. In an industrial plant, many cells are connected in series. Originally, diaphragm cells used asbestos as the diaphragm, but today special polymers are used. The reaction taking place in the diaphragm cell can be represented as: $2NaCl_{(aq)} + 2H_2O_{(l)} \rightarrow Cl_{2(g)} + H_{2(g)} + 2NaOH_{(aq)}$. The anode and cathode are made of graphite or steel alloys. The brine is a saturated 25% NaCl solution from which calcium and magnesium ions have been separated. The brine level in the anode chamber is elevated higher to allow the brine to pass through the diaphragm. Chlorine is oxidized at the anode and hydrogen and dilute sodium hydroxide are oxidized (mixed with unreduced sodium chloride) at the cathode. Cells referred to as membrane cells use a selective membrane in place of the diaphragm. The selective membrane allows the passage of positive ions (Na^+) from the anode to cathode chamber, increasing efficiency.

Contemporary mercury cells are based on the original designs of Castner and Kellner. Mercury is used as the cathode. Sodium produced forms an amalgam, which passes to a water chamber. Here graphite is used to catalyze the dissociation of sodium from the mercury. The sodium reacts with water to produce sodium hydroxide according to the reaction: $2Na_{(s)} + 2H_2O_{(l)} \rightarrow 2NaOH_{(aq)} + H_{2(g)}$.

Sodium hydroxide is sold commercially as anhydrous flakes or pellets or as 50% or 73% aqueous solutions. It has countless industrial uses and is one of the top 10 chemical in terms of production and use on a global scale. Approximately 15 million tons of sodium hydroxide is used annually. Its largest use, consuming about half of its production, is as a base in producing other chemicals. It is used to control pH and neutralize acids in chemical processes. The paper industry makes extensive use of sodium hydroxide in the pulping process. Sodium hydroxide is used to separate fibers by dissolving the connecting lignin. It is used in a similar fashion in the production of rayon from cellulose. Sodium hydroxide is a key chemical in the soap industry. In the saponification process, triglycerides obtained from animal and plants are heated in a basic solution to give glycerol and soap:

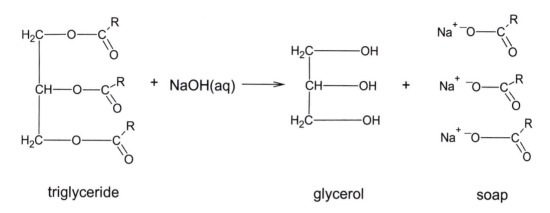

triglyceride   glycerol   soap

Sodium hydroxide is used in the textile industry for bleaching and treating textiles to make them dye more readily. The petroleum industry uses sodium hydroxide in drilling muds and as a bactericide. Sodium hypochlorite (NaOCl) is used extensively for cleaning and as a disinfectant. Common household bleach consists of about 5% sodium hypochlorite solution. Sodium hypochlorite is prepared by reacting chlorine with sodium hydroxide: $Cl_{2(g)} + 2NaOH_{(aq)} \rightarrow NaOCl_{(aq)} + NaCl_{(aq)} + H_2O_{(l)}$. Sodium hydroxide is used in the food industry for cleaning and peeling fruits and vegetables. Sodium hydroxide is a minor ingredient in many common household products, but in a few it may constitute more than half of the product. Drano crystals contain between 30% and 60% sodium hydroxide and some drain cleaners can consist of 100% sodium hydroxide.

84. Sodium Hypochlorite

CHEMICAL NAME = sodium hypochlorite
CAS NUMBER = 7681–52–9
MOLECULAR FORMULA = NaOCl
MOLAR MASS = 74.4 g/mol
COMPOSITION = Na(30.9%) Cl(47.6%) O(21.5%)
MELTING POINT = −6°C (5% solution)
BOILING POINT = decomposes at 40°C (5% solution)
DENSITY = 1.1 g/cm³

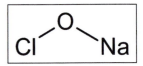

Sodium hypochlorite exists as an aqueous solution from 5–15% NaOCl and is commonly called bleach. Household bleach is typically a 5.25% solution, and industrial bleach is sold as a 12% solution. When sodium hypochlorite is used in this entry, it is assumed to be the aqueous solution, which is clear, slightly yellow, corrosive, and has a distinctive chlorine smell. Chorine gas was discovered by Carl Wilhelm Scheele (1742–1786) in 1774 and known initially as depholgisticated salt spirit. In 1787, the French chemist Claude Louis Berthollet (1749–1822) experimented with aqueous solution of chlorine gas as bleaching agents. Based on Berthollet's work, the Javel Company located on the outskirts of Paris began to produce bleaches in 1788. Chlorine gas was dissolved in a solution of soda potash (potassium carbonate) to obtain a product called *liqueur de Javel*, which was potassium hypochlorite. Potash treated with chlorine gas was also used to produce bleaching powders. In 1798, Charles Tennant (1768–1838) substituted cheaper slaked lime (calcium hydroxide, Ca(OH)$_2$) for potash to produce *liqueur de Javel*, but he was unable to patent his process. The next year he patented a bleaching powder using chlorine with slaked lime to produce calcium hypochlorite, Ca(ClO)$_2$. In 1820, Antoine Germaine Labarraque (1777–1850), an apothecary, substituted cheaper soda ash (sodium carbonate) for potash to produce *Eau de Labarraque* or Labarraque solution, which was sodium hypochlorite. *Eau de Labarraque* was used as a disinfectant and to bleach paper.

Bleaching powders, borax, lye, and blueing were used as bleaches throughout the 19th century. In the second half of the 19th century, the acceptance of Louis Pasteur's (1822–1895) germ theory of disease emphasized the need for sanitary conditions and human hygiene. Although hypochlorite solutions and bleaching powders were used as general disinfectants in the first half of the 19th century, the routine use of chlorine and chlorine compounds in public water systems did not begin until the end of the century. The first wastewater treatment plants to use chlorine opened in 1893 in Hamburg, Germany and, in 1894, in Brewster, New York. The first drinking water plant in the United States to use chlorine was opened in Chicago in 1908.

The increased production of sodium hypochlorite for bleaching and disinfection provided stimulus for large-scale industrial production of the compound. The method used to produce sodium hypochlorite involved absorbing chlorine gas in a sodium hydroxide solution: $Cl_{2(g)} + NaOH_{(aq)} \rightarrow NaOCl_{(aq)} + NaCl_{(aq)} + H_2O_{(l)}$. One of the first large-scale producers of bleach was Dow Chemical Company. This company was started by Herbert Henry Dow (1866–1930) in Canton, Ohio. Dow was a student at Case Institute in Cleveland who studied the characteristics of salt brines acquired from wells around the Great Lakes. Dow was determined to discover methods to extract chemicals from the salt brine. Rather than use the standard distillation method of his day to obtain chemicals, Dow used electrolysis to separate bromides, chlorides, and sodium hydroxide. He began Midland (Michigan) Chemical Company in 1890, returned to the Cleveland area and formed Dow Process Company in 1895, and finally returned to Midland for good in 1897, where he founded Dow Chemical Company. Dow Chemical began producing sodium hypochlorite in 1898.

Clorox, the largest producer of household bleach in the United States, was founded in 1913 by five businessmen in Oakland, California as The Electro-Alkaline Company; it changed its name to Clorox (coined from chlorine and hydroxide) in 1922. The original Clorox bleach solution was 21% sodium hypochlorite (NaOCl) and was delivered in five-gallon returnable buckets to dairies, breweries, and laundries that used the solution for disinfecting equipment. Trying to increase their market, Clorox's general manager William C. R. Murray packaged a 5.25% solution in 15-ounce bottles, which was given away in his grocery and sold door-to-door starting in 1916. Over the next several years free samples were provided to retailers to establish markets, and it became a useful household product. Clorox bleach remained the company's sole product until the company was acquired by Procter & Gamble in 1957.

Sodium hypochlorite's (and calcium hypochlorite's) disinfection property is due to its ability to form hypochlorous acid, HOCl. The hypochlorous acid oxidizes the cell walls and kills bacteria. Sodium hypochlorite generates hypochlorous acid according to the reaction: $NaOCl_{(aq)} + H_2O_{(l)} \rightarrow HOCl_{(aq)} + NaOH_{(aq)}$. The hypochlorite ion generated from NaOCl exists in equilibrium with water as represented by the equation: $OCl^- + H_2O \rightleftarrows HOCl + OH^-$.

The bleaching power of sodium hypochlorite and other bleaches results from their ability to disrupt the light-absorbing structures in organic molecules called chromophores. Chromophores are often associated with conjugated systems, which are structures with alternating single and double bonds. The electrons in conjugated systems are delocalized and have the ability to exist in different molecular orbitals. Electrons in one molecular orbital state can absorb energy and be promoted to a higher energy state. Electron transitions resulting from absorption of specific wavelengths creates a color that is the visual complement of the absorbed light. For example, absorption of red light of 500–520 nm light gives a substance a green appearance. The bleaching ability of sodium hypochlorite and other oxidizing bleaches breaks the double bonds in the chromospheres, which either destroys the chromospheres or changes it so that it no longer absorbs visible light.

Sodium hypochlorite is the primary hypochlorite used as a bleach and disinfectant, accounting for 83% of world hypochlorite use, with calcium hypochlorite accounting for the remaining 17%. Approximately 1 million tons of sodium hypochlorite was used globally in 2005, with about half this amount used in households for laundry bleaching and disinfection. The other half was used primarily for wastewater and drinking water treatment; other uses include pool sanitation, bleaching of pulp, paper, and textiles, and as an industrial chemical.

85. Strychnine

CHEMICAL NAME = strychnidin-10-one
CAS NUMBER = 57–24–9
MOLECULAR FORMULA = $C_{21}H_{22}N_2O_2$
MOLAR MASS = 334.4 g/mol
COMPOSITION = C(75.4%) H(6.6%) N(8.4%) O(9.6%)
MELTING POINT = 275°C-285°C
BOILING POINT = decomposes
DENSITY = 1.36 g/cm³

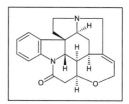

Strychnine is a white, odorless, toxic crystalline powder with a bitter taste. Strychnine is a well-known poison, but it has also been used in medications. It was one of the first alkaloids to be isolated in pure form when Joseph Bienaimé Caventou (1795–1877) and Pierre Joseph Pelletier (1788–1842) extracted it from Saint Ignatius beans (*Strychnos ignatii*) in 1818. Caventou and Pelletier subsequently obtained it from its main source *Strychnos nux vomica*; it is also found in other *Strychnos* species. *Strychnos nux vomica* is an evergreen tree native to east India, Southeast Asia, and northern Australia. The seeds of the plant are the main source of strychnine and several other alkaloids (brucine), but it is also obtained from the bark and roots. The seeds are heated and then ground to a powder that contains strychnine. The brownish-gray powder obtained from the ground seeds extracted from the nut is called *nux vomica*. In commercial production during the 1800s, the seeds were softened by boiling in a dilute sulfuric acid solution and then dried and ground to a powder. Strychnine was extracted from the powder using alcohol and separation techniques to remove impurities. The structure of strychnine was determined by Robert Robinson (1886–1975) in 1946; the next year Robinson received the Nobel Prize in chemistry for his work on alkaloids. Robert Burns Woodward (1917–1979) performed the first complete synthesis of strychnine in 1953.

Strychnine was used in its impure powder form centuries before it was isolated. The nuts that yielded the seeds were given names such as poison nut or vomit nuts. Natives prepared poison arrows using the seeds and excretions of *Strychnos* species, and Europeans imported seeds to use as medicine and poison during the Middle Ages. Valerius Cordus (1515–1544) described the properties of *Strychnos nux vomica* seeds in the 16th century. Early workers on its properties warned about its poisonous properties and recommended against it internal use. Strychnine as *nux vomica* was available in apothecaries throughout the 19th century, and strychnine salts such as strychnine sulfate, strychnine nitrate, and strychnine phosphate were a popular medicine. It was sold as pills, powders, and tinctures. Small quantities of strychnine were added to tonics to serve as stimulants. Because its bitter taste stimulates saliva and gastric production, strychnine was used to counteract the loss of appetite produced by various diseases. Hypodermic injections of strychnine along with digitalis were prescribed for treating acute heart failure in the early 20th century. Other conditions for which strychnine was prescribed included paralysis, impotence, alcoholism, and drug addiction.

Today strychnine as *nux vomica* is still available as an herbal ingredient for herbal and homeopathic remedies. The action of *nux vomica* is attributed to strychnine and brucine. It is recommended for stomach ailments such as atonic dyspepsia, constipation, heart burn, and indigestion. It is reported to relieve nausea during pregnancy. Chinese herbalists use *nux vomica* externally to treat tumors, headaches, and paralysis. It is recommended for the treatment of Bell's palsy. Strychnine has some uses in modern traditional medicine. It has been used to treat the genetic disorder nonketotic hyperglycinaemia (NKH). NKH is related to glycine metabolism; it can lead to high levels of the inhibitory receptor glycine resulting in seizures. Strychnine is used in neurological research by applying it to areas of the brain or spinal column and observing how the nervous system responds. This method is called strychnine neuronography. Strychnine has been used in conjunction with antivenoms to treat poisonous snake bites. Strychnine is sometimes cut into cocaine, heroin, and other illegal substances in the production of designer drugs.

Although *strychnine* and *nux vomica* has been used for several hundred years medicinally, it is best known as a powerful poison. The use of poison for murder has been used for thousands of years, but several events in the 19th century contributed to strychnine's notoriety as a poison. These included the development of the science of toxicology and practical pathological methods to identify the use of specific poisons, sensationalizing high-profile murder cases that used strychnine, and the use of strychnine as an exotic poison in fictional accounts.

Strychnine's poisoning effects are due to its ability to act as a central nervous system stimulant. Evidence indicates that strychnine binds to inhibitory glycine receptors in mammals, resulting in overproduction of synaptic signals; this can produce convulsions and other symptoms. The symptoms depend on mode of delivery, dose, and an individual's genetics. Symptoms typically appear within an hour, but as quickly as several minutes. Nonlethal doses can produce muscle spasms, fever, liver damage, kidney damage, rigidity of the limbs, and breathing difficulty. In lethal doses, death is typically due to asphyxiation as a result of spasms and paralysis of the brain's respiratory center. The lethal dose (LD_{50}) for humans is approximately 1–2 mg per kilogram of body weight. Therefore just a few tenths of a gram would be sufficient to kill a human. There is no antidote for strychnine poisoning. Because of its rapid action, quick medical attention is required to relieve symptoms. It is treated with barbiturates and muscle relaxants.

Strychnine was once used liberally as a pesticide, especially to control vertebrates. It is used as a salt in granular bait formulations. In the United States its use has been progressively limited over the years. In 1972, an executive order and the Environmental Protection Agency (EPA) action banned it use for predator control on federal lands or in federal programs. In 1978, its use in pesticide formulations was limited to 0.5%. In the 1980s, its use was curtailed for above-ground applications to reduce toxicity to nontarget species, especially birds; by the end of the decade, its use was limited to below ground. Currently, strychnine is only registered for use by the EPA as a below-ground bait pesticide for use on pocket gophers.

86. Styrene

CHEMICAL NAME = styrene
CAS NUMBER = 100–42–5
MOLECULAR FORMULA = C_8H_8
MOLAR MASS = 104.2g/mol
COMPOSITION = C(92.3%) H(7.7%)
MELTING POINT = −31°C
BOILING POINT = 145°C
DENSITY = 0.91 g/cm³

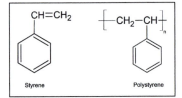

Styrene is a clear, colorless, flammable liquid with a sweet smell that is used in thousands of products. Styrene occurs naturally in plants. It was first isolated from a resin called storax obtained from the inner bark of the Oriental sweet gum tree (*Liquidambar orientalis*) by Bonastre. In 1839, the German apothecary Eduard Simon prepared styrene by distilling it from storax and called it styrol. Simon observed it solidified into a rubbery substance after being stored and believed it had oxidized into styrol oxide. Subsequent analysis showed the solid did not contain oxygen and it was renamed metastyrol. This was the first written record of polymerization in chemistry. In 1845, the English chemist, John Blyth, and the German chemist, August Wilhelm von Hofmann (1818–1892), observed that styrene was converted to polystyrene by sunlight and that styrene could be polymerized to polystyrene by heating in the absence of oxygen. It took another 70 years for the polymerization of styrene to be described by Hermann Staudinger (1881–1965) in the 1920s. This laid the foundation for the commercial polystyrene industry that developed in the 1930s.

Styrene is made by dehydrogenation of ethylbenzene at high temperature using metal catalysts: $C_6H_5CH_2CH_{2(g)} \xrightarrow{\Delta, metal} C_6H_5CH = CH_{2(g)} + H_{2(g)}$. This is called the EB/SM (ethylbenzene/styrene monomer) process. Styrene can also be made by PO/SM (propylene oxide/styrene monomer) process). This process starts by oxidizing ethylbenzene ($C_6H_5CH_2CH_2$) to its hydroperoxide ($C_6H_5CH(OOH)CH_3$), which is then used to oxidize propylene ($CH_3CH = CH_2$) to produce propylene oxide (CH_3CH_2CHO) and phenylethanol ($C_6H_5CH(OH)CH_3$). The phenylethanol is then dehydrated to give styrene and water. Styrene can also be synthesized by reacting benzene and ethylene or natural gas.

Styrene readily polymerizes into polystyrene when heated or using free radical polymerization. The first commercial polystyrene plant was opened in 1931 by BASF in Germany. Dow started producing polystyrene in 1938. Pure polystyrene is a colorless, hard, brittle plastic. One of the early challenges during commercial polystyrene production was controlling polymerization. The polymerization is highly exothermic, and heat generated must be removed to prevent thermal degradation of the polymer. To prevent premature polymerization of styrene, inhibitors were added to the monomer. Advances in the commercial production of styrene and polystyrene were accelerated during World War II. Styrene production was a top priority of the Allies who needed to produce synthetic rubber (see Isoprene) to replace natural sources controlled by the Japanese. During this period, continuous production methods replaced batch methods and new processes were developed.

During the early 1940s, the development of cellular polystyrene took place. Styrofoam was discovered accidentally by Ray McIntire (1918–1996), a Dow chemical engineer in 1944. McIntire was trying to make an artificial rubber for electrical insulation. He was combining isobutene with polystyrene when the isobutene formed bubbles in the styrene, resulting in a light cellular structure. Dow registered the trademark Styrofoam in 1954, but the name is now used generically for foam cellular insulation.

World demand for styrene monomer in 2006 is approximately 25 million tons. Styrene has numerous uses. The homopolymer, which is hard and clear, is used for plastic eating utensils, CD/DVD cases, and plastic hobby models. The most common forms of polystyrene is expanded polystyrene (EPS) and extruded expanded polystyrene (XEPS). EPS is produced using a mixture of polystyrene beads, pentane, and a blowing agent. The mixture is heated with steam, causing the beads to expand to 10 to 100 times their original volume as the pentane vaporizes. After this, the mixture is injected into a vacuum mold where heat and a partial vacuum cause further expansion. EPS can be molded into a variety of shapes. The pentane in the foam is replaced by air during the curing process. Extruded expanded polystyrene starts with polystyrene crystals. The crystals are mixed with additives and blowing agents in an extruder. In the extruder, the mix is heated under pressure into a plastic melt. This plastic melt expands through a die into foam. Extruded expanded polystyrene cannot be molded but is produced in sheets. At one time chlorofluorocarbons (CFC) were the preferred blowing agents used to produce expanded polystyrenes, but they have been replaced by hydrochlorofluorocarbons because of concerns about CFC's impact on the ozone layer (see Dichlorodifluoromethane). Styrofoam, produced by Dow, is extruded. Coffee cups and food packaging are technically not Styrofoam because Dow does not produce Styrofoam as molded expanded polystyrene. EPS and XEPS are used extensively for insulation in the construction industry. Polystyrene is used as a co-polymer with a number of other materials. Examples of co-polymers are acrylonitrile-butadiene-styrene, styrene-acrylonitrile, and styrene-butadiene rubber. Polystyrene is used in paints, coatings, adhesives, and resins.

Effects of styrene on human health depend on concentration, length of exposure, and individual genetics. Styrene vapor irritates the eyes, the nose, and the throat and can adversely affect the central nervous system. Health effects associated with breathing low concentrations of styrene over extended periods in the workplace include alterations in vision and hearing loss and increased reaction times. The Environmental Protection Agency has classified styrene as a potential human carcinogen.

87. Sucrose

CHEMICAL NAME = α-D-glucopyranosyl-(1→2)-
β-D-fructofuranoside
CAS NUMBER = 57–50–1
MOLECULAR FORMULA = $C_{12}H_{22}O_{11}$
MOLAR MASS = 342.3 g/mol
COMPOSITION = C(42.1%) H(6.5%) O(51.4%)
MELTING POINT = decomposes between 160°C and 186°C
BOILING POINT = decomposes between 160°C and 186°C
DENSITY = 1.59 g/cm³

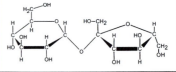

Sucrose is the white granulated compound referred to as sugar. Sucrose is a disaccharide made of glucose and fructose. The main sources of sucrose for the production of commercial sugar are sugarcane and sugar beets. Sugarcane is a tall perennial grass of the genus *Saccharum* native to Southeast Asia and the South Pacific. It has been consumed by chewing the stalk in areas where it grows for thousands of years. Sugarcane spread to India where it was processed to extract crude sugar as early as 2,500 years ago. Persian invaders discovered sugar after invading India and the plant and sugar production spread into the Middle East around 600 C.E. Europeans were introduced to sugar around 1100 C.E. when the first crusaders returned with knowledge of the sweet spice and the Arab Empire spread into Spain.

Sugar was an important trade commodity, and initially its limited availability made it expensive and use was limited to nobility and the wealthy. Venice became Europe's main entry point for sugar. As European countries colonized areas suitable for sugarcane, sugar plantations were established. Columbus brought sugarcane plantings to the Caribbean during his second voyage in 1493. Sugarcane quickly spread throughout the Caribbean and South America as Europeans established a sugar trade industry. The rise of sugar as a trade commodity had economic, social, and political relevance. Taxes were imposed on sugar and its increased supply and subsequent drop in price made it available to the general population. The growth of sugarcane and its processing in sugar mills were labor intensive. To supply the labor needed to work sugar plantations, African slaves were obtained in the triangle trade. Sugar was a main trade item in this triangle where sugar was exported to Europe and then traded for merchandise that could be transported to Africa and exchanged for humans. These slaves were then brought to the West Indies and the American colonies to work on sugar plantations.

The use of sugar beet to obtain sugar began when the German chemist Andreas Sigismund Marggraf (1709–1782) extracted sucrose from sugar beets using alcohol. The amount of sucrose obtained by Marggraf did not warrant commercial use of beets as a sucrose source. During the late 18th century, Franz Karl Archard (1753–1821), a student of Marggraf, selectively bred beets to increase the sucrose content to 5–6% and developed a commercial method to extract sucrose. In 1801, British blockades of European ports during the Napoleonic Wars spurred sugar production using sugar beets. This established the use of sugar beets as a viable alternative to sugarcane. Although Europeans subsequently returned to the use of sugar cane after the war, the abolition of slavery along with advances in breeding of sugar beets to produce 15–20% sugar made sugar beets competitive with sugarcane. Sugar beets provided a sucrose source that could be grown in temperate climates.

The processing of sugarcane into refined sugar starts with the harvesting of the stalks. The stalks are sent to a mill where they are chopped and shredded and then pressed between rollers to extract the juice. The juice's pH is adjusted to a neutral value of 7.0 with lime. The juice then passes through a clarifier, allowing some impurities to settle out before it is concentrated in boilers under reduced pressure. The sugar syrup thickens in the boilers where evaporation concentrates the juice to syrup of approximately 60% sucrose. The syrup then passes into crystallizers where it is further concentrated and a mixture of crystals and syrup is produced. Separation of the sugar crystals from the syrup occurs by spinning the mixture in centrifuges. The centrifuge produces raw brown sugar and the remaining syrup is molasses. The raw brown sugar leaves the mill and passes to the refinery. Here the raw sugar is washed, filtered, and further processed to remove remaining impurities and decolorize it. The refined sugar is 99.9% sucrose and is the familiar white, granulated product called table sugar. Sugar crystallized from the colored syrup results in brown sugar and contains between 92% and 96% sucrose. Brown sugars retain some of the molasses from the syrup and can be darker or lighter, depending on the amount of molasses in the final product. Sugar beet processing is similar to that of sugarcane, but the beets are soaked in hot water to extract the sucrose juice.

Sugar is widely used in food processing for numerous purposes. Its most prevalent use is as a flavoring agent to impart a sweet pleasant taste to food and beverages. It masks the bitter taste of coffee and teas and sweetens the tartness of acidic fruits. Sugar is a natural preservative used in jams, jellies, and other preserved foods. It works by binding water, which is a necessary ingredient for the microorganisms that cause spoilage. Sugar is also used for various culinary processes. For example, yeast action on sugar creates carbon dioxide, causing baked goods to rise. It binds proteins to retard coagulation in eggs, enhances flavoring, and caramelizes when heated above its melting point to produce a sweet, hard coating.

In recent years the consumption of carbohydrates in the form of refined sugar has received significant attention from health professionals. The annual consumption of sugar in the United States is about 50 pounds per person or just under 1 pound per week. The adverse effects of excessive sugar in the diet include obesity, increased risk of heart disease, diabetes, tooth decay, and disruptive behavior such as hyperactivity in children. Because of these problems, the food industry has used a number of synthetic or artificial sweeteners in place of sugar. These artificial sweeteners may reduce the use of sugars, but they have also been linked to adverse health problems such as cancer. Saccharin is the oldest synthetic sweetener. It was discovered in 1879 and has been used since 1900. During the 1970s, rats fed large doses of this substance developed bladder tumors and the federal government proposed banning this substance. Public resistance prevented its ban, but foods containing saccharin must carry a warning label (see Saccharin). Aspartame, known commonly as *Nutra-Sweet*, is the predominant synthetic sweetener in use today. It consists of two amino acids: aspartic acid and phenylalanine. It is approximately 160 times sweeter than sucrose. The health issues associated with aspartame are associated with the disease phenylketonuria (PKU). PKU results when excess amounts of phenylalanine occur in individuals owing to the lack of the enzyme needed to catalyze the essential amino acid phenylalanine. This genetic condition, if not treated, causes brain damage resulting in mental retardation. Because aspartame contains phenylalanine, it is suggested that infants under two years old not ingest any foods containing synthetic sweeteners (see Aspartame). Cyclamates were once a popular sweetener used in this country, but they were banned in 1969 because of possible carcinogenic effects. Their use is permitted in Canada. Sucralose is a synthetic sweetener approved for use in the United States in 1998. Sucralose is a chlorinated substituted chemical prepared from sucrose itself (Figure 87.1). It is 600 times sweeter than sucrose, passes unmodified through the digestive system and, unlike aspartame, is heat stable so it can be used in cooking.

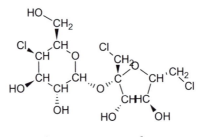

Figure 87.1   Sucralose.

Although the use of synthetic sweeteners is increasing annually, the quantity of sucrose is also increasing. This apparent contradiction seems to be related to the increased production of low-fat foods in recent years. The fat calories of many low-fat foods have been replaced by sugars. Questions persist regarding the safety of artificial sweeteners, but the consensus is that moderate intake of these substances do not pose a risk to human health. The federal government in approving synthetic sweeteners considers guidelines regarding safe daily limits. The acceptable daily intake (ADI) is considered the amount of a sweetener that can be consumed daily over a lifetime without adverse effects. For example, the ADI of aspartame is 50 mg per kilogram of body weight. No ADI has been established for saccharin, but the Food and Drug Administration limits the amount of saccharin per serving to 30 mg.

Sucrose is predominantly associated with the food industry, but it does have industrial uses in other areas. Sucrose fatty acid esters are a mixture of mono, di, and tri esters of sucrose with fatty acids. These are use in cosmetics, shampoos, resins, inks, paper processing, and pesticides. Sucrose benzoate is used as an emulsifier and in nail polishes. Sucrose has also been used in making glues and treating leather.

88. Sulfuric Acid

CHEMICAL NAME = sulfuric acid
CAS NUMBER = 7664–93–9
MOLECULAR FORMULA = H_2SO_4
MOLAR MASS = 98.08 g/mol
COMPOSITION = H(2.1%) S(32.7%) O(65.2%)
MELTING POINT = 10°C
BOILING POINT = 290°C
DENSITY = 1.84 g/cm³

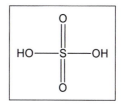

Sulfuric acid is a colorless, oily, dense liquid that is one of the most important industrial chemicals. Each year sulfuric acid tops the list of chemicals (ignoring water) used by industry, and it is claimed that a country's economic status can be gauged by the amount of sulfuric acid it consumes. More than 40 million tons are produced in the United States annually and approximately 170 million tons are produced globally. Sulfuric acid has a long history and was first produced by ancient alchemists. Its discovery is credited to the Persian physician Mohammad Ibn Zakariya al-Razi (Rhazes, 854–925), who produced sulfuric acid from the dry distillation of minerals. Dry distillation typically involves heating a substance in a closed container to limit oxygen and combustion. As the substance is heated, it decomposes and the volatile components can be captured. Because sulfuric acid was obtained from distilling minerals, it is called a mineral acid. The ancient method of sulfuric acid production involved heating either iron (II) sulfate heptahydrate ($FeSO_4 \bullet 7H_2O$), which was called green vitriol, or copper (II) sulfate pentahydrate ($CuSO_4 \bullet 5H_2O$), called blue vitriol. When minerals containing these compounds were heated, the products included sulfur trioxide (SO_3) and water. The combination of sulfur trioxide and water produced sulfuric acid: $SO_{3(g)} + H_2O_{(l)} \rightarrow H_2SO_{4(aq)}$. The production of sulfuric acid from natural minerals called vitriols and its oily appearance led to the common name "oil of vitriol" for sulfuric acid.

Sulfuric acid was one of the first chemicals to be produced industrially on a large scale. Until the early 1700s, sulfuric acid was produced in glass jars of several liters in which sulfur and potassium nitrate, KNO_3, were heated. Pyrite was often substituted for sulfur because of its availability and lower cost. In 1746, John Roebuck (1718–1794) substituted large lead-lined chambers for glass jars and thereby was able to increase the amount of sulfuric acid produced from a few liters or pounds at a time to tons. The lead chamber process was used throughout the 1800s to produce sulfuric acid. The reactions representing this process are:

$$6KNO_{3(s)} + 7S_{(s)} \xrightarrow{heat} 3K_2S + 6NO_{(g)} + 4SO_{3(g)}$$

$$SO_{3(g)} + H_2O_{(l)} \longrightarrow H_2SO_{4(aq)}$$

Although the lead-chamber processed increased the amount of sulfuric acid that could be produced, it relied on a source of nitrate that usually had to be imported. The process also produced nitric oxide gas (NO), which oxidized to toxic brown nitrogen dioxide (NO$_2$) in the atmosphere. To reduce the supply of nitrate required and the amount of nitric oxide produced, Joseph Louis Gay-Lussac (1778–1850) proposed that the nitric oxide be captured in a tower and recycled into the lead chamber. Although Gay-Lussac first proposed this modification to the lead-chamber method around 1830, it was not until the 1860s that John Glover (1801–1872) actually implemented Gay-Lussac's idea with the Glover tower.

The lead-chamber process supplied the world's need for sulfuric acid for a century and a half. In the late 19th century, the contact process replaced the lead-chamber process and is still used today to produce the world's supply of sulfuric acid. The contact process was first developed by Peregrine Phillips (1800–?), a British acid dealer, in 1831. The contact process used sulfur dioxide, SO$_2$, which was produced as a by-product when sulfur-bearing ores were smelted. The contact process was named because the conversion of sulfur dioxide to sulfur trioxide, SO$_3$, takes place on "contact" with a vanadium or platinum catalyst during the series of reactions:

$$S_{(s)} + O_{2(g)} \xrightarrow{heat} SO_{2(g)}$$

$$2SO_{2(g)} + O_{2(g)} \xrightarrow{heat,catalyst} 2SO_{3(g)}$$

$$SO_{3(g)} + H_2O_{(l)} \longrightarrow H_2SO_{4(l)}$$

The production of sulfuric acid was sufficient to meet world demand in the mid-18th century. At that time, sulfuric acid was used in producing dyes, for bleaching wools and textiles, and refining metals. Its demand greatly increased at the end of the 18th century when a method was discovered for preparing sodium carbonate, also known as soda ash or soda, using H$_2$SO$_4$ (see Sodium Carbonate).

Sulfuric acid is universally used in the chemical industry in the production of numerous products. Its largest use, accounting for approximately two-thirds of its total use, is in the production of fertilizer. Adding acid to phosphate rock produces phosphate fertilizer. Superphosphate fertilizer is produced by adding sulfuric acid to the phosphate rock fluorapate, [Ca$_3$(PO$_4$)$_2$]$_3$CaF$_2$, according to the reaction:

$$[Ca_3(PO_4)_2]_3CaF_2 + 7H_2SO_4 + 3H_2O \rightarrow 3[CaH_4(PO_4)_2 \bullet H_2O] + 7CaSO_4 + 2HF$$

Phosphate fertilizer may also be produced by reacting phosphate rock with phosphoric acid that has been produced from sulfuric acid. Phosphoric acid produced from sulfuric acid is called wet process phosphoric acid. In the production of fertilizer using this method, sulfuric acid reacts with the calcium phosphate, (Ca$_3$(PO$_4$)$_2$, in the rock to produce phosphoric acid

(H_3PO_4) and calcium sulfate ($CaSO_4 \cdot 2H_2O$, also known as gypsum). The phosphoric acid is then processed with the phosphate rock to produce triple superphosphate fertilizer. Triple superphosphate has three times the concentration of phosphorus as superphosphate. Another fertilizer produced from sulfuric acid is ammonium sulfate, $(NH_4)_2SO_4$. Ammonium sulfate is produced by reacting ammonia and sulfuric acid.

Because of its ability to form hydrates, sulfuric acid is used as a dehydrating agent. It is often used to remove water from organic compounds. When organic compounds are dehydrated, the hydrogen and oxygen are separated from carbon. This is observed when sulfuric acid is spilled on paper or textile, producing a dark burnt stain. The stain is similar in appearance to what would be produced if the substance were burned. Sulfuric acid's dehydrating properties are used to dehydrate and dry foods. Sulfuric acid will react with a metal to produce hydrogen gas and metal sulfate. Its reactions with metals make it useful in metal refining, metal cleaning, and metallurgical applications. For example, sulfuric acid is used in electroplating solutions. Sulfuric acid's electrolytic property makes it a common electrolyte used in car batteries (see Figure 88.1). Sulfuric acid is as a catalyst used in reactions in numerous industries such as petrochemicals, textiles, dyes, and pharmaceuticals. Sulfuric acid is also used to produce other acids such as nitric, hydrochloric, and phosphoric acids. For example, reacting salt, NaCl, with sulfuric acid produces hydrogen chloride gas, which, when dissolved in water, gives hydrochloric acid ($HCl_{(aq)}$): $NaCl_{(aq)} + H_2SO_{4(l)} \rightarrow NaHSO_{4(aq)} + HCl_{(g)}$

Lead Storage Battery

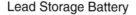

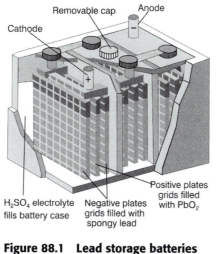

Figure 88.1 Lead storage batteries use sulfuric acid as an electrolyte.

Drawing by Rae Déjur.

Sulfuric acid plays a major role in air quality and is a primary contributor to acid deposition. The combustion of fossil fuels, which contain sulfur as an impurity, results in the production of sulfur oxides. Sulfur oxides react with water in the atmosphere to produce sulfuric acid, but they may also undergo other reactions leading to dry deposition. Clean air is slightly acidic, with a pH of approximately 5.6. The acidic conditions are primarily due to the presence of carbonic acid produced from the carbon dioxide present in the atmosphere.

Acid precipitation resulting from sulfuric acid (and nitric acid) has the ability to significantly lower the pH. The reactions forming sulfuric acid in the atmosphere often take several days; therefore acid precipitation can fall several hundred kilometers away from the pollution source. This has led to transnational negotiations and agreements in an attempt to control sulfur emissions, the precursor to acid precipitation.

Acid precipitation has both direct and indirect effects. Individual organisms can be harmed by the lowering of the pH in aquatic systems, or reproduction of organisms may be disrupted. Table 88.1 shows the effect of pH on several major groups. In addition to lowering the pH of the water body, acid precipitation can affect the surrounding watershed by mobilizing metals, such as aluminum, and leaching nutrients from the soil. Acid precipitation interferes with nutrient uptake in plants, causing plants to be stressed. This makes them more susceptible to diseases, climatic stresses, and insects, which ultimately leads to a decrease in natural plant and crop productivity and therefore a loss of agricultural productivity. Another major effect of acid precipitation is the deterioration of structures such as buildings, bridges, and artworks. Limestone and marble structures, which are made of calcium carbonate, such as statues are particularly vulnerable to the effects of acid precipitation. It is estimated that the economic effects of acid precipitation in the United States is more than $10 billion, but it is difficult to determine the exact costs of this problem.

Table 88.1   Critical pH Values for Several Types of Organisms

Group	pH
Frogs	4.0
Perch	4.5
Trout	5.0
Bass	5.5
Mayfly	5.5
Clams	6.0
Snails	6.0

In recent decades, countries have taken numerous measures to decrease acid precipitation. In the United States, sulfur emissions are regulated according to the Clean Air Act. Primary control methods to reduce sulfur emissions at its source include switching to low-sulfur coal, scrubbing coal to remove sulfur, substituting natural gas and oil for coal, using alternative fuels, and using scrubbers to remove sulfur oxides from stack gases. A technique used to combat acidification in lakes is to neutralize the acid by treating them with a base such as slaked lime, $Ca(OH)_2$.

89. Tetrafluoroethylene

CHEMICAL NAME = tetrafluoroethene
poly(tetrafluoroethene)
CAS NUMBER = 116–14–3 (TFE) 9002–84–0
(PTFE)
MOLECULAR FORMULA = C_2F_4
MOLAR MASS = 100.0 g/mol
COMPOSITION = C(24%) O(76%)
MELTING POINT = −142.5°C (TFE) 327°C (PFTE)
BOILING POINT = −76.3°C (TFE) 400°C (PFTE)
DENSITY = 0.0051 g/cm³ (3.9, air = 1) (TFE) 2.2 g/cm³ (PTFE)

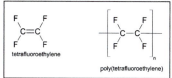

tetrafluoroethylene

poly(tetrafluoroethylene)

Tetrafluoroethylene (TFE), also known as perfluoroethylene, is a colorless, flammable, toxic gas. It is the monomer used for polytetrafluoroethylene (PTFE), which is sold under the DuPont tradename of Teflon. TFE is co-polymerized with other compounds to produce a variety of Teflons. TFE is produced by heating chlorodifluoromethane ($CHClF_2$, Freon-22) or trifluoromethane (CHF_3, Freon-23). TFE is used almost exclusively as a monomer in the production of PTFE. PTFE is a vinyl polymer, which means it is made from a monomer with carbon-carbon double bonds. PTFE is made from TFE by free radical polymerization.

The polymerized form of TFE was discovered accidentally by researchers at DuPont in 1938. During the mid-1930s, DuPont was investigating the development of new chloro-fluorocarbons (CFCs) for use as refrigerants. DuPont joined with General Motors to form a company called Kinetic Chemicals to combine their efforts in this area. Roy J. Plunkett (1910–1944) was one of the chemists assigned to the project. Plunkett was using TFE as a reactant to produce CFCs and stored it in pressurized canisters. Plunkett cooled the TFE canisters with dry ice to reduce the risk of explosion (TFE can explode owing to spontane-ous polymerization at high temperatures). On the morning of April 6, Plunkett and his laboratory technician, Jack Rebok, attempted unsuccessfully to deliver TFE from one of the cylinders. After discovering that the valve was not plugged and the gas had not escaped, Plunkett removed the valve and observed a white solid material. Plunkett and Rebok then sawed open several of the TFE storage containers and observed that they were coated with a white waxy substance. Plunkett noted that the TFE had polymerized. Over the next year Plunkett examined the properties of the substance. He discovered that the substance was inert to other chemicals, had a high melting point, and was very slippery. Plunkett also worked on duplicating the conditions necessary to produce it from TFE. He applied for a patent for polytetrafluoroethylene polymers in 1939, which was granted in 1941 (U.S. Patent 2230654). He assigned the patent to Kinetic Chemicals. DuPont registered the term Teflon in 1945 and introduced Teflon products the next year.

PTFE's unique physical properties are due to its chemical structure. PTFE consists of long chains of carbon atoms surrounded by fluorine atoms. The fluorine atoms act as a protective barrier that shields the carbon-carbon bond from chemical attack. The fluorine atoms repel other atoms, making it difficult for anything to stick to PTFE. PTFE resins have very low coefficients of friction (< 0.1). The strong fluorine-to-carbon bonds and high electronegativity of fluorine make PTFE very stable. The long chains of PTFE pack closely together to give a dense crystalline solid. The packing, which can be compared to stacking boards, produces little cross-linking.

Teflon is best known for its use in cookware, but its use in this area followed original industrial applications in gaskets, sealers, tape, and electrical insulation. These applications were a direct result of the use of PTFE for military purposes during World War II.

The use of Teflon in cookware was stimulated by a French inventor named Marc Grégoire (1905–1996). Grégoire used Teflon to coat his fishing gear to prevent tangles. His wife, Colette, thought Teflon could be applied to her frying pans and Grégoire was successful in this endeavor. He was granted a patent for the process in 1954 and began selling coated pans out of his home in the 1950s. In 1956, he formed the company Tefal for the production of Teflon-coated cookware and made plans to build a factory. In the late 1950s, Thomas G. Hardie, a businessman from the United States, met Grégoire in Paris and became convinced that Teflon cookware could be marketed in the United States. Upon returning to the United States, Hardie tried to convince American cookware producers to adopt Grégoire's technology, but he met with resistance. This led Hardie to import Teflon goods from Grégoire's French factory and market them under the name T-fal. Hardie met with DuPont executives and tried to accelerate their Teflon cookware plans. DuPont continued to conduct research on Teflon's safety in cookware and sought approval from the Food and Drug Administration (FDA) for its use in this application. Meanwhile, Hardie's T-fal sales soared in the early 1960s, leading him to build his own plant for production in the United States. The FDA approved Teflon's use in cookware in 1962. As Hardie's success with T-fal continued, other manufacturers started to produce Teflon cookware, which flooded the market with Teflon cooking items.

Teflon's marketing success for a non-stick cooking items was accompanied by concerns about the safety of cooking with Teflon products. These concerns continue today and have intensified in recent years. The concerns involve workers who produce Teflon, as well as general public use of Teflon cookware. A larger issue involves the use of perfluorooctanoic acid, $C_8HF_{15}O_2$ (PFOA). This chemical, also called C8, is used in small amounts during the polymerization of PTFE, and trace amounts of it may remain in the final product. PFOA (and its salt ammonium perfluorooctanoate) is a persistent chemical and likely human carcinogen. At high enough temperatures, residual PFOA may be released from Teflon products in fumes or can migrate into foods. Birds are particularly susceptible to PFOA. The exact temperature where

PFOA is mobilized is unknown; environmental groups and manufacturers cite temperatures that can vary by more than 100°C. Leading cookware manufacturers have recommended a maximum temperature of 260°C. Recommended frying temperature is 190°C. DuPont reached a $16.5 million settlement with the Environmental Protection Agency in 2005 over allegations that it covered up the health risks of PFOA. DuPont officials claimed the settlement was done to avoid protracted litigation and they had done nothing illegal. Class action lawsuits filed on behalf of consumers against DuPont are seeking billions of dollars in damages. These suits claim that DuPont concealed information about the health risks of Teflon. In response to the concern about C8, DuPont and cookware manufacturers plan to phase out the use of PFOA in Teflon over the next several years.

PTFE is used in hundreds of applications in addition to cookware. PTFE-based powders are use in inks, plastics, coatings, and lubricants. PTFE resins can be molded into gaskets, seals, bearings, gears, and other machine parts. Films of PTFE materials are used as liners, insulation, membranes, and adhesives. Teflon tape is commonly used in plumbing work. PTFE is used to produce rainproof garments. Gore-Tex, a PTFE material, was developed by Wilbert Gore (1912–1986) and his wife, Genevieve, in their basement. Gore was a DuPont engineer who worked on Teflon and started his business in 1958 after failing to persuade DuPont executives to pursue PTFE fabrics. The Gores mortgaged their home to start their business, and Gore-Tex outdoor clothing became a well-known product beginning in the 1980s.

90. Tetrahydrocannabinol (THC)

CHEMICAL NAME = delta-9-tetrahydrocannabinol
CAS NUMBER = 1972–08–3
MOLECULAR FORMULA = $C_{21}H_{30}O_2$
MOLAR MASS = 314.5 g/mol
COMPOSITION = C(80.2%) H(9.6%) O(10.2%)
MELTING POINT = 80°C
BOILING POINT = 200°C
DENSITY = not reported

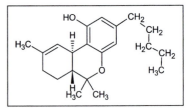

Tetrahydrocannabinol (THC) is the main active compound in marijuana. It comes from the plant *Cannabis sativa* (cannabis), which is a dioecious (monoecious varieties do exist) annual herb naturally found in many tropic and temperate regions of the world. Many varieties of cannabis exist, and two related species (*Cannabis indica* and *Cannabis ruderalia*) are main sources of THC. *Cannabis sativa* is also known as hemp, although this name is not unique to the species; its stem is a source of fiber that has been used throughout history for hundreds of applications including rope, twine, paper, and cloth. Hemp seeds are edible and high in protein. The seeds are also a source of fatty oil that can be used for food, cosmetics, medicines, and as a fuel source. *Cannabis* contains chemicals called cannabinoids; of the 60 cannabinoids found in *Cannabis*, one is THC, , which is the psychoactive ingredient in marijuana. Marijuana is produced from the leaves and flowers of cannabis, and hashish is a resin collected from the female flowers. The THC content, which determines the effect of cannabis drugs, varies with plant structure, variety, and preparation. Buds and flowers specifically cultivated for drug use have greater THC content than leaves. THC content may vary from a few tenths of a percent to more than 10%, but good quality marijuana has a THC content of approximately 10%, and good hashish and hashish oils generally have THC contents between 30% and 80%.

Cannabis use by humans dates from prehistoric times. It was used as a food source, medicine, fuel source, for fiber, and as a recreational drug. Although it is not known how *Cannabis* was first used, fibers and seeds have been found in Neolithic archaeological sites. Pottery recovered from Taiwan archaeological sites dating from more than 10,000 years ago contain cannabis fiber. The earliest evidence of the medicinal use of *Cannabis* comes from the Chinese Emperor and herbalist Shen-Nung, who lived in the 28th century B.C.E. Shen-Nung wrote a work on herbal medicines in which he recommended *Cannabis* for malaria, beri-beri, gout, constipation, rheumatic pains, and female menstrual problems. The first use of *Cannabis* as a psychoactive agent is unknown, but it is assumed that knowledge of its mind-altering effect was a natural consequence of its medicinal use. Indian and Chinese writings as early as 2000 B.C.E. refer to the psychedelic effect of *Cannabis,* and it was used in religious rituals in both regions. The Greek historian Herodotus (484–425 B.C.E.) wrote about Scythians inhaling *Cannabis* smoke for its euphoric effect. Medieval alchemists used *Cannabis* vapors to clear the mind, and hemp oil was distilled from flowers for use in various formulations.

THC was first isolated from hashish in 1964 by Raphael Mechoulam (1930–) and Yehiel Gaoni at the Weizmann Institute. Mechoulam had obtained 5 kg hashish from Israeli police officials and the earliest scientific work on THC and cannabinoids used this source. In the early 1990s, the specific brain receptors affected by THC were identified. These receptors are activated by a cannabinoid neurotransmitter called arachidonylethanolamide, known as anandamide. Anandamide was named by Mechoulam using *ananda,* which is the Sanskrit word for ecstasy. Anandamide is thought to be associated with memory, pain, depression, and appetite. THC is able to attach to and activate anandamide receptors. These receptors are actually called THC receptors rather than anandamide receptors because researchers discovered that THC attaches to these receptors before anandamide was discovered. The areas of the brain with the most THC receptors are the cerebellum, the cerebral cortex, and the limbic system. This is why marijuana affects thinking, memory, sensory perception, and coordination.

THC in the form of marijuana is the most widely used and available illicit drug in the United States. The potency of marijuana is directly related to its THC content, which averages between 5% and 10% for commercial-grade marijuana. During the last several decades, the THC content of seized marijuana has steadily increased. In 1985, the THC content averaged 3.5%; in 2005, the average was 8%. It is difficult to determine the amount of marijuana produced because of its illegal status. Production in different countries is highly variable depending on growing conditions, weather, and enforcement. Mexico, the chief U.S. supplier, has averaged approximately 10,000 tons of production annually over the last several years, although not all of this is exported to the United States. It is estimated that between 15,000 and 25,000 tons of marijuana are available in the United States, with Americans spending approximately $10 billion annually on the drug.

Cannabis has been used as a medicine for thousands of years, and there are many proponents for the use of medical marijuana. Some common medicinal uses include alleviating nausea associated with cancer therapies, treating glaucoma, alleviating suffering in AIDS patients, and helping multiple sclerosis patients control muscles. Medical marijuana is a controversial issue. Opponents claim that medical marijuana will make the drug more readily available, there are suitable alternatives, it has limited medical efficacy, it is difficult to control, and it has harmful side effects. Proponents claim it relieves suffering, is not addictive, is safe, and is easily administered. The American Medical Association has taken a neutral view and called for more controlled research on the topic. The federal government's view that marijuana use is illegal for all purposes is in direct conflict with several states that have passed medical marijuana laws legalizing limited medical usage. In 2005, the Supreme Court ruled against state medical marijuana laws and supported the federal government view that users in states where it was allowed medicinally could be prosecuted.

Medical marijuana remains a controversial topic, but synthetic THC, dronabinol, marketed under the trade name Marinol, has been available by prescription since 1986. The dronabinol analog nabilone is another THC prescription drug marketed under the name Cesamet. Marinol and Cesamet, taken as capsules, have Food Drug Administration approval as an antinausea agent and appetite stimulant (for AIDS patients), but they are also prescribed for depression and muscle spasms. In 2005, Canada was the first country to approve Sativex, a cannabis spray that relieves pain in people with multiple sclerosis.

91. *Thymine.* See 29. Cytosine, Thymine, and Uracil

92. *Trinitrotoluene (TNT)*

CHEMICAL NAME = methyl-2,4,6-trinitrobenzene
CAS NUMBER = 118–96–7
MOLECULAR FORMULA = $C_7H_5N_3O_6$
MOLAR MASS = 227.1 g/mol
COMPOSITION = C(37.0%) H(2.2%) N(18.5%) O(42.3%)
MELTING POINT = 80.1°C
BOILING POINT = unstable, explodes at 240°C
DENSITY = 1.65 g/cm³

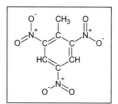

TNT is the abbreviation of the aromatic nitrated aromatic compound 2,4,6-trinitrotoluene. It is a pale-yellow crystalline solid that was first synthesized in 1863 by the German chemist Joseph Wilbrand (1811–1894), but it was not immediately used as an explosive. TNT is made by nitrating toluene using nitric acid, sulfuric acid, and oleum (a mixture of sulfuric acid and SO_3). Nitration of toluene occurs in stages, with the nitro units added sequentially in a stepwise process as the reaction proceeds. The last nitro unit is accomplished by using oleum (SO_3 dissolved in sulfuric acid). After nitration, unused acids are recycled, and the product is washed with sodium sulfite and water to remove impurities.

TNT is one of the most common explosives. Unlike nitroglycerin, TNT will not explode when subjected to significant shock and friction. It is classified as a secondary explosive, which means it requires an initiating explosive to detonate. The Germans began production of TNT in the last decade of the 19th century, and it was used in the mining industry. Military engineers adapted mining explosives for use in warfare, and TNT started to be incorporated in munitions in 1902. The first widespread use of TNT occurred during World War I. It had several advantages over picric acid, which had been used widely in munitions during the latter half of the 19th century and the first part of the 20th century. Unlike TNT, picric acid was much more likely to detonate when disturbed. Furthermore, picric acid could react with the metal in artillery shells, producing explosive picrate compounds. TNT did not react with metal and its low melting point meant it could be melted using steam and poured into shells. TNT's ability to withstand shock was an advantage in penetrating the armor of ships, tanks, and other metal-plated objects. TNT shells equipped with an appropriate fuse could penetrate objects and cause maximum damage on detonation. In contrast, picric acid shells would explode after striking an object. The British made extensive use of picric acid shells in World War I, which was called lyddite after the town of Lydd in England where it was manufactured; the Germans made greater use of TNT. Although picric acid had more explosive power, the Allies began to use more TNT because of its ability to penetrate armor without detonating prematurely.

Between World War I and II, TNT replaced picric acid as the explosive of choice in munitions. It was also mixed with other compounds to produce more powerful explosives with unique characteristics. Amatol is a mixture containing between 40% and 80% ammonium nitrate and TNT. Pentolite is a mixture of PETN (pentaerythritol tetranitrate) and TNT. Another common explosive mixture is RDX (cyclotrimethylenetrinitramine) and TNT. RDX is an abbreviation for Royal Demolition Explosive.

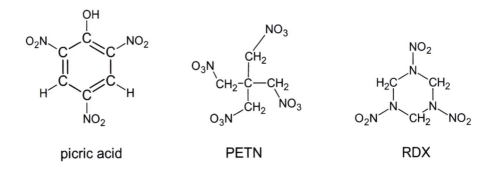

picric acid   PETN   RDX

TNT's explosive power results from the quick formation of stable gaseous products when TNT is detonated. The explosive reaction of TNT can be represented as: $2C_7H_5N_3O_{6(s)} \rightarrow 3N_{2(g)} + 7CO_{(g)} + 5H_2O_{(g)} + 7C_{(s)}$. On reaction, the volume of the products is a thousand times that of the TNT reactant. The rapid increase of volume causes the explosion. TNT's explosive power is used as a standard for energy, which is especially useful in rating explosives. Because the precise amount of energy release is difficult to calculate, 1 kiloton of TNT is defined as releasing 10^{12} calories of energy. Based on this equivalency, a thousand tons, 1 kiloton, would release 4.2×10^{12} joules; and a million tons, a megaton, releases 4.2×10^{15} joules. Nuclear weapons are rated in terms of megaton. The atomic bombs used in World War II were approximately 20 kilotons. Modern nuclear warheads have ratings on the order of several megatons.

TNT has limited use as a chemical intermediate in pharmaceuticals and for photographic chemicals. It is used to produce other nitrated compounds. Removing the methyl group from TNT produces 1,3,5-trinitrobenzene, and removing methyl and a nitro group produces 1,3-dinitrobenzene (1,3-DNB). Both trinitrobenzene and dinitrobenzene can be used as explosives. Trinitrobenzene is more powerful than TNT but less sensitive to impact. Dinitrobenzene has been used in the production of nitrocellulose, which is used for smokeless gunpowder and guncotton.

93. Toluene

CHEMICAL NAME = toluene
CAS NUMBER = 108–88–3
MOLECULAR FORMULA = C_7H_8
MOLAR MASS = 92.1 g/mol
COMPOSITION = C(91.3%) H(8.7%)
MELTING POINT = –95°C
BOILING POINT = 110.6°C
DENSITY = 0.87 g/cm³

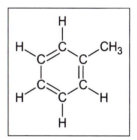

Toluene is a clear, flammable, aromatic hydrocarbon liquid with a smell similar to benzene. It is also called methylbenzene, indicating that a methyl group has been added to one of benzene's carbon atoms. Toluene was first isolated by Pierre-Joseph Pelletier (1788–1842) and Philippe Walter (1810–1847) in 1837. The name toluene comes from the South American tree *Toluifera balsamum.* Henri-Etienne Sainte-Claire Deville (1818–1881) isolated toluene from the tree's gum, *Tolu balsam,* in 1841.

The main source of toluene is from the catalytic reforming of naphthas during petroleum processing. During this process cycloalkanes are dehydrated, forming aromatics such as toluene and xylene along with hydrogen. Toluene can also be obtained from the pyrolysis of gasoline. It is a by-product when styrene is produced and can also be produced from coal tar, which was its main source in the first half of the 20th century.

Toluene has numerous applications in the chemical and petroleum industry, with approximately 6 million tons used annually in the United States and 16 million tons used globally. The major use of toluene is as an octane booster in gasoline. Toluene has an octane rating of 114. Toluene is one of the four principal aromatic compounds, along with benzene, xylene, and ethylbenzene, that are produced during refining to enhance gasoline's performance. Collectively, these four compounds are abbreviated as BTEX. BTEX is a major component of gasoline, forming about 18% by weight of a typical blend. Although the proportion of the aromatics is varied to produce different blends to meet geographic and seasonal requirements, toluene is one of the major components. A typical gasoline contains approximately 5% toluene by weight.

Toluene is a primary feedstock used to produce various organic compounds. It is used to produce diisocyanates. Isocyanates contain the functional group $-N = C = O$, and diisocyanates contain two of these. The two main diisocyanates are toluene 2,4-diisocyanate and toluene 2,6-diisocyanate. The production of diisocyanates in North America is close to a billion pounds annually. More than 90% of toluene diisocyanate production is used for making polyurethanes foams. The latter are used as flexible fill in furniture, bedding, and cushions. In rigid form it is used for insulation, hard shell coatings, building materials, auto parts, and roller skate wheels.

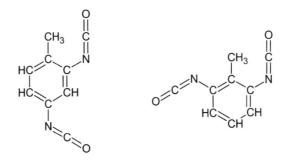

toluene 2,4-diisocyanate   toluene 2,6-diisocyanate

Benzene is produced from toluene through a process called hydrodealkylation. In this process, toluene reacts with hydrogen in the presence of a chromium, platinum, or molybdenum catalysts at temperatures of several hundred degrees Celsius and pressures of about 50 atmospheres: $C_6H_5CH_3 + H_2 \rightarrow C_6H_6 + CH_4$. Toluene can also be used to produce phenol, (C_6H_5OH), benzoic acid (C_6H_5COOH), and benzaldehyde (C_6H_5CHO). Nitrated forms of toluene produce explosive compounds; the most common of these is TNT (See Trinitrotoluene).

Toluene is insoluble in water but is a useful organic solvent. It is used in paint, paint thinner, shellacs, varnishes, dyes, and inks. It is a common component of cosmetics, nail polish, stain removers, and glues. A small amount of toluene occurs in cigarette smoke. In recent years toluene has replaced benzene in a number of products because of benzene's carcinogenetic properties. Toluene is not classified as a carcinogen, but it does pose a number of health hazards. Humans are exposed by breathing fumes from consumer products containing toluene. The main effect of toluene is on the central nervous system. Headaches and drowsiness may result at concentrations of 50 parts per million (ppm). Symptoms intensify with increasing concentrations. At 100 ppm dizziness and fatigue result; at 200 ppm a drunken state can result. As concentrations increase, nausea and disorientation take place. Death can result with concentrations of several thousand ppm. Toluene exposure can cause liver and kidney damage at high exposures. Toluene is responsible for some of the adverse effects associated with sniffing glues and other consumer products.

94. Triuranium Octaoxide

CHEMICAL NAME = triuranium (V) octaoxide
CAS NUMBER = 1344–59–8
MOLECULAR FORMULA = U_3O_8
MOLAR MASS = 842.1 g/mol
COMPOSITION = U(84.8%) O(15.2%)
MELTING POINT = 1,150°C
BOILING POINT = decomposes to UO_2 at 1300°C
DENSITY = 8.4 g/cm³

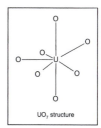

UO_7 structure

Uranium occurs in nature as a mixture of numerous uranium oxides. Triuranium octaoxide, U_3O_8, is the most stable and common chemical form of uranium oxide found naturally; uranium dioxide (UO_2) and uranium trioxide (UO_3) are also commonly found in uranium ores. Triuranium octaoxide, which is a complex oxide composed of the oxides U_2O_5 and UO_3, is a dark-green to black solid most commonly found in the mineral pitchblende. Uranium dioxide produces U_3O_8 when oxidized: $3UO_2 + O_2 \rightarrow U_3O_8$. Uranium trioxide is reduced to U_3O_8 when heated above 500°C: $6UO_3 \rightarrow 2U_3O_8 + O_2$. The structure of U_3O_8 is pentagonal bipyramidal, containing repeating UO_7 units.

Uranium is best known as a fuel for nuclear power plants. To prepare this fuel, uranium ores are processed to extract and enrich the uranium. The process begins by mining uranium-rich ores and then crushing the rock. The ore is mixed with water and thickened to form a slurry. The slurry is treated with sulfuric acid and the product reacted with amines in a series of reactions to give ammonium diuranate, $(NH_4)_2U_2O_7$. Ammonium diuranate is heated to yield an enriched uranium oxide solid known as yellow cake. Yellow cake contains from 70–90% U_3O_8 in the form of a mixture of UO_2 and UO_3. The yellow cake is then shipped to a conversion plant where it can be enriched.

Natural uranium consists of different isotopes of uranium. Natural uranium is 0.7% U-235 and 99.3% U-238. Uranium-238 is nonfissionable, and therefore naturally occurring uranium must be enriched to a concentration of approximately 4% to be used as fuel for nuclear reactors or 90% for weapons-grade uranium. Yellow cake is shipped to conversion plants for enrichment. The process involves dissolving yellow cake in nitric acid to produce uranyl nitrate hexahydrate, $UO_2(NO_3)_2 \cdot 6H_2O$. The uranyl nitrate solution is purified and heated to extract UO_3, which is then reduced to UO_2 with H_2: $UO_{3(s)} + H_{2(g)} \rightarrow UO_{2(s)} + H_2O_{(g)}$. To enrich uranium, the solid uranium oxide is fluorinated to put it into a gaseous phase by reacting with hydrogen fluoride: $UO_{2(s)} + 4HF_{(g)} \rightarrow UF_{4(s)} + 4H_2O_{(g)}$. Uranium tetrafluoride, UF_4, is combined with fluorine gas to yield uranium hexafluoride, UF_6: $UF_{4(s)} + F_{2(g)} \rightarrow UF_{6(g)}$. Uranium hexafluoride is a white crystalline solid at standard temperature and pressure, but it sublimes to a gas at 57°C. The U-235 in uranium hexafluoride can be enriched by several methods based on the difference in masses of the uranium isotopes. Two common methods are gaseous diffusion and gas centrifuge.

Enriched UF_6 is processed into UO_2 powder at fuel fabrication facilities using one of several methods. In one process uranium hexafluoride is vaporized and then absorbed by water to produce uranyl fluoride, UO_2F_2, solution. Ammonium hydroxide is added to this solution and ammonium diuranate is precipitated. Ammonium diuranate is dried, reduced, and milled to make uranium dioxide powder. The powder is pressed into fuel pellets for nuclear reactors.

Traditional nuclear power involves using the heat generated in a controlled fission reaction to generate electricity (Figure 94.1). The reactor core consists of a heavy-walled reaction vessel several meters thick that contains fuel elements consisting of zirconium rods filled with the UO_2 pellets. The reactor is filled with water for two purposes. One purpose is to serve as a coolant to transport the heat generated from the fission reaction to a heat exchanger. In the heat exchanger the energy can be used to generate steam to turn a turbine for the production of electricity. Water also serves as a moderator. A moderator slows down the neutrons, increasing the chances that the neutrons react with uranium, causing it to split. Interspersed between the hundreds of fuels rods are control rods. Control rods are made of cadmium and boron. These substances absorb neutrons. Raising and lowering the control rods control the nuclear reaction.

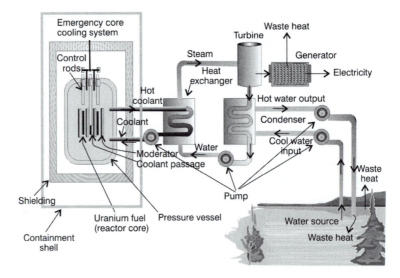

Figure 94.1   Schematic of nuclear power plant. Drawing by Rae Déjur.

The basic design of most nuclear reactors is similar, but several types of reactors are used throughout the world. In the United States most reactors use plain water as the coolant. Reactors using ordinary water are called light water reactors. Light water reactors can be pressurized to approximately 150 atmospheres to keep the primary coolant in the liquid phase at temperatures of approximately 300°C. The heat from the pressurized water is used to heat secondary water to generate steam. In a boiling water reactor, water in the core is allowed to boil. The steam produced powers the turbines directly. Heavy water reactors use water in the form of D_2O as the coolant and moderator. Heavy water gets its name from the fact that the hydrogen in the water molecule is the isotope deuterium, D ($_1^2H$), rather than ordinary hydrogen. The use of heavy water allows natural uranium rather than enriched uranium to be used as the fuel.

Breeder reactors were developed to use the 97% of natural uranium that occurs as nonfissionable U-238. The idea behind a breeder reactor is to convert U-238 into a fissionable fuel material. A reaction to breed plutonium is:

$$^{238}_{92}U + ^1_0n \rightarrow ^{239}_{92}U \rightarrow ^{239}_{93}Np + ^0_{-1}e \rightarrow ^{239}_{94}Pu + ^0_{-1}e$$

The plutonium fuel in a breeder reactor behaves differently than does uranium. Fast neutrons are required to split plutonium. For this reason water cannot be used in breeder reactors, as it moderates the neutrons. Liquid sodium is typically used in breeder reactors, and the term liquid metal fast breeder reactor (LMFBR) is used to describe it. One of the controversies associated with the breeder reactor is the production of weapon-grade plutonium and nuclear arms proliferation.

95. Uracil. See 29. Cytosine, Thymine, and Uracil

96. Urea

CHEMICAL NAME = Urea
CAS NUMBER = 57–13–6
MOLECULAR FORMULA = CH_4N_2O
MOLAR MASS = 60.06 g/mol
COMPOSITION = C(20.0%) H(6.7%) N(46.7%) O(26.6%)
MELTING POINT = 133°C
BOILING POINT = Decomposes
DENSITY = 1.32 g/cm³

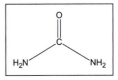

Urea has the distinction of being the first synthesized organic compound. Until the mid-18th century, scientists believed organic compounds came only from live plants and animals. They reasoned that organisms possessed a vital force that enabled them to produce organic compounds. The first serious blow to the theory of vitalism, which marked the beginning of modern organic chemistry, occurred when Friedrich Wöhler (1800–1882) synthesized urea from the two inorganic substances, lead cyanate and ammonium hydroxide: $Pb(OCN)_2 + 2NH_4OH \rightarrow 2(NH_2)_2CO + Pb(OH)_2$. Wöhler's discoveries on urea occurred while he was studying cyanates; he was attempting to synthesize ammonium cyanate when he discovered crystals of urea in his samples. He first prepared urea in 1824, but he did not identify this product and report his findings until 1828. In a note written to the famous chemist Jöns Jakob Berzelius (1779–1848) he proclaimed: "I must tell you that I can make urea without the use of kidneys, either man or dog. Ammonium cyanate is urea." Wöhler's synthesis of urea signaled the birth of organic chemistry.

Urea is a colorless, odorless crystalline substance discovered by Hilaire Marin Rouelle (1718–1779) in 1773, who obtained urea by boiling urine. Urea is an important biochemical compound and also has numerous industrial applications. It is the primary nitrogen product of protein (nitrogen) metabolism in humans and other mammals. The breakdown of amino acids results in ammonia, NH_3, which is extremely toxic to mammals. To remove ammonia from the body, ammonia is converted to urea in the liver in a process called the urea cycle. The urea in the blood moves to the kidney where it is concentrated and excreted with urine.

In the urea cycle, two molecules of ammonia combine with a molecule of carbon dioxide to produce a molecule of urea and water. The overall cycle involves a series of biochemical reactions dependent on enzymes and carrier molecules. During the urea cycle the amino acid ornithine ($C_5H_{12}N_2O_2$) is produced, so the urea cycle is also called the ornithine cycle. A number of urea cycle disorders exist. These are genetic disorders that result in deficiencies in enzymes needed in one of the steps in the urea cycle. When a urea cycle deficiency occurs, ammonia cannot be eliminated from the body and death ensues.

Urea is an important industrial compound. The synthesis of urea was discovered in 1870. Commercial production of urea involves the reaction of carbon dioxide and ammonia at high pressure and temperature to produce ammonium carbamate. Ammonium carbamate is then dehydrated to produce urea (Figure 96.1). The reaction uses a molar ratio of ammonia to carbon dioxide that is approximately 3:1 and is carried out at pressures of approximately 150 atmospheres and temperatures of approximately 180°C.

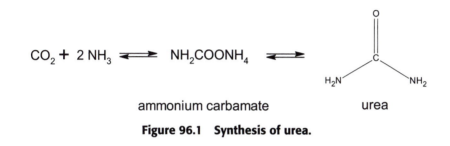

ammonium carbamate   urea

Figure 96.1   Synthesis of urea.

More than 100 million tons of urea is produced annually worldwide. The primary use of urea is as a nitrogen source in fertilizers, with about 90% of the urea production being used for this purpose. Urea's high nitrogen content (46%) makes it a concentrated source for adding fixed nitrogen to soils. It can be applied to the soil alone, but its high nitrogen content can stress plants and impact the soil negatively, so it is often blended with other nutrients. Blending also reduces the nitrogen content of the fertilizer. For example, blending with ammonium nitrate, NH_4NO_3, in different proportions produces fertilizers with various nitrogen contents. Urea in the soil is converted to ammonium nitrogen and taken up by plants. It can be applied in solid granule form or dissolved in water and used as a spray. Urea is also used agriculturally as a supplement in livestock feeds to assist protein synthesis.

Another use of urea is for resins, which are used in numerous applications including plastics, adhesives, moldings, laminates, plywood, particleboard, textiles, and coatings. Resins are organic liquid substances exuded from plants that harden on exposure to air. The term now includes numerous synthetically produced resins. Urea resins are thermosetting, which means they harden when heated, often with the aid of a catalyst. The polymerization of urea and formaldehyde produces urea-formaldehyde resins, which is the second most abundant use of urea. Urea is dehydrated to melamine, which, when combined with formaldehyde, produces melamine-formaldehyde resins (Figure 96.2). Melamine resins tend to be harder and more heat-resistant than urea-formaldehyde resins. Melamine received widespread attention as the primary pet food and animal feed contaminant causing numerous cat and dog deaths in early 2007. Melamine is a nitrogen-rich compound. Because the protein content of grain is tested by measuring its nitrogen content, some individuals speculate that melamine was deliberately added to wheat and rice products produced in China to falsely produce higher protein content for these products.

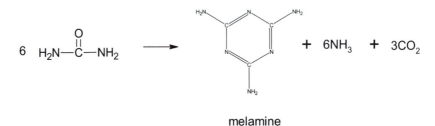

melamine

Figure 96.2   Dehydration of urea produces melamine, ammonia, and carbon dioxide.

Another use of urea is in pharmaceuticals. On December 4, 1864, Adolf von Baeyer (1835–1917) synthesized a compound from urea and malonic acid called barbituric acid. Baeyer derived the name barbituric from the name Barbara and urea. December 4, the date of Baeyer's synthesis, is Saint Barbara Day. Saint Barbara is the patron saint of individuals associated with military bombardment and firefighters. There was no immediate medicinal use for barbituric acid, but in 1903, Emil Fischer (1852–1919) and Joseph von Mehring (1849–1908) synthesized diethylbarbituric acid. In dietylbarbituric acid two ethyl groups replace two hydrogens in barbituric acid. Fischer and Mehring named their compound veronal, related to the city of Verona, Italy. They also discovered that veronal was effective in inducing sleep, which led to its use as a drug to help people sleep. This prompted the use of derivatives of barbituric acid in sleeping pills and sedatives, hypnotics, and anesthetics. These drugs came to be known as barbiturates.

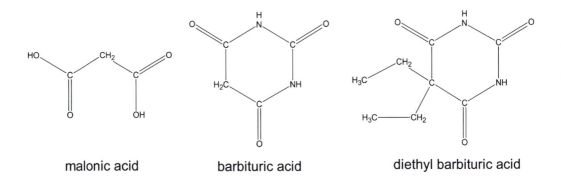

malonic acid   barbituric acid   diethyl barbituric acid

Other minor uses of urea are as rehydrating lotion, diuretics, deicers, and cold-compresses.

97. Vanillin

CHEMICAL NAME = 4-hydroxy-3-methoxy-benzaldehyde
CAS NUMBER = 121–33–5
MOLECULAR FORMULA = $C_8H_8O_3$
MOLAR MASS = 152.1 g/mol
COMPOSITION = C(63.1%) O(31.6%) H(5.3%)
MELTING POINT = 81°C
BOILING POINT = 285°C
DENSITY = 1.06 g/cm³

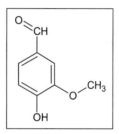

Vanillin is the primary chemical compound responsible for the flavoring commonly called vanilla. In pure form it exists as a white crystalline substance. Natural vanilla is extracted from the seed pods (wrongly referred to as beans) of several orchid vines that grow throughout the tropics. Although there are more than 50 species of vanilla orchids, only a few are used commercially: *Vanilla planifolia* (also called *Vanilla fragrans*), *Vanilla tahitensis,* and *Vanilla pompona.* The plant is native to southeast Mexico and Guatemala, but it is now cultivated in the tropics throughout the world. The word vanilla comes from the old Spanish word for the plant, *vainilla.* *Vainilla* comes from the diminutive of *vaina,* which is derived from the Latin word *vagina.* In Latin, *vagina* was the word for sheath or scabbard, so *Vainilla* was named for the small sheathlike pods on the plants. The vanilla orchid flower blooms for only a brief period, perhaps a few hours to a day, and, if pollinated, a seed pod ripens over nine months. The seed pods are green, several inches long, and resemble large green beans; they encase the vanilla seeds and oil.

It is not known how long vanilla has been used as a spice, but it dates back at least 1,000 years. The first known cultivators of vanilla were the Totonac people in the Veracruz region of Mexico, who regarded vanilla as a sacred plant and used it as a deodorant. The use of vanilla was acquired by the Aztecs after their invasion and interaction with the Totonacs. The Aztecs called vanilla *tlilxochitl,* which translates as black flower, a reference to the dark brown-black color of the dried pods after curing. Aztec royalty used vanilla to sweeten the bitter taste of their cocoa drink *xocolatl* and for medicinal purposes. The Spanish explorer Hernando Cortez

(1485–1547) introduced vanilla (as an ingredient in cocoa drink) to Europeans in the 1520s after conquering the Aztecs. The vanilla-flavored cocoa drink was enjoyed by European aristocracy, but in 1602 Queen Elizabeth's apothecary Hugh Morgan (1530–1613) recommended that vanilla be used as a flavoring apart from its use with cocoa. Vanilla's use as a flavoring spread, with the French especially using it increasingly in recipes, for example French vanilla ice cream. Thomas Jefferson was introduced to it during his time in France and helped to introduce it to the United States.

Attempts to cultivate vanilla orchids outside of Mexico were unsuccessful until the 1800s, allowing the Spanish to establish a monopoly on vanilla as a trade item. Individuals smuggled cuttings and plants out of Mexico in unsuccessful attempts to establish the orchid in other regions. Smuggled plants survived and flowered but did not produce seed pods. Meanwhile, vanilla remained an expensive commodity enjoyed mainly by the aristocracy. In 1836, the Belgian botanist Charles François Antoine Morren (1807–1858) discovered that the vanilla orchid depended on a tiny bee from the genus *Melipona,* which was found only in Mexico, for pollination. Morren produced seed pods using artificial pollination and published his work in 1837. He used scissors to cut the plant's rostellum, separating the stigma and anther in an intricate procedure. In 1841, Edmond Albius (1828–1880), a slave on the French island of Reunion (formerly called Bourbon) in the western Indian Ocean east of Madagascar, found a quick and easy method of pollinating orchids using a small pointed stick or thorn; this method is still used today. Albius pealed back the rostellum and used the stick to press the anther and stigma together. Albius's discovery broke the Spanish monopoly on Mexican vanilla and led to vanilla plantations in many tropical areas. Today, Madagascar leads the world in vanilla production, annually producing between 50% and 60% of the world output of vanilla beans. Indonesia and Comoros are also large producers. In 2005, world production was roughly 2,500 tons of cured beans with a yield of approximately 50 tons of natural vanillin.

When vanilla pods are picked, they are green and lack flavor. They are subjected to an extensive curing process to promote biochemical reactions that develop the characteristic flavor. Curing involves several stages. The first of these is killing the pods to terminate ripening and promote enzymatic conversion of glucosides to aromatic phenolic compounds, primarily glucovanillin to vanillin. The traditional killing method was to wilt the beans in the hot sun. Sun drying is still used, but modern methods involve scalding in hot water, oven drying, and freezing. The second stage is sweating, during which pods are kept at an elevated temperature and humidity for several days to weeks. Sweating promotes flavor development as biochemical reactions occur in the pod. The third and fourth stages consist of drying and conditioning the pods, respectively. Conditioning occurs over several months where the dried pods are aged in closed dark containers to develop flavor.

Cured vanilla pods are sorted and graded and used in a variety of forms: extract, paste, powder, flavorings, vanilla sugars, and vanilla beans. The most commonly used form is vanilla extract, which can be made by soaking vanilla beans in a water-ethyl alcohol solution (the Food and Drug Administration specifies at least 35% alcohol) for several months. Extract is typically produced by percolating the alcohol solution through macerated vanilla pods, similar to brewing coffee. Natural vanilla is a mixture of several hundred different compounds, but the main compound responsible for its characteristic flavor and smell is vanillin or 4-hydroxy-3-methoxy-benzaldehyde. Vanillin was first isolated in 1858 by the French chemist Nicolas-Theodore Gobley (1811–1876). German chemists Ferdinand Tiemann (1848–1899) and Wilhelm Haarmann (1847–1931) deduced vanillin's structure and synthesized vanillin in 1874 by oxidizing coniferin ($C_{16}H_{22}O_8$), a glucoside extracted from the sap of coniferous trees. The synthesis of vanillin was patented, and Haarman and Tiemann teamed with another German chemist, Karl Reimer (1845–1883), to form the Haarman and Reimer Company in 1874, which produced vanillin fragrances and flavorings. A number of laboratory methods were found to synthesize vanillin. In 1876, Reimer and Tiemann published an article describing the synthesis of phenolic aldehydes by the reaction of phenol and chloroform in a basic medium. Guaiacol ($C_7H_8O_2$), made using the Reimer-Tiemann reaction, was used to prepare vanillin. In 1896, vanillin was synthesized from the eugenol extracted from clove oil. Eugenol was isomerized to isoeugenol, which was then oxidized to vanillin. This became a commercially viable process for producing vanillin for several decades. In the 1930s, a method was found for making vanillin from lignin obtained from the waste pulp of paper production. Heating pine lignin obtained from the Kraft process in the presence of active alkali oxidizes lignin to vanillin and a number of other compounds. This method was used liberally to make vanillin for 50 years, and in the mid 20th century, 60% of the world's supply of synthetic vanillin came from one Ontario paper plant. This method has been abandoned in the last several decades owing to environmental and economic concerns. Today vanillin (metyl vanillin) and ethyl vanillin are synthesized from petrochemicals such as guaiacol or glyoxylic acid (HOC-COOH). The catechol method of vanillin synthesis starts with benzene conversion to phenol, which is then converted to catechol. Catechol is then converted to guaiacol and then vanillin:

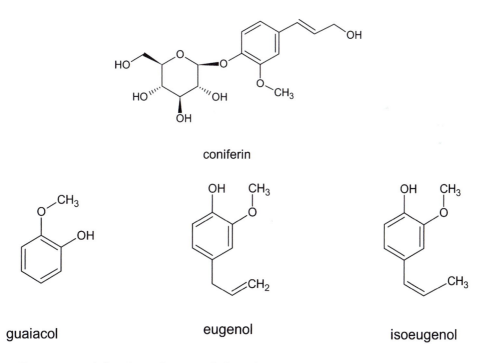

coniferin

guaiacol   eugenol   isoeugenol

Recent research has focused on catechol production using a green biosynthetic process that involves glucose. Ethyl vanillin is also used as a vanilla flavoring. It has the ethyl group bonded to the singly bonded oxygen rather than the methyl group as in methy vanillin (4-hydroxy-3-methoxy-benzaldehyde). Ethyl vanillin has a stronger vanillin flavor.

Synthetic vanillin formulations dominate global use of vanillin. As noted, approximately 50 tons of natural vanillin is used annually; global use of synthetic vanillin is approximately 15,000 tons. More than 90% of the vanilla used in the United States is synthetic (imitation), and imitation vanilla meets more than 97% of the world demand. The primary use of vanillin is as a flavoring ingredient; the largest demand is for ice cream. It is also used regularly in soft drinks (vanilla colas), candies, baked goods, coffees, and teas. Vanillin is used in perfumes, deodorants, and personal care products. The chemical industry uses vanilla to mask offensive odors of certain products. Vanillin is used as an intermediate in pharmaceutical and fine chemical production.

CHEMICAL NAME = chloroethylene
CAS NUMBER = 75–01–4 (vinyl chloride)
  9002–86–2 (polyvinyl chloride)
MOLECULAR FORMULA = C_2H_3Cl
MOLAR MASS = 62.5 g/mol
COMPOSITION = C(38.4%) H(4.8%)
  Cl(56.7%)
MELTING POINT = –153.8°C decomposes
  (polyvinyl chloride)
BOILING POINT = –13.4°C decomposes (polyvinyl chloride)
DENSITY = 2.9 g/L (vapor density = 2.2, air = 1)   1.41 g/cm³ (polyvinyl chloride)

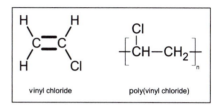

vinyl chloride   poly(vinyl chloride)

At room temperature vinyl chloride is a colorless, highly flammable gas with a faint sweet odor. Vinyl chloride is the monomer used to produce the polymer polyvinyl chloride, also called PVC. The first documented production of vinyl chloride was by the French scientist Henri-Victor Regnault (1810–1878) around 1835. Regnault used an alcoholic solution of potassium hydroxide to treat the compound ethylene dichloride ($C_2H_4Cl_2$), which, in Regnault's time, was known as Dutch oil: $C_2H_4Cl_{2(l)} + KOH \rightarrow C_2H_3Cl_{(g)} + KCl + H_2O$. Regnault observed that when vinyl chloride was exposed to sunlight, it polymerized into a white powder. In 1872, the German chemist Eugen Baumann (1846–1896) also noted that vinyl chloride could be polymerized into polyvinyl chloride and is often credited as its inventor. The first uses for polyvinyl chloride were described in a 1913 patent by Friedrich Heinrich August Klatte (1880–1934).

Polymerized vinyl chloride as a homopolymer is hard and brittle, making it difficult to work and impractical as a commercial material. In 1926, Waldo Lonsbury Semon (1898–1999) was working for B. F. Goodrich searching for a synthetic rubber that could adhere to metal objects. Semon examined vinyl chloride and found that when polyvinyl chloride powder was mixed in certain solvents, he obtained a stiff gel that could be molded into a plastic material. The material's hardness and pliability depended on the mix of solvent and polyvinyl chloride. Semon had discovered that polyvinyl chloride could be plasticized with additives, but plasticized polyvinyl chloride would still not bind to metal. Goodrich developed several applications for Semon's polyvinyl chloride, but these were not economically viable for large-scale production. Goodrich was about to discontinue production when Semon got the idea to coat fabrics with polyvinyl chloride to produce a waterproof material. Semon successfully convinced company officers to develop the product and in 1931, Goodrich introduced products using the material. It was initially used for raincoats, shower curtains, and waterproof linings. Semon was granted the U.S. patent for polyvinyl chloride in 1933 as a continuation of an original patent he had submitted in 1928 and assigned it to Goodrich (U.S. Patent Number 1929453). Goodrich named its polyvinyl material Koroseal. The registered trademark Koroseal is now owned by RJF International and numerous products such as wall coverings, liners, and flooring are produced from it. Semon expanded on the patent in 1940, which provided additional applications for molding PVC without heat (U.S. Patent Number 2188396).

Vinyl chloride monomer (VCM) is produced from ethylene dichloride, which is synthesized by the direct chlorination of ethylene: $CH_2 = CH_{2(g)} + Cl_{2(g)} \rightarrow CH_2ClCH_2Cl$ or oxychlorination using hydrogen chloride: $2CH_2 = CH_{2(g)} + HCl_{(g)} + O_{2(g)} \rightarrow 2CH_2ClCH_2Cl_{(g)} + H_2O_{(g)}$. The thermal dechlorination of ethylene dichloride at a temperature of approximately 500°C yields vinyl chloride and hydrogen chloride: $CH_2ClCH_2Cl_{(l)}$ $CH_2 = CHCl_{(g)} + HCl_{(g)}$. The hydrogen chloride produced in this reaction is used to produce more ethylene dichloride by the oxychlorination reaction. Vinyl chloride can also be produced from the hydrochlorination of acetylene: $HC \equiv CH(g) + HCl(g) \rightarrow CH_2 = CHCl_{(g)}$. Approximately 8.5 million tons of VCM are produced annually in the United States and about 40 million tons globally. Almost all (98%) of the monomer is used to produce polyvinyl chloride, which is second only to polyethylene as a thermoplastic in terms of volume produced.

Polyvinyl chloride is processed into a number of forms by including additives. Additives are used to vary the properties of PVC so that it can be made soft and flexible or hard and rigid. Additives are also used to inhibit decomposition as a result of exposure to sunlight, ozone, and chemicals. Plasticizers are the primary additive included in PVC materials. Di(2-ethylhexyl) phthalate (DEHP) and a host of other phthalates are the most common plasticizers. Plasticizers impart flexibility, thermal stability, strength, and resilience to PVC compounds. PVCs without plasticizers are classified as UPVC; the letters stand for unplasticized polyvinyl chloride. UPVC is rigid and used for conduit, containers, gutters, and floor tiles. Other common PVC additives are biocides, lubricants, and pigments.

Polyvinyl chloride has several advantages as a material: it is weather resistant; resists corrosion; is highly chemically inert; is inexpensive, durable, and fire resistant; and has a high strength-to-weight ratio. It is also a good thermal and electrical insulator. Approximately 32 million tons of PVC were used worldwide in 2005. PVC is used in countless applications, but the largest category of these is construction materials where it is used extensively for vinyl siding, floors, and pipes. It is also used for interior wall coverings, roof membranes, door and window casements, and gutters. Because of its insulating properties, PVC is used as sheathing for wires and cables. Other common uses of PVC are plastic furniture, imitation leather commonly called vinyl, cling film for food wraps, housings for electronics, computers, cell phones, and credit cards.

Although PVC use during the last 70 years has grown tremendously, health and environmental problems have been associated with its use. Vinyl chloride was identified as a carcinogen in the 1960s. Short-term exposure to high levels of vinyl chloride may result in damage to the central nervous system, resulting in dizziness, drowsiness, and headaches. Long-term exposure to vinyl chloride by workers in VCM and PVC manufacturing plants has resulted in liver damage. Vinyl chloride exposure has been shown to increase the risk of angiosarcoma of the liver. In addition to health problems to workers exposed to VCM in the chemical industry, many of the additives used in producing PVC have adverse health effects. Of particular concern are phthalate plasticizers used in PVC. Some of these may outgas and are believed to be endocrine disruptors. In 2007, greater restrictions on the use of phthalate plasticizers commonly used in PVC toys, which children may chew on, took effect throughout the European Union. Diisononyl phthalate (DINP), which is the most common plasticizer, and diisodecyl phthalate (DIDP) and di-n-octyl phthalate (DNOP) can only be used in infant items that cannot be placed in the mouth. Di(2-ethylhexyl) phthalate (DEHP), di-n-butyl phthalate (DBP), and butylbenzyl phthalate (BBP) are no longer permitted children articles. A general environmental concern deals with the amount of PVC waste produced and the formation of toxic substances when it is incinerated, especially in medical, municipal, and industrial incinerators. When PVC is burned it has the potential to produce dioxins. Dioxins are compounds characterized by the dioxin structure. They have been shown to cause birth defects and cancers in tests with laboratory animals. Many governing bodies have legislated laws intended to reduce PVC from waste that will be incinerated. Many environmental groups have called for the reduction of PVC waste, and private companies have voluntarily substituted other materials for PVC packaging. For example, Microsoft phased out PVC packaging in 2005, eliminating several hundred thousand pounds of PVC waste. Another way to reduce PVC waste is through recycling. PVC material is designated with the recycling number 3.

dioxin structure

CHEMICAL NAME = water
CAS NUMBER = 7732–18–5
MOLECULAR FORMULA = H_2O
MOLAR MASS = 18.0 g/mol
COMPOSITION = H(11.2%) O(88.8%)
MELTING POINT = 0.0°C
BOILING POINT = 100.0°C
DENSITY = 1.00 g/cm³ (liquid)   0.92 g/cm³ (ice)

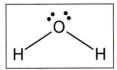

Water is the tasteless, odorless, clear liquid essential for all life. Water's importance for life is so critical that the question of extraterrestrial life is based to a large extent on the presence of water. The amount of water in the human body varies with gender, age, and body type, but humans generally contain between 50% and 70% water. A person can go weeks without food, but only several days without water. Seventy percent of the earth's surface is covered by water. Water is the only natural substance that can be found in all three states: liquid, solid (ice), and gas (steam) under normal conditions. In addition to its physical and biological importance, water has a spiritual quality, as witnessed by its use in many diverse religions. Many religions view water's cleansing ability as symbolic for cleaning the spirit and ridding the body of impurities.

Water played a key role in the development of modern chemistry. Thales of Miletus (624–560 B.C.E.) considered water to be the basic element of matter. The philosopher Empedocles of Agrigentum (495–435 B.C.E.) proposed that water was one of the four basic elements along with earth, air, and fire that made up matter. Water was considered an element until the development of modern chemistry in the 18th century. Henry Cavendish (1731–1810) discovered that inflammable air, which was hydrogen, and dephlogisticated air (oxygen) produced water when ignited. Cavendish's experiments were repeated by other chemists such as Joseph Priestley (1733–1804) and Antoine Lavoisier (1743–1794). Lavoisier decomposed water into oxygen and inflammable air, demonstrating that water was not an element, but a compound.

Water has a number of unique properties among substances. It has the highest latent heat of fusion of any natural liquid, highest heat of vaporization, highest surface tension, and the highest specific heat except for ammonia. Its boiling point is extremely high compared to other compounds with similar molar masses, and ice is less dense than liquid water at 0°C. Water's unique physical properties are due to its structure. Water is a polar molecule with an uneven distribution of electron density. The hydrogen part of the molecule has a partial positive charge, and the oxygen part of the molecule a partial negative charge (Figure 99.1). The electrostatic attraction between the positively charged hydrogen atoms on one water molecule and the negatively charged oxygen atom on another water molecule results in hydrogen bonding. Hydrogen bonds are relatively strong intermolecular forces that give water its unique physical properties. The traditional view of water's structure is that a water molecule hydrogen bonds to four other hydrogen molecules in a tetrahedral structure (Figure 99.1). This occurs because a water molecule's oxygen has two lone pairs of electrons that form hydrogen bonds with two water molecules, whereas its two hydrogen atoms form hydrogen bonds to another two water molecules. This tetrahedral arrangement is currently being challenged, with research demonstrating that a liquid water molecule may only hydrogen bond to two other water molecules, producing linear or ring arrangements.

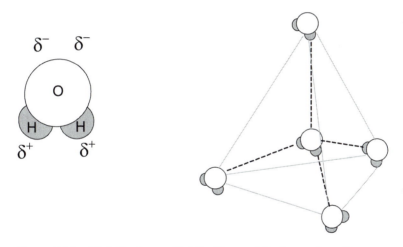

Figure 99.1   Water's charge distribution and hydrogen bonding tetrahedral structure in water.

Hydrogen bonding is an attractive force between water molecules that must be overcome when it changes phases from solid to liquid and liquid to gas. For this reason water has a high latent heat of fusion and latent heat of vaporization. The attractive force provided by hydrogen bonding is also responsible for water's high boiling point, high specific heat, high surface tension, and other unique properties.

Although almost all materials expand when heated, water contracts as it is heated just above its melting point. This results in the abnormal property of ice being less dense than liquid water. As the temperature of ice increases to its melting point of 0°C, hydrogen bonding provides enough attractive force to maintain the approximate tetrahedral structure of ice. Increasing the temperature above water's freezing increases kinetic energy of molecules, causing thermal expansion, and this would normally lead to an increase in volume and a corresponding decrease in density. Water actually becomes denser because heating water to 4°C provides energy for some of the water molecules to overcome the intermolecular attraction provided by the hydrogen bonding that produces the crystalline structure. These water molecules occupy void space in the residual crystalline structure, producing a tighter packing of the water molecules. Because more molecules of water occupy the same volume, the density increases. The tighter packing continues until the maximum density is reached at 4°C. At this point, the trapping of additional water molecules in the void space is not great enough to overcome the kinetic thermal expansion effect that lowers the density; therefore solid ice is less dense than liquid water.

Water exists in three phases under normal temperatures and pressures that exist on Earth. A phase diagram summarizes the relationship between temperature, pressure, and phases for a substance. The phase diagram for water is shown in Figure 99.2. Water can change from a solid to gas or gas to solid in processes called sublimation and deposition, respectively. At low temperature and pressure, water exists as a gas and a solid, but not as a liquid. At a temperature slightly above 0 and a pressure of 4.58 torr, water exists in all three phases. This point on the phase diagram is called the triple point. As temperature and pressure increase, a point is reached called the critical point. At the critical point and beyond, the substance cannot be liquefied and exists as a dense gaseous fluid called a supercritical fluid. Supercritical fluids have properties of both liquids and gases. The boundary between solid and liquid in the phase diagram for water has a slight negative slope. This means that as pressure is applied to ice, its melting point decreases and solid water is converted to liquid. This process takes place when an ice skater glides across the ice. The pressure the skater exerts on the ice is concentrated under the blades of the skates. This pressure is sufficient to momentarily melt the ice as the skater glides across the ice. The water quickly solidifies back to ice when the pressure is reduced after the skater passes.

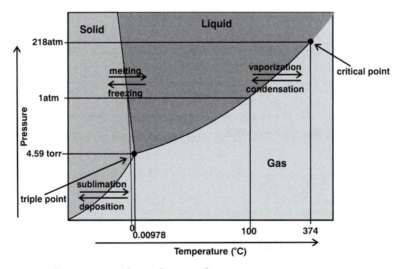

Figure 99.2 Phase diagram for water. Drawing by Rae Déjur.

Water is often referred to as the universal solvent, but the expression "oil and water don't mix" defies this label. Many substances dissolve in water to varying degrees. It is an excellent solvent for ionic compounds (salts) and polar molecules. Ionic and polar solutes interact electrostatically with water. The electrostatic attraction between the electronegative oxygen atom in water and a positive cation metal in a salt, along with the attraction of water's electropositive hydrogen atoms for the negative anion, is sufficient to overcome the strong ionic attraction between a salt's metal cation and anion. Nonpolar substances dissolve poorly in water because electrostatic attractions between water and nonpolar substances are low.

Pure water can be considered both an acid and a base because water dissociates slightly to produce hydrogen and hydroxide ions: $H_2O_{(l)} \rightleftarrows H^+_{(aq)} + OH^-_{(aq)}$. This equation depicts water as an Arrhenius acid and base. An Arrhenius acid is any substance which when dissolved in water produces hydrogen ions, H^+. An Arrhenius base is a substance that produces hydroxide ions, OH^-. Treating water in terms of the Brønsted-Lowry theory, a more appropriate reaction would be: $H_2O_{(l)} + H_2O_{(l)} \rightleftarrows H_3O^+_{(aq)} + OH^-_{(aq)}$, where H_3O^+ is the hydronium ion. At 25°C, the concentrations of hydronium and hydroxide ions in water are both equal to 1.0×10^{-7} M. To put this concentration into perspective, consider that about two of every billion water molecules dissociates into ions. The equilibrium constant for this reaction is known as the ion product constant of water, symbolized by K_w, and is equal to the product of the H^+ and OH^- molar concentrations: $K_w = [H^+][OH^-] = (1.0 \times 10^{-7})(1.0 \times 10^{-7}) = 1.0 \times 10^{-14}$. The equation for K_w applies to both pure water and aqueous solutions.

Because the ion concentrations are small and the negative exponents make them tedious to work with, Soren Peer Lauritz Sorenson (1868–1939), a Danish biochemist, devised the pH concept in 1909 to express the hydrogen ion concentration. The abbreviation pH comes from the French *pouvoir hydrogène* meaning power of hydrogen. The pH of a solution is given by the equation: $pH = -\log[H^+]$. The brackets are used to signify the molar concentration of a substance. This equation states that pH is equal to the negative log of the hydrogen ion molar concentration, but it should be remembered that it is really the hydronium ion concentration in solution that is being measured. In a similar fashion, the pOH of a solution can be defined as: $pOH = -\log[OH^-]$.

Using these definitions, the pH and pOH of a neutral solution at 25°C are both equal to 7. The expression for K_w shows that $[H^+]$ and $[OH^-]$ are inversely related, and consequentially pH and pOH are inversely related. The product of the hydrogen and hydroxide concentrations is equal to 1.0×10^{-14}, and the sum of the pH and pOH is equal to 14. In an acidic solution the hydrogen ion concentration increases above 1.0×10^{-7}, the hydroxide concentration decreases, and the pH value gets smaller. The relationship between the type of solution, pH, pOH, and ion concentrations is shown in Table 99.1.

Table 99.1   Ion Concentrations and pH

Solution	pH	pOH	[H⁺]	[OH⁻]
Neutral	7	7	1.0×10^{-7}	1.0×10^{-7}
Acid	<7	>7	$>1.0 \times 10^{-7}$	$<1.0 \times 10^{-7}$
Base	>7	<7	$<1.0 \times 10^{-7}$	$>1.0 \times 10^{-7}$

The pHs of a number of common substances are presented in Table 99.2. Because the pH and pOH scales are based on logarithms, a change in one pH or pOH unit represents a change in ion concentration by a factor of 10. Coffee with a pH of 5 has approximately 100 times the hydronium ion concentration as tap water, with a pH of 7.

Table 99.2 The pH of Some Common Substances

Battery acid	0.5	Coffee	5.0	Blood	7.4
Stomach acid	1.5	Rain	5.6	Seawater	8.3
Lemon juice	2.3	Urine	6.0	Baking soda	8.3
Vinegar	2.9	Milk	6.6	Toothpaste	10
Cola	3.0	Saliva	6.8	Milk of magnesia	10.5
Apple	3.5	Tap water	7.0	Ammonia cleaner	12

Daily water usage in the United States is estimated to be approximately 400 billion gallons each day. Major uses can be divided into several broad areas: agriculture, industry, power generation, and residential. Almost half of all water used in the United States is for power generation. Large volumes are used for steam generation and for cooling water. The second highest water usage is for irrigation water, with about one-third of the daily demand. Public water supply accounts for about 10% of daily usage and industry uses another 5%. Domestic water, which consists of drinking, food preparation, bathing, washing clothes and dishes, watering lawns, and flushing toilets, accounts for less than 1% of daily usage.

CHEMICAL NAMES =
  1,2 dimethylbenzene,
  1,3 dimethylbenzene,
  1,4 dimethylbenzene
CAS NUMBERS = 95–
  47–6, 108–38–3,
  106–42–3
MOLECULAR FORMULA

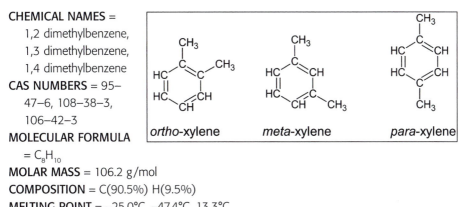

ortho-xylene	meta-xylene	para-xylene

 = C_8H_{10}
MOLAR MASS = 106.2 g/mol
COMPOSITION = C(90.5%) H(9.5%)
MELTING POINT = –25.0°C, –47.4°C, 13.3°C
BOILING POINT = 144.4°C, 139.1°C, 138.4°C
DENSITY = 0.88 g/cm³, 0.86 g/cm³, 0.86 g/cm³

Xylene is benzene to which two methyl groups have been added to two carbon atoms in the benzene ring. The addition of two methyl groups gives three isomers of xylene labeled according to the relative positions of the methyl groups. Ortho-xylene has methyl groups on consecutive carbons in the ring, meta-xylene's metyl groups are separated by a single carbon bonded to hydrogen atoms, and para-xylene has the methyl groups on carbon atoms on opposite sides of the ring. The three xylene isomers are abbreviated using o-,m-, p- for ortho, meta, and para, respectively. Xylene is used both as a mixture, where it is referred to as xylenes or xylol, and as individual isomers. Because their boiling points are close, separation using distillation is diffi-cult. Therefore isomers are separated using techniques such as recrystallization and adsorption. Xylenes are flammable, colorless liquids with a pleasant odor. Xylene was first isolated from coal tar in the mid-19th century. The name xylene comes from the Greek word for wood *xulon* because xylene was obtained from the distillation of wood in the absence of oxygen.

Xylenes are produced from the reformulation of naphthas during petroleum refining in a process that also produces benzene, toluene, and ethylbenzene. Collectively, these are abbreviated as BTEX. A portion of the xylene made through reformulation is used to boost the octane content of gasoline. Meta- and para-xylenes have octane ratings of approximately 135; that of ortho-xylene is about 115. Mixed xylenes are used extensively as organic solvents similar to toluene. It is used as a thinner in paints and other coatings, and as a degreaser and cleaning agent. It is also used in dyes, pigments, cosmetics, pharmaceuticals, and pesticides.

Xylene is used as a chemical feedstock in the chemical industry. Xylenes can undergo oxidation where the side methyl groups are oxidized to give a carboxyl group (COOH) yielding a carboxylic acid. The particular acid produced depends on the isomer oxidized. When o-xylene is oxidized phthalic acid is produced, and when p-xylene is oxidized terephthalic acid results (Figure 100.1). Terephthalic acid is one of the main feedstocks in making polyesters. Terephthalic acid reacts with ethylene glycol to form the ester polyethylene terephthalate (PET). PET is one of the most common plastics used as food and beverage containers. PET containers contain the recycling symbol with a number 1. PET is marketed using a number of commercial names; the most generic of these is polyester. It is also the material known as Dacron. Mylar is PET in the form of thin films. Although all three isomers of xylene are used as chemical feedstocks, the greatest demand is for para-xylene to produce terephthalic acid. The smallest demand is for meta-xylene. Approximately 30 million tons of xylenes are used annually worldwide.

Xylene, similar to toluene, acts as a neurotoxin. Most exposure results from inhalation of fumes from products that contain xylene. Xylene concentrations in the air of 100 ppm or less can irritate the nose, throat, and eyes. As concentrations increase, xylene produces headaches, dizziness, and disorientation. Exposures at high concentrations for an extended time may result in paralysis or death. Xylene may cause liver and kidney damage.

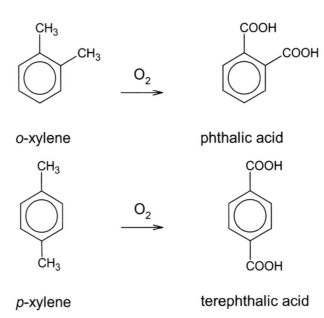

Figure 100.1   Oxidation of xylene produces carboxylic acid.

A closely related compound to the xylenes is ethylbenzene. Ethylbenzene is a structural isomer with the xylenes, but its properties distinguish it from the three xylene isomers. Ethylbenzene is part of xylene mixtures and is also one of the principal aromatic components in BTEX associated with petroleum products.

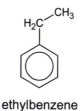

ethylbenzene

Common and Ancient Names of Substances

Common Name	Chemical Name	Formula
Alumina	aluminum oxide	Al_2O_3
Aqua fortis	nitric acid	HNO_3
Baking soda	sodium hydrogen carbonate	$NaHCO_3$
Bleach	sodium hypochlorite	$NaClO$
Borax	sodium tetraborate decahydrate	$Na_2B_4O_7 \cdot 10H_2O$
Brimstone	sulfur	S_8
Chalk	calcium carbonate	$CaCO_3$
Chilean saltpeter	sodium nitrate	$NaNO_3$
Cream of tartar	potassium hydrogen tartrate	$KHC_4O_4H_6$
Dephlogisticated air	oxygen	O_2
Epsom salt	magnesium sulfate heptahydrate	$MgSO_4 \cdot 7H_2O$
Fixed air	carbon dioxide	CO_2
Grain alcohol	ethanol	C_2H_5OH
Gypsum	calcium sulfate dihydrate	$CaSO_4 \cdot 2H_2O$
Inflammable air	hydrogen	H_2
Laughing gas	nitrous oxide	N_2O
Lime (quicklime)	calcium oxide	CaO
Lime (slaked)	calcium hydroxide	$Ca(OH)_2$
Limestone	calcium carbonate	$CaCO_3$
Lye	sodium hydroxide	$NaOH$
Magnesia	magnesium oxide	MgO
Marble	calcium carbonate	$CaCO_3$
Marsh gas	methane	CH_4
Milk of lime	calcium hydroxide	$Ca(OH)_2$
Milk of magnesia	magnesium hydroxide	$Mg(OH)_2$
Muriatic acid	hydrochloric acid	$HCl_{(aq)}$
Natron	hydrated sodium carbonate	$Na_2CO_3 \cdot 10H_2O$
Nitrous air	nitric oxide	NO
Oil of vitriol	sulfuric acid	$H_2SO_{4(aq)}$
Phlogisticated air	nitrogen	N_2
Plaster of Paris	calcium sulfate	$CaSO_4$

(Continued)

(Continued)

Common Name	Chemical Name	Formula
Potash	potassium carbonate	K_2CO_3
Quicksilver	mercury	Hg
Sal ammoniac	ammonium chloride	NH_4Cl
Saltpeter	potassium nitrate	KNO_3
Talc	magnesium silicate	$Mg_3S_4O_{10}(OH)_2$
Vinegar	acetic acid	CH_3COOH
Washing soda	sodium carbonate	Na_2CO_3
Water	water	H_2O
Wood alcohol	methanol	CH_3OH

acetate ester. Ester derived from acetic acid.

acid hydrolysis. Splitting of a molecule by reacting with water under the action of an acid.

acylation. Process resulting in the substitution of an acyl group into a molecule.

acyl group. Functional group characterized by $R-\overset{|}{C}=O$

alkaloid. A nitrogen containing compound obtained from plants, for example, caffeine, nicotine.

alkene. An acyclic hydrocarbon that contains at least one carbon-carbon double bond.

alkylate. Product of alkylation, typically associated with petroleum industry.

alkylation. Chemical process in which an alkyl group is introduced into an organic compound.

alkyl group. The group left when a hydrogen atom is removed from an alkane; for example, removing a hydrogen atom from methane produces the methyl group

alkyne. An acyclic hydrocarbon that contains at least one carbon-carbon triple bond.

alpha cells. Cells in pancreas that make and secrete glucagon.

alpha receptor. Site in nervous system activated by norepinephrine and epinephrine, causing various physiological responses.

amine. Organic compounds that result when one or more hydrogen atoms in ammonia are replaced by organic radicals.

amino acid. Organic acids that contain both an amino group, NH_2, and a carboxyl group, COOH, the building blocks of proteins.

analgesic. Substance that relieves pain.

angina. Chest pain associated with lack of blood supply to the heart.

angiosarcoma. Cancerous tumor that develops in blood vessel lining.

anneal. To heat a metal or glass to a high temperature and slowly cool, reducing brittleness of material.

antipyretic. Substance that reduces fever.

arenes. Compounds that contain both aromatic and aliphatic units.

azeotrope. Mixture of two or more liquids in which the distillate has the same composition as original mixture and therefore cannot be separated by ordinary distillation procedures.

beta cells. Cells in the pancreas that make and release insulin.

beta receptor. Site in nervous system activated by norepinephrine and epinephrine causing various physiological responses.

biomagnification. An increase in the concentration of a chemical moving up the food chain.

calcination. Slow heating of a substance without causing it to melt.

calorie. Unit of energy equivalent to the energy required to raise the temperature of 1 gram of water from 14.5°C to 15.5°C, equivalent to 4.18 Joules.

carbonyl. A carbon atom and oxygen atom joined by a double bond $C = O$, present in aldehydes and ketones.

carbonylation. Process in which a carbonyl group is introduced into a molecule.

carboxyl. Carbonyl with an OH group attached, -COOH.

carboxylation. Process in which a carboxyl group is added to a molecule.

carcinogenic. A substance that causes cancer.

catalytic cracking. Breaking a compound into a smaller compounds with the use of catalysts.

catecholamine hormone. Biological compounds derived from tyrosine that include epinephrine, norepinephrine, and dopamine.

chelate. Coordinate or complex compound consisting of a metal atom bonded to ligand.

chelation. Holding of a metal atom between two atoms in a single molecule.

chiral center. Atom in a tetrahedral arrangement with four different groups attached to it.

chlorination. Addition or substitution of chlorine to an organic compound; the addition of chlorine to water or wastewater for the purpose of disinfection.

chromophore. Structure that leads to color in many organic substances.

coagulant. Substance that causes precipitation or separation from a dispersed state.

coal tar. Viscous hydrocarbon mixture produced in the destructive distillation of bituminous coal.

coke. Carbon product produced by heating coal in the absence of air to remove volatile components.

condensation reaction. Combination of two or more reactants with the elimination of water.

conjugated system. Organic structure characterized by alternating single and double carbon-carbon bonds.

co-polymers. Polymers containing two or more structural units.

cracking. Process by which a compound is broken down into simpler substances typically used in petroleum industry to break carbon-carbon bonds.

cultural eutrophication. Accelerated aging of water body as a result of human influence and water pollution.

cycloalkane. Unsaturated monocyclic hydrocarbon with the general formula C_nH_{2n-2}.

cyclopentadienyl ligand. $C_5H_5^-$ complex.

denature. To make unsuitable for human consumption.

dielectric fluid. Substance with high resistivity that acts as a good electrical insulator.

diisocyanates. Class of chemical compounds containing the isocyanate functional group: $-N = C = O$.

dimer. A chemical molecule consisting of the combination of two similar units.

dimerization. Chemical process resulting in dimers.

electron transport chain. Series of biochemical reactions involving transfer of electrons resulting in the production of adenosine triphosphate (ATP).

enantiomer. One of a pair of non superimposable mirror image forms of a compound.

epoxidation. Reaction resulting in the formation of an epoxide generally by reacting an alkene with a peracid.

essential amino acid. An amino acid that cannot be produced by our bodies and must be ingested.

ester. Class of organic compounds that results from the reaction of carboxylic acid and an alcohol.

euthrophication. The natural aging of a water body.

extractive distillation. Distillation using a solvent to separate azeotropic mixtures.

extrusion. Manufacturing process to produce elongated materials with a specific cross sectional area.

feedstock. A process chemical used to produce other chemicals or products.

free base. Basic, uncombined form of a substance or alkaloid as opposed to the form combined with an acid such as HCl to produce a salt.

free radical. A species possessing at least one unpaired electron.

free radical polymerization. Formation of polymers by formation of free radicals in a chain-reaction process until process is terminated.

galvanizing. A method used to protect metals by plating them with another metal, for example, coating iron with zinc.

glucosides. Glycocide derived from glucose.

glyceride. Esters formed from acids and glycerol.

glycocide. Molecule containing a sugar and nonsugar unit.

glycolysis. Oxidation of glucose.

Grignard ether. Grignard reagent dissolved in ethyl ether solvent.

Grignard reagent. An organic magnesium halide compound used in a variety of organic syntheses.

homopolymer. Polymer formed from polymerization of a single structural unit called a monomer.

hormone. Substance produced in cells in one part of the body that move to cells in another part of the body acting as chemical messengers.

hydrocarbon. Organic compound consisting only of carbon and hydrogen atoms.

hydrochlorination. Reaction involving hydrogen chlorine compounds such as alkenes reacting with hydrogen chloride

hydrodealkylation. Chemical reaction resulting in the formation of a simple aromatic compound by the use of hydrogen gas and an appropriate catalyst.

hydrogen bond. Intermolecular force formed between hydrogen of one molecule and highly electronegative atom of nitrogen, oxygen, or fluorine on another molecule.

hydrolysis. Decomposition or change of a substance by its reaction with water.

hydrolyze. Breaking a bond by the addition of water, to undergo hydrolysis.

hydroxylation. Chemical process in which hydroxyl groups, –OH, are introduced into a molecule.

intermediate. Compound or substance used or formed in synthesis of a desired product.

joule. SI unit for energy and work equivalent to 0.24 calorie.

ketone. Class of organic compounds containing a carbonyl group bonded to two other carbon atoms.

Kraft process. Process used in paper industry to extract lignin from pulp using sodium hydroxide and sodium sulfide.

Le Châtelier's principle. Principle that describes when a system is at equilibrium and a change is imposed on the system that the system will shift to reduce the change.

ligand. Ion, molecule, or group bonded to the central metal ion in a chelate or coordinate compound.

metallocene. Organometallic compounds that consist of a metal ion sandwiched between two cyclopentadienyl (Cp⁻) ligands.

methine. CH, carbon with two single bonds with one double bond;

mineral acid. Acid derived from an inorganic substance such as hydrochloric acid, sulfuric acid, and nitric acid.

mitochondria. Organelles found in cells involved in energy production.

molar. Unit to express the concentration of a solution in moles of solute per liter of solution.

monomer. The basic unit that repeats to make a polymer.

naphtha. Liquid, volatile hydrocarbon mixtures obtained during petroleum distillation.

nitriding. Absorption of nitrogen into steel to harden it.

nonessential amino acid. An amino acid that can be produced by our bodies.

olefin. Hydrocarbons with one carbon-carbon double bond such as ethylene.

oxidation. Process involving the loss of electrons.

oxychlorination. Reaction in which a substance is reacted with oxygen gas and hydrogen chloride.

phenyl. A benzene ring with a hydrogen atom removed, C_6H_5.

phosphorylation. Introduction of a phosphate group into an organic molecule.

piperidine. Cyclic compound with the formula $C_5H_{11}N$.

polyester. Polymer with an ester functional group, COOR, as the repeating unit.

polymer. Giant molecule formed by the linking of simple molecules called monomers.

polymerization. Process used to produce polymers.

porphyrin. Structure of four pyrrole rings linked on opposite sides by four methine bridges.

pyridine ring. Benzene with a nitrogen atom substituted for one of the hydrogen atoms, C_6H_5N.

pyrimidine. Heterocyclic aromatic compound,

pyrolysis. Heating in low-oxygen atmosphere to break down a compound.

pyrrole. C_4H_5N,

pyrrolidine. C_4H_9N,

quarternary carbon. A carbon atom bonded to four other carbon atoms by single bonds.

racemate. A mixture of equal amounts of right-handed and left-handed enantiomers of a chiral compound.

racemic mixture. Compound consisting of equal mixtures of levorotatory and dextrorotatory.

reduction. Process involving the gain of electrons.

reforming. Process in which naphthas are converted into higher octane isoparaffins and aromatics.

resin. Sticky liquid organic substance exuded from plants that harden on exposure to air.

scrubber. Device used to remove air pollutants from stacks and industrial exhausts.

steam reforming. Method used to produce hydrogen by reacting methane with steam.

stereo isomers. Molecules of the same compound with different spatial arrangements.

synaptic neurons. Nerve cells that transmit information; information moves from presynaptic neuron to postsynaptic neuron.

tachycardia. Abnormal rapid heart rate.

tacticity. Arrangement of units along a chain in macromolecule or a polymer.

thermal cracking. Use of high temperature (and pressure) to break hydrocarbons into smaller compounds.

transition metal. An element with an incomplete d subshell or one that forms a cation with an incomplete d subshell

Ziegler-Natta catalyst. Organometallic compounds used in vinyl polymerization to control tacticity.

Selected Bibliography

Aftalion, Fred. *A History of the International Chemistry Industry*, 2nd ed. Philadelphia: Chemical Heritage Foundation, 2001.

Atkins, Peter William. *Molecules*. New York: Scientific American Library, 1987.

Baird, Colin, and Michael Cann. *Environmental Chemistry*, 3rd ed. New York: W. H. Freeman and Company, 2004.

Baker, David A., and Robert Engel. *Organic Chemistry*. St. Paul: West Publishing Company, 1992.

Brock, William H. *The Norton History of Chemistry*. New York: W. W. Norton and Company, 1993.

Cambridge Soft Corporation. ChemFinder.com <http://chemfinder.cambridgesoft.com/> The Web site is a searchable database giving structure, physical data, and links to other sites on compounds.

Chang, Raymond. *Chemistry*, 7th ed. Boston: McGraw-Hill, Inc., 2002.

Chemical & Engineering News. *Top Pharmaceuticals*. 83(25): 44–139. Special issue published on June 20, 2005 devoted to reports on 46 of the most important drugs for human society.

Chemical Heritage Foundation. <http://www.chemheritage.org/> This organization is devoted to preserving the history of chemistry.

Chemical Land21.com. <http://www.chemicalland21.com/> An industrial supplier that provides a comprehensive list of compounds with their structures, selected physical properties, and a general description.

Chemical Technology, An Encyclopedic Treatment. New York: Barnes and Noble Books, 1972.

Chenier, Philip J. *Survey of Industrial Chemistry*, 3rd ed. New York: Kluwer Academic/Plenum Publishers, 2002.

Downing, Ralph C. *Fluorocarbon Refrigerants Handbook*. Upper Saddle River, NJ: Prentice Hall, 1988.

Drug Enforcement Administration, Department of Justice. *Drugs of Abuse*. <http://www.usdoj.gov/dea/pubs/abuse/index.htm> Provides general background information on illegal drugs.

European Fertilizer Manufacturers Association. Best Available Techniques for Pollution Prevention and Control in the European Fertilizer Industry. <http://www.efma.org/publications/index.asp> 2000 A comprehensive site with a wealth of information on ammonium, sulfuric acid, nitrates, and other compounds involved in making fertilizer.

The Formaldehyde Council, Inc. <http://www.formaldehyde.org/> Devoted to the uses, health and safety, and facts about formaldehyde.

Furniss, Brian S., Anthony J. Hannaford, Peter W. G. Smith, and Austin R. Tatchell. *Vogel's Textbook of Practical Organic Chemistry*, 5th ed. Harlow: Longman Scientific & Technical, 1989.

Garfield, Simon. *Mauve*. London: Faber and Faber Limited, 2000.

Giunta, Carmen. *Selected Classic Papers from the History of Chemistry*. <http://webserver.lemoyne.edu/faculty/giunta/papers/html> A site to access classical papers in the history of chemistry.

Guyton, Arthur C., and John E. Hall. *Human Physiology and Mechanisms of Disease,* 6th ed. Philadelphia: W. B. Saunders, 1977.

Hopp, Vollrath, and Ingo Hennig. *Handbook of Applied Chemistry*. Washington: Hemisphere Publishing Corporation, 1983.

Hounshell, David A., and John Kenly Smith, Jr. "The Nylon Drama." *Invention and Technology,* Fall, 1988: 40–55.

Idhe, Aaron J. *The Development of Modern Chemistry*. New York: Harper and Row, 1964.

IMA Europe. <http://www.ima-eu.org/index.html> European site devoted to industrial minerals such as gypsum, calcium carbonate, and so forth.

International Directory of Company Histories. Detroit: St. James Press, 1988.

Kent, James A. *Riegel's Handbook of Industrial Chemistry,* 10th ed. New York: Kluwer Academic/Plenum, 2003.

Kirk-Othmer Encyclopedia of Chemical Technology, 5th ed. New York: John Wiley & Sons Inc., 2004.

Kress, Henriette. *King's American Dispensatory* by Harvey Wickes Felter and John Uri Lyold, 1898. February 8, 2002. <http://www.henriettesherbal.com/eclectic/kings/index.html> A comprehensive work describing medicinal plants used at the beginning of the 20th century. This pharmacopeia gives the history of natural compounds used from antiquity until the 20th century.

Landenburg, A. *Lectures on the History of the Development of Chemistry Since the Time of Lavoisier*. Translated by Leonard Dobbin. Chicago: University of Chicago Press, 1911.

LeCouteur, Penny, and Jay Burreson. *Napolean's Buttons*. New York: Jeremy P. Tarcher, 2003.

May, Paul. *The Molecule of the Month*. <http://www.chm.bris.ac.uk/motm/motm.htm> Information and links on over 100 chemical compounds.

MayoClinic. <http://www.mayoclinic.com/health/site-map/smindex> Clinic's Web site provides background information on various diseases, conditions, and pharmaceuticals.

The Merck Index, 12th ed. Whitehouse Station, NJ: Merck & Co. Inc., 1996.

Molecules. <http://www.mdpi.org/molecules/> Links to a journal of synthetic organic chemistry and natural products. Provides scholarly articles on various compounds.

Morrison, Robert Thornton, and Robert Neilson Boyd. *Organic Chemistry*. Boston: Allyn and Bacon, Inc., 1959.

Multhauf, Robert P. *Neptune's Gift*. Baltimore: The John Hopkins Press, 1978.

Myers, Richard L. *The Basics of Chemistry*. Westport, CT: Greenwood Press, 2003.

National Institute of Health. *Third Report of the National Cholesterol Education Program*. September 2002. <http://circ.ahajournals.org/cgi/reprint/106/25/3143> Comprehensive findings on research on cholesterol and health.

National Lime Association. <http://www.lime.org/> Chemistry, uses, and production of various forms of lime.

National Toxicology Program, Department of Health and Human Services. *11th Report on Carcinogens.* <http://ntp-server.niehs.nih.gov> Overview of numerous carcinogenic compounds giving uses, production, exposure, and properties of these compounds.

The Nobel Foundation. <http://nobelprize.org/nobel_prizes/> Acceptance speeches and background information on Nobel laureates.

Parker, Sybil P., Ed. *McGraw-Hill Dictionary of Chemical Terms.* New York: McGraw Hill Book Company, 1985.

Plastics and Chemical Industries Association. *Chemical Fact Sheets.* <http://www.pacia.org.au/index.cfm?menuaction = mem&mmid = 004&mid = 004.009.001> Physical data, chemistry, and production processes of numerous chemical compounds.

Raven, Peter H., and George B. Johnson. *Biology,* 4th ed. Dubuque, IA: Wm. C. Brown Publishers, 1996.

Reusch, William. *Virtual Textbook of Organic Chemistry.* <http://www.cem.msu.edu/~reusch/VirtualText/intro1.htm> 1999 A comprehensive online text of organic chemistry.

Schatz, Paul W. "Indigo and Tyrian Purple-In Nature and in the Lab." *Journal of Chemical Education* 78 (November, 2001):1442–43.

Sidgwick, N. V. *The Chemical Elements and Their Compounds* Volume I and Volume II. Oxford: Clarendon Press, 1930.

Snyder, Carl H. *The Extraordinary Chemistry of Ordinary Things,* 3rd ed. New York: John Wiley & Sons Inc., 1998.

The Soap and Detergent Association. <http://www.cleaning101.com/>

Technology Information Group. Chemical Market Reporter, *Chemical Profiles* http://www.the-innovation-group.com/chemprofile.htm Production statistics and data on producers of various industrial compounds.

3DChem.Com. Chemistry, Structures and 3D Molecules. <http://www.3dchem.com/index.asp> Comprehensive site on compounds giving the top 50 pharmaceuticals, molecule of the month, and structures on more than 1,600 compounds.

Tobin, Allan J., and Jennie Dusheck. *Asking About Life.* Fort Worth, TX: Saunders College Publishing, 1998.

United States Department of Energy. *Depleted UF$_6$ Management Information Network.* <http://web.ead.anl.gov/uranium/index.cfm> Information on uranium compounds.

United States Energy Information Administration. <http://www.eia.doe.gov/environment.html> Gives information on alternative fuels such as ethanol, methanol, and pollutants.

United States Environmental Protection Agency. <http://www.epa.gov/> Information on air pollutants such as nitrogen oxides, greenhouse gases, CFC's, and toxic chemicals.

United States Geological Survey. *Minerals Yearbook Volume I*—Metals and Minerals. <http://minerals.usgs.gov/minerals/pubs/commodity/myb/> Background information and production statistics on various mineral compounds such as soda ash, salt, gypsum, etc.

Voet, Donald, Judith G. Voet, and Charlotte W. Pratt. *Fundamentals of Biochemistry Upgrade Edition.* New York: John Wiley & Sons, 2002.

Walsh, Christopher. *Antibiotics: Actions, Origins, Resistance.* Herndon, VA: ASM Press, 2003.

Index

About the Author

RICHARD L. MYERS is Professor of Environmental Science at Alaska Pacific University. He has taught chemistry, oceanography, meteorology, and physics classes and conducted research on urban environmental issues. His work has appeared in the *Journal of Environmental Health, The Northern Engineer, Journal of Chemical Education,* and *Journal of College Science Teaching.* His teaching awards include the Carnegie Foundation Alaska Professor of the Year, Higher Education Award from the United Methodist Church, President's Forum Teaching Award, and Ohaus Award for Innovations in College Science Teaching. He teaches undergraduate and graduate courses in chemistry, statistics, and environmental science. Dr. Myers' published research includes work on air quality, water quality, science education, and science and the humanities.